Ethics and Regulation of Clinical Research

Second Edition

Entia non sunt multiplicanda praeter necessitatem

William of Occam

Ethics and Regulation of Clinical Research

Robert J. Levine

Professor of Medicine
Yale University, School of Medicine
New Haven, Connecticut

Second Edition

Yale University Press
New Haven and London

First published 1986 by Urban & Schwarzenberg.
Published 1988 by Yale University Press.

Copyright ©1986 by Urban & Schwarzenberg.

Printed in the United States of America

Library of Congress cataloging card number: 87–51614
International standard book number: 0–300–04288–4

The paper in this book meets the guidelines for
permanence and durability of the Committee on
Production Guidelines for Book Longevity of the
Council on Library Resources.

10 9 8 7 6 5 4

To the Memory of
Henry K. Beecher and Franz J. Ingelfinger
and for

John, Elizabeth, and Jeralea,
My touchstones to validity

Contents

Preface to the First Edition, Abridged and Edited

Scope

This book is a survey of the ethical and legal duties of clinical researchers. The term "clinical" is derived from the Greek *klinikos*, meaning 1) of a bed, or 2) a physician who attends bedridden patients. The term "clinical research" is used in a less strict sense to mean research involving human subjects that is designed either to enhance the professional capabilities of individual physicians or to contribute to the fund of knowledge in those sciences that are traditionally considered basic in the medical school setting—e.g., biochemistry, physiology, pathology, pharmacology, epidemiology, and cognate behavioral sciences; my conceptualization of sciences that are "basic" to medicine is elaborated in an earlier publication (399). Although the chapter on deception deals with some of the ethical problems of social scientists, most of my attention is focused on the relevance of these problems to the conduct of clinical research. Social policy research, administrative sciences, and other disciplines that may affect profoundly the capabilities of groups of physicians to provide care for patients are mentioned only in passing.

By "legal duties" I mean obligations imposed by the law on clinical researchers. In this book I am concerned primarily with federal regulations for the protection of human research subjects. The most generally applicable passages of the Code of Federal Regulations (CFR) are Title 45, Part 46, the Regulations of the Department of Health and Human Services (DHHS), and Title 21, those of the Food and Drug Administration (FDA). These are discussed in detail throughout this book, and the most important passages are reprinted as Appendices. Some other federal regulations are discussed in relation to specific topics—e.g., in Chapter 7 there is a discussion of the United States Public Health Service (USPHS) Regulations (Title 42) on privacy and confidentiality. References to regulations appear in two forms: DHHS, Section 46.107a or 45 CFR 46.107a both mean Title 45 (DHHS Regulations), Part 46, Section 107, paragraph a; FDA, Section 50.132k means the same as 21 CFR 50.132k. When titles other than 45 and 21 are cited, I always use the style "28 CFR 22." DHHS regulations or proposals that were published before 1980, when this agency was separated from the Department of Health, Education, and Welfare, are cited as DHEW.

Occasionally, I discuss other types of law—e.g., federal statutes and the jurisprudence of American courts—usually to elaborate points on which the regulations seem unclear.

The idea of ethical duties is elaborated in Chapters 1 and 2. At its best, the law reflects its society's concepts of ethical duties, both positive ("thou shalt") and negative ("thou shalt not"). Also at its best, the law is concerned only with duties that should apply to all under its jurisdiction, establishing appropriate limits to its society's tolerance of moral pluralism. Moreover, the law should confine itself to establishing duties that are so important that the interests of society require that those who deviate be punished by the government. This book reflects my judgment that, in several important respects, the law governing clinical research is not at its best. I identify several specific problems in almost every chapter and in Chapter 14 I present a proposal for radical reform.

DHHS Regulations point out that they are not intended to be an ethical code. In fact, they require that each institution covered by them must provide a "statement of principles governing the institution in . . . protecting the rights and welfare of" human research subjects. The statement adopted by most institutions that conduct clinical research is the Declaration of Helsinki. This Declaration is derived from the Nuremberg Code; both documents, which are reprinted as Appendices, are discussed and interpreted extensively in this book. Some of the flaws in these documents are also identified and analyzed. For guidance on various points on which these codes and the regulations are silent, I occasionally turn to ethical codes of various other professional and academic organizations.

A final comment on the scope of this book. As a survey, there is no pretense that it is comprehensive; I shall not write all the wrongs. Although I have attempted at least to identify the major themes in the field, there are many ethical codes and regulations that are not even mentioned. Because FDA Regulations alone are substantially longer than this book, perhaps the reader will forgive me. Similarly, I have not attempted a comprehensive survey of the literature; there are over 4000 articles published on the topic of informed consent alone. Rather, in this survey I have generally cited recent reviews of each topic, when such are available, in order to afford the interested reader an adequate portal of entry into the literature.

Audience

I hope that this book will provide something of value to all persons having a serious interest in the ethics and regulation of clinical research. This, of course, is not a particularly exclusive class of persons. As I write, I have in mind that I am speaking to a person who has just been appointed to serve on a medical school or hospital committee that is charged to "safeguard the rights and welfare of human research subjects"—an Institutional Review Board (IRB). There are certain facts, opinions, judgments, and arguments that I am urging this novice to consider so that he or she might become a competent IRB member. There are certain other arguments, analyses, and theoretical works that I suggest for further study for those who wish to go beyond competence to an enriched appreciation of what they are doing and why. I acknowledge that one can be a perfectly competent academic administrator without having read *Magister Ludi*, just as one can manage a tight budget without having read *Walden* or play baseball without having read *Catcher in the Rye*. A competent clinical researcher need never have heard of Doctors Sydenham, Bichat, Bernard, Osler, or even Beecher, not to mention Aristotle, Immanuel Kant, Talcott Parsons, or Benjamin Cardozo. But there are those who are pleased that they have.

The novice IRB member might be a physician, pastor, philosopher, physiologist, lawyer, student, administrator, or community representative. Thus, I presume no special knowledge of any technical or professional jargon. Occasionally, I let a bit of jargon slip in without definition. The IRB member-reader should have very easy access to colleagues who can define and explain these words; they can, for example, ask a physician to explain "cholera" or a lawyer to explain "fiduciary." Readers who have no consultants will find no terms that cannot be understood easily with the aid of a standard medical dictionary and a standard general dictionary.

Motivation

In 1973, DHEW published its first set of proposed regulations on the protection of human research subjects. On May 30, 1974, clinical researchers were presented with final regulations in the form of 45 CFR 46. By this time, DHEW had already published proposed regulations designed to provide special protections for persons having "limited capacities to consent." Almost overnight the clinical researcher was presented with a bewildering array of new and potential legal duties.

In July 1974, Congress passed the National Research Act (Public Law 93–348), which established the National Commission for the Protection of Human Subjects of Biomedical and Behavioral Research; I shall call this the Commission. Among the charges to this Commission were:

> Sec. 202(a)(1)(A) The Commission shall (i) conduct a comprehensive investigation and study to identify the basic ethical principles which should underlie the conduct of biomedical and behavioral research involving human subjects, (ii) develop guidelines which should be followed in such research to assure that it is conducted in

accordance with such principles, and (iii) make recommendations to the Secretary (I) for such administrative actions as may be appropriate to apply such guidelines to biomedical and behavioral research conducted or supported under programs administered by the Secretary. . . .

From 1975 through 1978, the Commission published a series of reports on various aspects of research involving human subjects. Each of these reports presented recommendations to the DHEW Secretary; some also contained recommendations for congressional or presidential action. These recommendations formed the bases for the development of new DHEW and DHHS Regulations, some of which are still in the proposal stage.

Although the various regulations and ethical codes provide statements of ethical norms and procedures, they tend not to explain their purpose. By contrast, the Commission's reports refer to the fundamental ethical principles these norms and procedures are meant to uphold. Each report contains a section entitled "Deliberations and Conclusions" that shows how the Commission derived its recommendations and what the recommended regulations are expected to accomplish. These reports are excellent resources; I discuss them extensively in this book.

Early in 1978, during the final months of the Commission's existence, I began to write this book as a chapter for another book. The other book—which has not been published—had as its intended audience clinical researchers. My purpose at the time was to explain to clinical researchers what the regulations were and what new regulations they could anticipate. I also planned to draw on the Commission's reports to explain the purpose of the regulations. I hoped that clinical researchers who read my chapter would be better able than they had been to perform intelligently within a complicated regulatory framework.

Shortly after the original chapter was completed, FDA, through its regulatory proposal of August 8, 1978, notified all concerned of its intention to disregard both the letter and the spirit of the Commission's recommendations on IRBs (409). At this time the Commission had already announced its intention to dissolve at its September 1978 meeting. Congress had prolonged its life on several previous occasions, but in the Spring of 1978 the Commission protested. As one Commissioner put it, "Even a Commission is entitled to die with dignity."

Subsequently, and quite promptly, DHHS began to violate both the letter and the spirit not only of the Commission's recommendations but also of the National Research Act's clear directions as to how and when it was to respond to the Commission's recommendations. And Congress has yet to respond to the Commission's recommendations for congressional action.

Early in 1980, I decided that my purpose in writing the chapter could not be accomplished. I could have chosen to rewrite the chapter to try to explain the rationale—the ethical grounding—for the new regulations; in fact, this book began as an attempt to do precisely that. But as I began to write I began to sense a need to say something more, some things in addition, to an audience more diverse than clinical researchers. In response to the Commission's recommendations, DHHS has established a set of regulations that are far more rational than those it had proposed in 1973–1974. However, both DHHS and Congress have missed a golden opportunity to do much better; they could have adhered much more faithfully to the Commission's recommendations.

Now, I do not think that the Commission did a perfect job. But I watched very carefully as 11 dedicated women and men spent a lot of their time and the taxpayer's money responding to their Congressional charge. They studied, they investigated (with the aid of many qualified consultants), they deliberated and concluded. They argued. They strove mightily as adversaries do at court and they ate and drank together as friends. I will never forget David W. Louisell living as a model of responsiveness to the principle of respect for persons, defending, for example, his dissent from the report on *Research on the Fetus*; on what he considered the most crucial issue he was outvoted 9 to 1. But I digress; there are too many stories to tell.

The Commission's publications reflect rather well the moral pluralism of the United States as it could be perceived by members of the upper middle class in the 1970s. Consequently, I draw heavily on these publications in this book, disagreeing with them only when I find it necessary.

Acknowledgements

As a physician I am enjoined by Hippocrates to "hold my teachers in this art" as equal to my own parents. Some of my most important teachers present themselves bearing other titles: patients who teach me what problems merit a physician's attention, students who never leave me alone and thereby compel my learning, and colleagues who share with me their knowledge and their wisdom.

I must also acknowledge gratefully the many research subjects with whom I have worked. Because all my interactions with them were in institutions of my choosing, I have had from the outset an awareness that these interactions were being watched. Consequently, in my dealings with them, I practiced my devices to avoid misusing or abusing them, trying to use them as means only to mutually agreeable ends. Particularly in the early years, when I did not anticipate writing about these devices, I could not disclose all possible consequences of our interactions. What can I do now but thank them anonymously for having allowed themselves to be used as means to my unforeseen ends. To those who are capable of learning of this, I hope you find these ends agreeable.

Finally, and above all, I am thankful to and for all persons who argue with me, or who praise or blame me, when such is my due. If Joseph V. Brady, Karen A. Lebacqz, Kenneth J. Ryan and Eliot Stellar were not such persons, this book would not exist. Such persons serve me as models and examples and as means to mutually agreeable ends.

Robert J. Levine
New Haven, Connecticut
May 19, 1981

Preface to the Second Edition

There was a time I would look back on what I had written 5 years ago and say, "How could you have been so stupid?" Now I say instead, "Look at how we have grown!" Changing in this way the description of my experiences does not, of course, change the experience. It does, however, serve to diminish duodenal disturbance.

We, the community of persons having a serious interest in the ethics and regulation of clinical research, have grown indeed in the past 5 years. In 1981, federal regulations in this field were new; special protections for children were not to be promulgated until 1983. All these regulations have since been the object of extensive critical analysis and interpretative commentary by scholars, practitioners, administrators, and others. Through such writings we have begun to grasp the meaning of the rules.

A particularly important and authoritative voice was that of the President's Commission for the Study of Ethical Problems in Medicine and Biomedical and Behavioral Research; given the brevity of its existence (1980–1983), its accomplishments must be regarded as prodigious. In this book I refer to it as the "President's Commission" to distinguish it from the National Commission for the Protection of Human Subjects of Biomedical and Behavioral Research, which I continue to call the "Commission."

Various events have directed our attention to problems presented by advancing technology (e.g., *in vitro* fertilization with cryopreservation and gene therapy) and new diseases (e.g., Acquired Immune Deficiency Syndrome (AIDS)) and accentuated our concern with old problems (e.g., academic misconduct, including fraud and plagiarism, and senile dementia of the Alzheimer's type).

In my view, the greatest of all resources has been and continues to be case studies, which are articles describing problems confronted in the real world by Institutional Review Boards (IRBs) and investigators and showing how they interpreted applicable ethical codes and regulations to reach judgments about particular research protocols. In these publications, the necessarily vague abstractions of Nuremberg, Helsinki, and federal regulations are given practical meaning. What is a reasonable balance of harms and benefits, a minor increment above minimal risk, an adequate justification for waiver or alteration of some or all of the elements of informed consent? Case studies of the sort published frequently in *IRB: A Review of Human Subjects Research* and *Clinical Research* are beginning to yield a common sense of the answers to these and similar questions.

The second edition of this book reflects the growth of our understanding of ethics and regulation of clinical research as I see it. Otherwise, its scope, intended audience, and purpose are as presented in the preface to the first edition. As in the first edition, I try to demonstrate the controversial nature of many of the topics. In some cases I take sides in the arguments. When I do not, it is not an oversight; it means that I remain to be persuaded to one of the competing positions.

Also as in the first edition, I take very seriously the reasoning and recommendations of such deliberative bodies as the Commission and the President's Commission. In consideration of the way the commissioners were selected, the resources available to them, and the circumstances under which they conducted their debates and wrote their reports (1,334,337), I think those who would disagree with them assume a weighty burden of proof. When I cite a judgment of one of the Commissions as authority on a point, I often do not repeat the reasoning that led to this judgment; I do, however, provide references so that those who are so inclined may decide for themselves whether to accept the Commission's reasoning.

About the Author

Because this book is concerned primarily with values that cannot or should not be quantified, the reader is entitled to know something about the author, particularly those things that have been most influential in shaping my relevant perspectives and intuitions. I am or have been each of the following: an academic physician specializing in internal medicine; chief of a university section of clinical pharmacology; a laboratory researcher in biochemical pharmacology; a research subject; and a clinical investigator doing research involving patients with various rare and common diseases, normal volunteers, children, students, elderly persons, mentally retarded persons and their families, and two dead fetuses. I have taught various courses ranging from pharmacology to physical diagnosis to medical ethics. For many years I have been active in the policy arena, serving in various capacities such agencies as the Commission, the President's Commission, the National Institutes of Health, and the Commission on the Federal Drug Approval Process, as well as various other international, national, state, and Yale University agencies, committees, working groups, task forces, and advisory boards.

One constant in my professional life has been a fondness for words, written or spoken. In the organizations I take seriously, the positions to which I tend to be elected or appointed are secretary or editor. I once declined a nomination for presidency of a national research organization in favor of remaining editor of its journal, stating on that occasion that it matters less what the president says than what the editor or secretary records as his or her statement. Among my major assignments have been Associate Editor of *Biochemical Pharmacology*, Editor of *Clinical Research*, and Secretary of the Eastern Section of the American Federation for Clinical Research. I am now Editor of *IRB: A Review of Human Subjects Research*, Secretary as well as Chair of the Publications Committee of the Connecticut Humanities Council, and Secretary of Medicine in the Public Interest.

Words are perhaps my most important professional tools; they are also among my favorite toys. As much as I love puns, twists, double-entendres, and the like, I try to keep them out of my professional writings; occasionally, I fail.

If ever in my life I possess simultaneously sufficient arrogance and energy to write an autobiography, I think I shall follow W. H. Auden's example and write a commonplace book, a collection of passages from my readings that are important in shaping the way I see and think about the world. This approach, of course, acknowledges the necessity of omiting some phenomena and noumena that cannot be described. For example, when I write, the background music is often Brahms or Liszt; what can one say about the experience of listening to music? Words set to music are another matter; when writing about autonomy I might reflect on Kris Kristofferson's "Freedom's just another word for nothing left to lose."

If I were to begin my commonplace book today, I would have to find places for the following entries:

"Like all young men I set out to be a genius, but mercifully laughter intervened."

Lawrence Durrell

"The madman shouted in the market place. No one stopped to answer him. Thus it was confirmed that his thesis was incontrovertible."

Dag Hammarskjöld

"And were an epitaph to be my story
I'd have a short one ready for my own.
I would have written of me on my stone:
I had a lover's quarrel with the world."

Robert Frost

"That was the second year of the Third World War,
The one between Us and Them. . . .
Who started it? Oh, they said it was Us or Them
and it looked like it at the time."

Stephen Vincent Benet

"Oh, those who are my generation!
We're not the threshold, just a step.
We're but the preface to a preface,
a prologue to a newer prologue!"

Yevgeny Yevtushenko

"It was your vision of the pilot
confirmed my vision of you: you said, *He keeps*
on steering headlong into the waves, on purpose
while we crouched in the open hatchway
vomiting into plastic bags
for three hours. . . ."

Adrienne Rich

"The difference between the two ways of seeing is the difference between walking with and without a camera. When I walk with a camera, I walk from shot to shot, reading the light on a calibrated meter. When I walk without a camera, my own shutter opens, and the moment's light prints on my own silver gut. When I see this second way I am above all an unscrupulous observer."

Annie Dillard

"All persons, living and dead,
are purely coincidental
and should not be construed."

Kurt Vonnegut, Jr.

Acknowledgments

I wish to thank Edythe M. Sheridan and Kathryn P. Potter, who carefully and cheerfully typed the original and final drafts of each chapter. I also wish to thank my friends and colleagues who read and criticized drafts of one or more chapters: Nancy Rockmore Angoff, Margaret Farley, John N. Forrest, Jr., Jay Katz, Carol Levine, Maurice J. Mahoney, and Kathleen A. Nolan. I owe special thanks to Angela Roddey Holder, who not only read and criticized very carefully each and every chapter before I wrote the final drafts, but also gave generously of her advice and assistance in developing other components of the book, such as the index.

Further acknowledgments are in the preface to the first edition.

Robert J. Levine
New Haven, Connecticut
February 18, 1986

Chapter 1

Basic Concepts and Definitions

> Then the Lord came down to see the city and tower which mortal men had built, and he said, "Here they are, one people with a single language, and now they have started to do this; henceforward nothing they have a mind to do will be beyond their reach. Come, let us go down there and confuse their speech, so that they will not understand what they say to one another." . . . That is why it is called Babel, because the Lord there made a babble of the language of all the world. . . .
>
> Genesis 11:5–9

In the twentieth century discourse on law and ethics, men and women, not the Lord, are building a Tower of Babel. They are doing this not by creating diverse languages, but rather by confounding the meanings of words and phrases within the English language. While most are doing this inadvertently, some are doing it deliberately.

I shall now identify some of the more common obfuscations. Stipulated definitions often create serious misunderstandings. Consider, for example, the word "fetus." All dictionaries I have consulted provide a more or less definite gestational age at which the product of human conception is first called a fetus; the earliest is during the seventh week and the latest is the end of the third month of gestation; no dictionary definition accommodates an extrauterine existence for the fetus. Yet, Department of Health and Human Services, (DHHS) regulations state " 'fetus' means the product of conception from the time of implantation

/ 1

. . . until a determination is made, following expulsion or extraction of the fetus, that it is viable" (Section 46.203c). Various bills introduced in state and federal legislative bodies offer the following definition: "A child exists as a living and growing person from the moment of fertilization. . ." (391). In these bills, there is no time at which the developing human is called a fetus.

Many authors stipulate their own definitions for such words as "fetus." They then proceed to refer to the writings of another who has stipulated a different definition, creating the incorrect impression that the other writer either supports or opposes their position. The prudent reader will be wary whenever he or she encounters a stipulated definition.

Meanings are distorted frequently through incautious use of language. Consider, for example, this excerpt from a judge's opinion: "The court finds that there is clear and convincing evidence beyond a reasonable doubt that (the patient) made an informed and rational and knowing decision to forego dialysis treatment. . . . (His) decision was based on a compelling desire to escape the constant and severe pain caused by his multiple debilitating, irreversible and terminal conditions" (439). Imagine having multiple terminal conditions. DeBakey has published a fascinating collection of such abuses of the English language (180).

Some authors distort meanings deliberately as a rhetorical device. For an excellent discussion of this problem, see Jonsen and Yesley (337).

Many authors use language that is heavily laden with symbolic meaning. Use of such language often conveys unintended impressions and evokes misunderstandings that the authors may not anticipate. Current discussions of cardiopulmonary resuscitation are rich in examples. What do we mean by "heroic" therapy? Who is the hero? What is "aggressive" therapy? Who is the aggressor? What does it mean to "call a code?" Are we trying to reinforce an aura of mystery? In general, rational discussion can be facilitated by avoiding the use of such terms (439, 442).

It is impossible to overestimate the importance of terminological precision and conceptual clarity in satisfactory discussions of the ethics and regulation of research involving human subjects. In several essays I have called attention to the confusion that can be attributed to inattentiveness to such details (391, 401, 412, 439); some examples will be repeated in this chapter.

This chapter provides an account of some of the basic terms and concepts used in this book, e.g., definitions of "research" and "practice" and an introduction to the fundamental ethical principles identified by the National Commission for the Protection of Human Subjects of Biomedical and Behavioral Research, (referred to throughout this book as the Commission) as those that should underlie the conduct of research involving human subjects. Definitions that have somewhat less global application are presented in subsequent chapters.

Research and Practice

Until recently, most regulations and ethical codes did not define such terms as "research" and "practice." Although distinguishing research from practice might not seem to present serious problems, in the legislative history of PL 93–348, the act that created the Commission, we find that some prominent physicians regarded this as a very important and exceedingly difficult task (347). Jay Katz identified ". . . drawing the line between researh and accepted practice . . .(as) the most difficult and complex problem facing the Commission." Thomas Chalmers stated, "It is extremely hard to distinguish between clinical research and the practice of good medicine. Because episodes of illness and individual people are so variable, every physician is carrying out a small research project when he diagnoses and treats a patient."

Chalmers, of course, was only echoing the views of many distinguished physicians who had spoken on this issue earlier. For example, in *Experimentation with Human Subjects*, Herrman Blumgart stated, "Every time a physician administers a drug to a patient, he is in a sense performing an experiment," (63). To this Francis Moore added, ". . . every (surgical) operation of any type contains certain aspects of experimental work" (509). Although these statements are true, they tend to obfuscate the real issue; they tend to make more difficult the task of distinguishing research from practice. The following definitions are compatible with those adopted by the Commission (526) and incorporated subsequently in federal regulations (477).

The term "research" refers to a class of activities designed to develop or contribute to generalizable knowledge. Generalizable knowledge consists of theories, principles, or relationships (or the accumulation of data on which they may be based) that can be corroborated by accepted scientific observation and inference.

The "practice" of medicine or behavioral therapy refers to a class of activities designed solely to enhance the well-being of an individual patient or client. The purpose of medical or behavioral practice is to provide diagnosis, preventive treatment, or therapy. The customary standard for routine and accepted practice is a reasonable expectation of success. The absence of validation or precision on which to base such an expectation, however, does not in and of itself define the activity in question as research. Uncertainty is inherent in therapeutic practice because of the variability of the physiological and behavioral responses of humans. This kind of uncertainty is, itself, routine and accepted.

There are two additional classes of activities performed by physicians that require definition; these are "nonvalidated practices" and "practice for the benefit of others." These are less familiar terms than "research" and "practice."

For reasons I shall present, some of the ethical norms and procedures designed for research might, in some situations, be applicable to these classes of activities.

Nonvalidated Practices

Early in the course of its deliberations, the Commission defined a class of activities as "innovative therapy" (394):

> Uncertainties may be introduced by the application of novel procedures as, for example, when deviations from common practice in drug administration or in surgical, medical, or behavioral therapies are tried in the course of rendering treatment. These activities may be designated innovative therapy, but they do not constitute research unless formally structured as a research project. . . . There is concern that innovative therapies are being applied in an unsupervised way, as part of practice. It is our recommendation that significant innovations in therapy should be incorporated into a research project in order to establish their safety and efficacy while retaining the therapeutic objectives.

In the context of the Commission's deliberations, it is clear that diagnostic and prophylactic maneuvers are included under this rubric. Thus, a more descriptive title for this class of activities is "innovative practices." Subsequently, it has become clear that novelty is not the attribute that defines these practices; rather, it is the lack of suitable validation of the safety of efficacy of the practice. Therefore, the best designation for this class of activities is "nonvalidated practices." A practice might be nonvalidated because it is new; i.e., it has not been tested sufficiently often or sufficiently well to permit a satisfactory prediction of its safety or efficacy in a patient population. It is equally common for a practice to merit the designation "nonvalidated" because in the course of its use in the practice of medicine there arises some legitimate cause to question previously held assumptions about its safety or efficacy. It might be that the practice was never validated adequately in the first place (e.g., implantation of the internal mammary artery for the treatment of coronary artery disease or treatment of gastric ulcers with antacids) or because a question has been raised of a previously unknown serious toxicity (e.g., renal failure with some sulfa drugs). At the time of the first substantial challenge to the validity of an accepted practice modality, the proposition that subsequent use of that modality should be considered nonvalidated should be taken seriously. For purposes of ethical justification, all nonvalidated practices may be considered similarly. As the Commission has suggested for innovative therapy, these practices *should* be conducted in the context of a research project designed to test their safety or efficacy or both; however, the research should not interfere with the basic therapeutic (or diagnostic or prophylactic) objectives.

As we shall see subsequently, the ethical norms and procedures that apply to to nonvalidated practices are complex. Use of a modality that has been classified as a nonvalidated practice is justified according to the norms of practice. However, the research designed to develop information about the safety and efficacy of the practice is conducted according to the norms of research.

The term used by the Food and Drug Administration (FDA) for regulated test articles (e.g., drugs and medical devices) that have not yet been approved for marketing is "investigational." Because I find this term acceptable, when discussing test articles I shall often use the terms "investigational" and "nonvalidated" interchangeably.

In the routine practice of medicine, many therapeutic modalities have not been validated in the strict sense. For example, many drugs and devices approved for use by the FDA are prescribed for uses that are not listed on the FDA-approved package label. This does not mean that all such uses must be made the object of a formal study designed to establish safety and efficacy. At Yale, for example, formal studies that must be reviewed and approved by the Institutional Review Board (IRB) are required when they are mandated by FDA regulations (425). In general, this means that the industrial sponsor of an approved drug intends to collect data to support addition of the new indication to the FDA-approved label.

In situations in which the regulations do not give clear guidance, Yale's policy is in accord with the Commission's conclusion on this point (526, pp. 3–4). If a drug or other therapeutic modality is to be studied systematically according to a formal plan with the aim of developing generalizable knowledge about the drug, then the physician is required to prepare a protocol for review by the IRB. The status of the proposed use of the drug as approved or nonapproved by the FDA is immaterial. We require, for example, IRB approval of plans to conduct a randomized clinical trial designed to compare the safety and efficacy of two drugs even when both drugs are being administered according to the advice provided on the FDA-approved package label.

Physicans who propose to use drugs or other therapeutic modalities in ways that differ substantially from their use in customary medical practice are encouraged—and occasionally required—to develop a protocol designed to show that the drug is truly safe and effective for the new indication or in the new dosage schedule. In such cases, it is necessary to exercise judgment as one determines whether the degree of departure from the standards of customary medical practice is sufficient to impose the requirement for IRB review. This is not accomplished by comparing the physician's plans with the FDA-approved package label. If so, it would be necessary for the IRB to review approximately 80% of all prescriptions written by pediatricians (Chapter 10).

FDA Classification of Drug Studies   FDA regulations refer to various phases of drug studies (222, pp. 6–7). Phases II and III conform to the definition of nonvalidated practices. In general, Phase I studies do not, because the purpose of drug administration is to develop information on drug toxicities, metabolism, and dynamics rather than to enhance the well-being of an individual patient or client. Although the recipients of drugs in Phases II and III are always patients for whom the drug is intended to provide a therapeutic effect, in Phase I they are most commonly "normals." When Phase I studies are performed using patients for whom a therapeutic effect is intended, as in the development of most cancer chemotherapeutic agents, they may at times also be considered nonvalidated practices (cf. Chapter 2). In Phase IV, the drugs may be either validated or nonvalidated, as in cases in which marketed drugs are being tested for safety and efficacy for new indications or new populations, such as children.

Phase I Clinical Pharmacology is intended to include the initial introduction of a drug into man. It may be in the usual "normal" volunteer subjects to determine levels of toxicity and, when appropriate, pharmacologic effect and be followed by early dose-ranging studies in patients for safety and, in some cases, for early evidence of effectiveness.

Alternatively, with some new drugs, for ethical or scientific considerations, the initial introduction into man is more properly done in selected patients. When normal volunteers are the initial recipients of a drug, the very early trials in patients which follow are also considered part of Phase I.

Drug dynamic and metabolic studies, in whichever stage of investigation they are performed, are considered to be Phase I clinical pharmacologic studies. While some, such as absorption studies, are performed in the early stages, others, such as efforts to identify metabolites, may not be peformed until later in the investigations.

Phase II Clinical Investigation consists of controlled clinical trials designed to demonstrate effectiveness and relative safety. Normally, these are peformed on closely monitored patients of limited number.

Phase III Clinical Trials are the expanded controlled and uncontrolled trials. These are performed after effectiveness has been basically established, at least to a certain degree, and are intended to gather additional evidence of effectiveness for specific indications and more precise definition of drug-related adverse effects.

Phase IV Postmarketing Clinical Trials are of several types:
1. Additional studies to elucidate the incidence of adverse reactions, to explore a specific pharmacologic effect or to obtain more information of a circumscribed nature.
2. Large scale, long-term studies to determine the effect of a drug on morbidity and mortality.
3. Additional clinical trials similar to those in Phase III, to supplement premarketing data where it has been deemed in the public interest to release a drug prior to acquisition of all data which would ordinarily be obtained before marketing.
4. Clinical trials in a patient population not adquately studied in the premarketing phase, e.g., children.

5. Clinical trials for an indication for which it is presumed that the drug, once available, will be used.

Practice for the Benefit of Others

As we examine the universe of professional activities of physicians, we find that there is one set that does not conform to the definitions of practice, research, or nonvalidated practice (405). It departs from the definition of practice only in that it is not ". . . designed solely to enhance the well-being of an individual." However, it does meet ". . . the customary standard for routine and accepted practice . . ." in that it has ". . . a reasonable expectation of success." In addition, the purpose of this activity is not ". . . to develop or contribute to generalizable knowledge." Thus, it does not conform to the definitions of either research or nonvalidated practice. Practice for the benefit of others is a class of activities in which interventions or procedures are applied to one individual with the expectation that they will (or could) enhance the well-being or convenience of one or many others.

Although practices for the benefit of others may yield direct health-related benefits to the individuals on whom they are performed, this is not necessarily the case. For example, one activity in this class, the donation of an organ (e.g., kidney) or tissue (e.g., blood), brings no direct health benefit to the donor; in this case, the beneficiary is a single other person who may or may not be related to the donor. In some activities, the beneficiary may be society generally as well as the individual patient (e.g., vaccination), while in others the only beneficiary may be society (e.g., quarantine). At times individuals are called upon to undergo psychosurgery, behavior modification, psychotherapy, or psychochemotherapy so as to be less potentially harmful to others; this is particularly problematic when the individual is offered the "free choice" between the sick role and the criminal role. In some cases the beneficiaries may include succeeding generations, as when persons are called upon to undergo sterilization because they are considered genetically defective or otherwise incompetent to be parents. There are also situations in which one beneficiary of therapy may be an institution; there may be serious disputes over the extent to which the purpose of therapy is to provide administrative convenience to the institution, e.g., heavy tranquilization of disruptive patients in a mental institution or treatment of hyperkinetic schoolchildren with various stimulant and depressive drugs.

Although practice for the benefit of others is clearly a subset of practice, it has one important attribute in common with research: persons are called upon to assume various burdens for the benefit of others. Therefore, it seems appropriate to apply some of the ethical norms and procedures designed for research to this class of activities (405).

Other Definitions

In this book the term "physician" will be used to refer to a professional who is performing the practice of medicine; his or her client will be referred to as a "patient." An "investigator" is an individual who is performing research; the investigator may or may not be a physician. The individual upon whom the investigator performs research will be called "subject;" the terms "volunteer," "participant," and "respondent" will be used as synonyms. At times, more complex constructions will be required. For example, in the conduct of research designed to prove the safety and efficacy of a nonvalidated practice, various roles might be performed by such individuals as "physician-investigators" and "patient-subjects."

DHHS regulations provide this definition (Section 46.102f):

> "Human subject" means a living individual about whom an investigator (whether professional or student) conducting research obtains (1) data through intervention or interaction with the individual, or (2) identifiable private information.

DHHS's definition of "research" in Section 46.102e is an abbreviated version of the one provided earlier because it, too, is derived from the definition of the Commission.

Unacceptable Terminology

There are some terms that are commonly used in the discussion of the ethics and regulation of research that will not be used in this book. These are "therapeutic research," "nontherapeutic research," and "experimentation." The confusion created through the use of such terms has been reviewed in detail elsewhere (372, 401, 412).

It is not clear to me when the distinction between therapeutic and nontherapeutic research began to be made in discussions of the ethics and regulation of research. The Nuremberg Code (1947) draws no such distinction. The original Declaration of Helsinki (1964) distinguishes nontherapeutic clinical research from clinical research combined with professional care. In the 1975 revision of this Declaration, "medical research combined with professional care" is designated "clinical research," while "nontherapeutic biomedical research" is also called "non-clinical biomedical research."

The problem with the distinction between therapeutic and nontherapeutic research is, quite simply, that it is illogical (372, 412). All ethical codes, regulations, and commentaries relying on this distinction have, therefore, contained serious errors. This is exemplified by placing a Principle of the Declaration of

Helsinki developed for clinical research (II.6) in immediate proximity with one developed for nonclinical research (III.2).

II.6. The doctor can combine medical research with professional care, the objective being the acquisition of new medical knowledge, only to the extent that medical research is justified by its potential diagnostic or therapeutic value for the patient.

III.2. The subjects should be volunteers—either healthy persons or patients for whom the experimental design is not related to the patient's illness.

This classification has several unfortunate and unintended consequences. First, many types of research cannot be defined as either therapeutic or nontherapeutic. Consider, for example, the placebo-controlled, double-blind drug trial. Certainly, the administration of a placebo for research purposes is not ". . . justified by its potential diagnostic or therapeutic value for the patient." Therefore, according to the Declaration of Helsinki, this is nontherapeutic, and those who receive the placebo must be ". . . either healthy persons or patients for whom the experimental design is not related to the patient's illness." This, of course, makes no sense. Another unfortunate consequence is that a strict interpretation of the Declaration of Helsinki would lead us to the conclusion that all rational research designed to explore the pathogenesis of a disease is to be forbidden. Because it cannot be justified as prescribed in Principle II.6, it must be considered nontherapeutic and therefore can be done only on healthy persons or patients not having the disease one wishes to investigate. Again, this makes no sense.

In recognition of these sorts of problems, the Commission abandoned the concepts of therapeutic and nontherapeutic research. Instead of "nontherapeutic research," the Commission simply refers to "research" which is, by definition, something other than therapeutic. More cumbersome language is employed by the Commission to convey the meaning intended most commonly by those who had used the expression "therapeutic research." For example, in *Research Involving Prisoners*, Recommendation 2 states in part, "Research on practices, both innovative and accepted, which have the intent and reasonable probability of improving the health or well-being of the individual may be conducted. . . ." The same concept is reflected in Recommendation 3 of *Research Involving Children*. It is made clear that the risks and benefits of therapeutic maneuvers are to be analyzed similarly, notwithstanding the status of the maneuver as either nonvalidated or standard (accepted). "The relation of benefit to risks . . . (should be) at least as favorable to the subjects as that presented by any . . ." available alternative. The risks of research maneuvers (designed to benefit the collective) are perceived differently. If they are ". . . greater than those normally encountered (by children) in their daily lives or in routine medical or psychological examinations . . ." review at a national level may be required.

It is worth underscoring the point that risks and benefits of therapeutic maneuvers are to be assessed similarly notwithstanding their status as either non-validated or standard. This point is particularly important in understanding the requirement of a null hypothesis in the ethical justification of a randomized clinical trial (RCT) (Chapter 8). Investigators who propose to begin a RCT must be able to defend a claim that there is no scientifically valid reason to predict that either of the two or more therapies to be compared will prove superior. In many RCTs, a new therapy (nonvalidated practice) is compared with an established standard therapy. Thus, what is called for is a formal claim of therapeutic equivalency between a nonvalidated modality and an accepted standard. This is consistent with the Commission's recommendations.

It should be noted that the Commission completed its report, *Research on the Fetus*, before it began to address the general conceptual charges in its mandate. Consequently, as we shall see in Chapter 13, the Commission's recommendations on research on the fetus and the DHHS regulations derived from them rely on the distinction between therapeutic and nontherapeutic research. Apart from this, federal regulations and, with occasional exceptions (114, 423), major public policy documents have been rid of these erroneous concepts. The Canadian Medical Research Council also rejected this distinction in developing its guidelines for research involving human subjects (102).

A third term that I shall not use is "experimentation." Experimentation is commonly and incorrectly used as a synonym for research. Although some experimentation is research, much of it is not. "Experiment" means to test something or to try something out. In another sense, an experiment is a tentative procedure, especially one adopted with uncertainty as to whether it will bring about the desired purposes or results. As noted earlier, much of the practice of diagnosis and therapy is experimental in nature. One tries out a drug to see if it brings about the desired result. If it does not, one either increases the dose, changes to another therapy or adds a new therapeutic modality to the first drug. All of this experimentation is done in the interests of enhancing the well-being of the patient.

When experimentation is conducted for the purpose of developing generalizable knowledge, it is regarded as research. One of the problems presented by much research designed to determine the safety and efficacy of drugs is that this activity is much less experimental than the practice of medicine. It must, in general, conform to the specifications of a protocol. Thus, the individualized dosage adjustments and changes in therapeutic modalities are less likely to occur in the context of a clinical trial than they are in the practice of medicine (cf. Chapter 8). This deprivation of the experimentation ordinarily done to enhance the well-being of a patient is one of the burdens imposed on the patient-subject is a clinical trial (441).

Fundamental Ethical Principles

In this discussion and, indeed, throughout this book, I follow the theoretical assumptions embodied in the Commission's *Belmont Report* (526). Because the Commission never articulated its concepts of ethical theory, the assumptions that follow are my own; they are based on my careful reading of all the Commission's publications as well as my participation in virtually all of the discussions that led to the writing of its reports. The four major assumptions I address at this point are relevant generally to all considerations in this book. I shall consider some others in connection with specific points in other chapters.

1. Both utilitarian and deontological reasoning have a place in ethical decision-making.

In determining the moral rightness of human actions, the deontologist (from the Greek, *deon*, meaning "duty") evaluates the action per se apart from any of its expected consequences. If a deontologist concludes it is always wrong to lie, then lies cannot be justified, not even to prevent serious harm to a person. By contrast, the utilitarian (from the Latin *utilitas*, meaning "useful") evaluates the moral rightness of human actions in terms of their consequences. The morally right act is that which is most likely to be useful, to contribute to the development of goods. The goods in question are intrinsically nonmoral. Classical utilitarians have defined the good in such terms as "happiness" or "pleasure." Thus, for example, a utilitarian who concludes that it is always wrong to lie argues that lying is always detrimental to the pursuit of happiness. My hypothetical deontologist will argue that deception in research must be forbidden simply because lying is wrong—intrinsically immoral in that it treats persons as merely means to researchers' ends (*infra*). My hypothetical utilitarian will, instead, point to undesirable consequences, such as that use of deception in research undermines trust in researchers and thus hampers the efforts of all researchers to contribute to the good of society through the development of new knowledge.

The Commission did not fail to choose among competing ethical theories. Rather, as explained well by Abram and Wolf (1), "Aware of Kantian (deontological), utilitarian, and Aristotelian traditions, for instance, the commission nonetheless refrained from relying on any one of them for the legitimacy of its conclusions. Agreement on a fundamental moral system was not sought or needed." Although these authors discuss the President's Commission, they make it clear that it was patterned after the National Commission, which similarly refrained from relying exclusively on any particular moral system.

2. Ethical reasoning is, and ought to be, hierarchical and deductive.

The hierarchy begins with basic ethical principles (e.g., respect for persons) from which we deduce ethical norms, which in turn represent requirements (e.g., for informed consent). The Commission thus appears to pay no attention to philosophers who hold differing viewpoints (e.g., that the traditional or received norms of society are most important and logically prior to the basic principles (101, 364). Those who hold this viewpoint claim that the purpose of ethical theory and basic principles is to explain and systematize ethical norms scientifically or inductively; in this view, the principles are induced from the norms.

Given that the Commission affirmed a heirarchical deductive system, it is fortunate that it did not choose between deontological and utilitarian reasoning. As Ladd has observed (364, p. 16),

> A head-on collision between adherents of two different foundational principles (e.g., between a utilitarian and a rights theorist) is bound to lead to an impasse. On a social level, the kind of blind dogmatism reflected in the categorical and unilateral assertion of controversial first principles leads sooner or later to violence: war and terrorism. What is seldom realized is that these moral, legal, and political outcomes are conceptually rooted in irresolvable conflicts of foundational first principles.

3. Ethical reasoning should begin with multiple basic ethical principles (in this case, three), each of which in the abstract has equal moral force.

In this regard, the Commission ignores the arguments of philosophers who claim that there should be a single overarching superprinciple, such as justice or *agape*. It further ignores those who claim that there is a need to assign priority to principles—to rank them in order of moral forcefulness—to resolve disputes engendered by conflicting requirements of two or more principles (588).

The Commission identified three ethical principles that, as we shall see, give rise to norms that in turn often create conflicting requirements. For example, the principle of justice, as articulated by the Commission, creates requirements that are incompatible with some of those created by its principle of respect for persons (cf. Chapter 12). Similarly, norms derived from the principle of beneficence inevitably engender conflicts with those arising from the principle of respect for persons. Implicitly, the Commission also endorsed the notion of *prima facie* rules; these are rules that are binding unless they are in conflict with other stronger rules or unless in specific situations there is ethical justification for overriding the rule's requirements (cf. Chapter 9). The Commission's specific recommendations for dealing with many of these conflicts are presented in every chapter of this book. However, the Commission did not attempt to identify all such conflicts. It left a good deal for investigators, IRB members, and others to think about; through its identification of three coequal ethical principles, it assured continuing productive argument.

4. Public policy should be addressed to the behavior, rather than the character, of the actors or agents.

In this regard, the Commission appears to ignore those who claim that the virtues of investigators are more important than rules governing their behavior in assuring that research subjects will be treated with due regard for their rights and welfare (457). This is one possible reading of the Commission. Another, which I prefer, is that the Commission was mindful of the importance of virtue and character and, for this reason, recommended a striking decrease in requirements for bureaucratic oversight and monitoring as compared with regulations that had been proposed by the Department of Health, Education, and Welfare (DHEW) shortly before the Commission was impanelled. Moreover, the Commission's recommendations acknowledge implicitly that, although behavior can be regulated, character cannot. The Commission's recommendations relating to IRBs also can be construed as reflecting its awareness of the importance of character (cf. Chapter 14).

These assumptions and several others that I shall mention later have been the target of harsh criticism by various authors; some of this criticism has been followed by rebuttals by Commissioners, Commission staff, and others not connected with the Commission. For an example, see the exchange between Annas (22) and Jonsen and Yesley (337). I also wish to call attention to Jonsen's important article on the grounds for respecting the authority of agencies such as the Commission and the President's Commission (334).

There is yet another charge frequently levelled against the Commission. In each of its publications, it seems to embrace an atomistic view of the person. The person is seen as a highly individualistic bearer of duties and rights; among his or her rights, some of the most important are to be left alone, not to be harmed, and to be treated with fairness. Except, perhaps, in its report on research involving children, there is little or no reference to persons in relationship to others or as members of communities. This perception of the person dominated the mainstream of medical ethics in the 1970s (241, 284, 363). Only recently have we seen signs of an important movement away from this individualism in the writings of Farley (212), Gilligan (261), Thomasma (681), Toulmin (686), Whitbeck (726), the President's Commission (576), and others (122). Ladd (362–364), Cassell (119, 121), and others (cf. Chapter 5) are among those who never embraced extreme individualism. I acknowledge the importance of the emphasis in the 1970s on individualism; among other things, this emphasis helped serve as a corrective to the highly paternalistic doctor-patient relationship (cf. Chapter 5). Now I am pleased to see philosophers and other writers adopting a new perspective on persons as beings who can exist only in relation to others and whose very being is defined by these relationships. One of these important relationships, which cannot be defined or described adequately solely in terms of rights and duties, is the doctor-patient relationship.

Having said this, I do not think the Commission's recommendations are obsolete. As I shall discuss in Chapter 5, it is usually quite appropriate to view investigator-subject relationships as relationships between strangers. Thus, in general, an individualistic ethic is appropriate. Also, in several chapters I shall discuss how the Commission's recommendations and the resulting federal regulations have been interpreted to satisfy the needs of complex relationships, such as those between physician-investigators and patient-subjects.

Let us now proceed with a consideration of ethics as envisioned by the Commission. Ethical discourse and reasoning may be conducted at various levels of abstraction and systematization. At the most superficial level, there are *judgments*. A judgment is a decision, verdict, or conclusion about a particular action. Particular judgments, if challenged, may be justified ethically by showing that they conform to one or more *rules* or *norms*. A rule is a more general statement that actions of a certain type ought (or ought not) to be done because they are right (or wrong). At an even more general and fundamental level, there are ethical *principles* that serve as foundations for the rules and norms. Finally, there are ethical *theories*; these consist of systematically related bodies of principles and rules.

In the practical world, discussions of ethics are usually conducted at the most superficial level acceptable to the participants in the discussion. In response to challenges or demands for justification, one moves to deeper levels of reasoning.

Here is an example.

Investigator I: "Before we treat Mr. Jones with investigational drug X, we tell him that drugs Y and Z are also likely to be effective for his condition." (Judgment)

Colleague C: Why *should* we tell him that?"

I: "Because the regulations state that we may not proceed without informed consent, of which one essential component is a disclosure of any appropriate alternative procedure that might be advantageous." (Rule)

C: "Why was that rule promulgated?"

I: "It is designed to be responsive to the principle of *respect for persons*, which requires that we treat individuals as autonomous agents. Persons cannot be fully autonomous without an awareness of the full range of options available to them." (Principle)

Readers who are interested in an excellent introduction to ethical theory are referred to Frankena (244) or, for a discussion of ethical theory as it relates to medicine and biomedical research, to Beauchamp and Childress (49). These books are selected because they present ethical theory compatible with my intepretation of the Commission. Considerations of ethical norms will be continued in Chapter 2 and examples of judgments are provided in most other sections. In this section we shall concentrate on the fundamental ethical principles.

The Commission defined a "basic ethical principle" as a "general judgment that serves as a basic justification for the many particular prescriptions for and evaluations of human actions'' (526, p. 4) By ''fundamental ethical principle,'' I mean a principle that, within a system of ethics, is taken as an ultimate foundation for any second-order principles, rules, and norms; a fundamental principle is not derived from any other statement of ethical values. The Commission has identified three fundamental ethical principles as particularly relevant to the ethics of research involving human subjects: *respect for persons, beneficence,* and *justice*. The norms and procedures presented in regulations and ethical codes are derived from and are intended to uphold these fundamental principles. As the Commission observed in an early draft of the Belmont Report (405):

> Reliance on these three fundamental underlying principles is consonant with the major traditions of Western ethical, political and theological thought presented in the pluralistic society of the United States, as well as being compatible with the results of an experimentally based scientific analysis of human behavior. . . .

Thus, in the view of the Commission, these principles pertain to human behavior in general; it is through the development of norms that they are made peculiarly relevant to specific classes of activities such as research and the practice of medicine.

The President's Commission also identified three basic principles that ''predominated'' in their analyses of the problems presented by medicine and research (576, pp. 66–71). These principles, respect, well-being, and equity correspond closely to the Commission's *respect for persons*, *beneficence*, and *justice*, respectively.

Respect for Persons

The principle of *respect for persons* was stated formally by Immanuel Kant: ''So act as to treat humanity, whether in thine own person or in that of any other, in every case as an end withal, never as a means only.'' However, what it means to treat a person as an end and not merely as a means to an end may be variously interpreted (378). The Commission concluded that (526):

> Respect for persons incorporates at least two basic ethical convictions: First, that individuals should be treated as autonomous agents, and second, that persons with diminished autonomy and thus in need of protection are entitled to such protections.

An autonomous person is ''. . . an individual capable of deliberation about personal goals and of acting under the direction of such deliberation'' (526). To show respect for autonomous persons requires that we leave them alone, even to the point of allowing them to choose activities that might be harmful (e.g., mountain climbing), unless they agree or consent that we may do otherwise. We are not to touch them or to encroach upon their private spaces unless such touching or encroachment is in accord with their wishes. Our actions should be designed to affirm their authority and enhance their capacity to be self-determining; we are not to obstruct their actions unless they are clearly detrimental to others. We show disrespect for autonomous persons when we either repudiate their considered judgments or deny them the freedom to act on those judgments in the absence of compelling reasons to do so.

Clearly, not every human being is capable of self-determination. The capacity for self-determination matures during a person's life; some lose this capacity partially or completely owing to illness or mental disability or in situations that severely restrict liberty, such as prisons. Respect for the immature or the incapacitated may require one to offer protection to them as they mature or while they are incapacitated.

Beneficence

The Commission observed that:

> The term, *beneficence*, is often understood to cover acts of kindness or charity that go beyond strict obligation. In this document, *beneficence* is understood in a stronger sense, as an obligation. Two general rules have been formulated as complementary expressions of beneficent actions in this sense: 1) Do no harm and 2) maximize possible benefits and minimize possible harms.

The principle of *beneficence* is firmly embedded in the ethical tradition of medicine. As Hippocrates observed in the *Epidemics*, "As to diseases, make a habit of two things—to help, or at least to do no harm" (332). Unfortunately, this is often oversimplified as "Do no harm (*primum non nocere*). Some authors point out that separation of these two obligations into two ethical principles, *beneficence* (do good) and *nonmaleficence* (do no harm), would tend to decrease confusion (49, 441).

Frankena identifies four obligations that derive from the principle of *beneficence* (244, p. 47); listed in decreasing order of ethical force, these are:

A. One ought not to inflict evil or harm.

B. One ought to prevent evil or harm.

C. One ought to remove evil.

D. One ought to do or promote good.

Statement A is a straightforward articulation of the principle of *nonmaleficence*. Few, if any, would argue that the injunction against inflicting harm or evil is not a very strong *prima facie* duty. Statement D, by contrast, is not regarded generally as a duty or obligation but rather, as the Commission pointed out, an exhortation to act kindly or charitably. Statement D becomes a duty in the strict sense when one consents or contracts to be bound by it. Physicians, for example, pledge themselves to act for the benefit of patients. Similarly, researchers who accept public support for their work assume a contractual obligation to promote good by contributing to the development of new knowledge. It is in this context that the Commission identified *beneficence* as a basic ethical principle.

The principle of *beneficence* is interpreted by the Commission as creating an obligation to secure the well-being of individuals and to develop information that will form the basis of our being better able to do so in the future (societal benefit, the goal of much research). However, in the interests of securing societal benefits, we should not intentionally injure any individual. Calling upon persons to agree to accept risks of injury in the interest of producing a benefit for themselves or for others is morally less offensive than taking an action that will certainly injure an identifiable individual. As I shall discuss in detail subsequently, research tends to present us with much more complex risk-benefit calculations than do most classes of medical practice.

Justice

Justice requires that we treat persons fairly and that we give each person what he or she is due or owed (47). *Justice* is either comparative or noncomparative. Comparative justice is concerned with determining what a person is due by weighing his or her claims against the competing claims of others. Noncomparative justice is concerned with identifying what persons are due without regard to the claims of others (e.g., never punish an innocent person). The concerns addressed by the Commission under the rubric of *justice* are exclusively concerns of distributive justice, a type of comparative justice. Consequently, in this book, unless specified otherwise, the term "justice" may be understood as meaning "distributive justice." The Commission also addressed the issue of compensatory justice; this is discussed in Chapter 6.

A formal statement of the principle of *comparative justice* is that generally attributed to Aristotle: "Equals ought to be treated equally and unequals unequally." *Distributive justice* is concerned with the distribution of scarce benefits where there is some competition for these benefits. If there is no scarcity, there is no need to consider just systems of distribution. *Distributive justice* is also concerned with the distribution of burdens, specifically when it is necessary to impose burdens on fewer than all members of a seemingly similar class of persons.

Justice requires a fair sharing of burdens and benefits; however, just what constitutes a fair sharing is a matter of considerable controversy (441). To determine who deserves to receive which benefits and which burdens, we must identify morally relevant criteria for distinguishing unequals. Various criteria have been proposed (552). Is it fair for persons to be treated differently on the basis of their needs? Their accomplishments? Their purchasing power? Their social worth? Their past record or future potential?

There are those who argue that the fairest distribution of burdens and benefits is precisely that which creates the most benefits for society at large. This is the classical utilitarian argument, which harmonizes the principles of *justice* and *beneficence* by stipulating that there is no conflict. To create good is to do *justice*; just institutions act so as to produce "the greatest good for the greatest number." The Commission rejected the utilitarian formulation because it does not accord either with Western concepts of the fundamental equality of persons (e.g., before the law) or with the very strong tradition that interprets fairness to require extra protection for those who are weaker, more vulnerable, or less advantaged than others. This latter interpretation is reflected in such disparate sources as the injunction in the Judeo-Christian tradition to protect widows and orphans, the Marxist dictum "from each according to ability; to each according to need," and, most recently, the contractual derivation of principles of *justice* of Rawls (588).

The Commission's interpretation of the requirements of *justice* is embodied in one of its statements on the relevance of this principle to the problem of selection of subjects (526, pp. 9–10):

> the selection of research subjects needs to be scrutinized in order to determine whether some classes (e.g., welfare patients, particular racial and ethnic minorities, or persons confined to institutions) are being systematically selected simply because of their easy availability, their compromised position, or their manipulability, rather than for reasons directly related to the problem being studied. Finally, whenever research supported by public funds leads to the development of therapeutic devices and procedures, justice demands both that these not provide advantages only to those who can afford them and that such research should not unduly involve persons from groups unlikely to be among the beneficiaries of subsequent applications of the research.

Chapter 2

Ethical Norms and Procedures

An ethical norm is a statement that actions of a certain type ought (or ought not) to be done. If reasons are supplied for these behavioral prescriptions (or proscriptions), they are that these acts are right (or wrong). Statements of ethical norms commonly include the words "should" or "ought," but in some cases there are stronger terms such as "must" or "forbidden." A typical statement of an ethical norm is: research should be conducted only by scientifically qualified persons. The behavior-prescribing statements contained in the various codes and regulations on research involving human subjects may be regarded as variants of five general ethical norms (441). There should be 1) good research design, 2) competent investigators, 3) a favorable balance of harm and benefit, 4) informed consent, and 5) equitable selection of subjects. In addition, a sixth general ethical norm has begun to appear in some guidelines; 6) there should be compensation for research-related injury. The purpose of ethical norms is to indicate how the requirements of the three fundamental ethical principles may be met in the conduct of research involving human subjects. These general ethical norms will be discussed in detail subsequently.

Because statements of the ethical norms in codes and regulations tend to be rather vague, they permit a variety of interpretations; it is difficult to know exactly how to apply them to particular cases. When faced with such uncertainty, it is generally helpful to look behind the norm to examine the fundamental ethical principle or principles it is intended to embody. Accordingly, in the discussion of each of the ethical norms, I shall call attention to the fundamental ethical principle or principles it is designed to serve. Examples of resolution of uncertainties in the application of norms to particular cases through referring to underlying principles are provided in most chapters of this book.

The codes and regulations contain, in addition to the ethical norms, descriptions of procedures that are to be followed to assure that investigators comply with the requirements of the ethical norms. The most important general procedural requirement that is relevant to all research involving human subjects conducted in the United States is review by an Institutional Review Board (IRB). Another general procedural requirement (general in that it is designed to assure compliance with all of the ethical norms) is concerned with the publication of the results of research that seems to have been done unethically; it is discussed in this chapter. Other specific procedural requirements such as documentation of informed consent are discussed in connection with the norms to which they are related.

Good Research Design

The experiment should be so designed and based on the results of animal experimentation and a knowledge of the natural history of the disease or other problem under study that the anticipated results will justify the performance of the experiment (Nuremberg 3).

Biomedical research involving human subjects must conform to generally accepted scientific principles and should be based on adequately performed laboratory and animal experimentation and on a thorough knowledge of the scientific literature (Helsinki I.1).

These are typical expressions of the generally accepted requirement that research must be sufficiently well designed to achieve its purposes; otherwise, it is not justified. The primary purpose of this norm is to uphold the principle of *beneficence*. If the research is not well designed, there will be no benefits; investigators who conduct badly designed research are not responsive to the obligation to do good or to develop knowledge that is sufficiently important to justify the expenditure of public funds, to impose upon human subjects risks of physical or psychological harm, and so on. The norm is also responsive to the principle of *respect for persons*. Persons who agree to participate in research as subjects are entitled to assume that something of value will come of their participation. Poorly designed research wastes the time of the subjects and frustrates their desire to participate in a meaningful activity.

Department of Health and Human Services (DHHS) regulations (Section 46.111a) charge the IRB to make determinations of "the importance of the knowledge that may reasonably be expected to result." Because determinations of the importance of the knowledge to be gained and the likelihood of gaining that knowledge require consideration of the adequacy of research design, this statement may be construed as a charge to the IRB to perform such consideration. In its report on IRBs, the Commission had recommended that DHHS make this obligation of the IRB more explicit:

Approval of such research shall be based upon affirmative determinations by the IRB that: The research methods are appropriate to the objectives of the research and the field of study . . . (Recommendation 4A).

In its reports on research involving those having limited capacity to consent, the Commission has generally charged the IRB with determining that all research proposals should be "scientifically sound." In its commentary on Recommendation 4A in the IRB report, it elaborates and interprets this requirement. It recognizes that equally rigorous standards are not suitable in all disciplines or necessarily appropriate for all research purposes. For example, not all research is designed to provide a definitive test of a hypothesis, and not all results need be subjected to statistical analyses of their significance.

It is customary for IRBs to perform rather superficial examinations of the scientific design of research protocols, particularly when IRB members are aware that the investigators are applying for funding from some agency outside their own institution. In most cases in which research is funded by the federal government or by a private philanthropy, such agencies provide a very careful review by a panel of experts in the field of specialization of the research. For example, at the National Institutes of Health (NIH), there are Initial Review Groups (IRGs), commonly known as Study Sections. Rigorous review by IRGs of the research design is, in general, the basis for their funding decisions. The IRB tends to review with much more care the scientific design of research proposals for which outside support is not being sought. The principle that seems to guide the IRB is that each protocol should have careful review to see to it that the requirements of each of the general ethical norms are given due consideration. However, when the IRB knows that review for compliance with one of the norms is being performed by another competent agency (often one having much greater competence than the IRB), it may assume that its obligations with regard to that norm are being discharged satisfactorily. In other cases in which such review may not be performed by some competent agency, the IRB tends to examine scientific design much more carefully, as for example research proposals for which outside support is not being sought and proposals to modify substantially ongoing research activities. I do not mean to say that the IRB does not review the scientific merits of research proposals. Rather, it is not competent to perform in-depth review with the same degree of expertise that is available to advise most funding agencies or the Food and Drug Administration (FDA).

FDA regulations require a statement from the sponsor of an "investigational drug" (Notice of Claimed Investigational Exemption for a New Drug; Form FD–1571) (IND) that includes a detailed description of planned investigations (Section 312.1a). If the description of the planned investigations seems inadequate, the FDA has the authority to forbid the sponsor to proceed. In addition, the FDA has published various documents that provide general considerations for the clinical evaluation of drugs, as well as guidelines for the clinical evaluation of various specific classes of drugs (222). The purpose of these guidelines is to provide assistance in the scientific design and ethical conduct of clinical trials.

The FDA has analogous regulations and guidelines covering the development of medical devices (230, 231). Instead of an IND, an IDE (Application for an Investigational Device Exemption) is required. Unlike the situation with drugs, however, an IDE is required only for "significant risk devices." The determination as to whether any article is a significant risk device is the duty of the IRB (21 CFR 812.65a). This determination is discussed further in Chapter 3.

In 1981, the FDA Commissioner requested information on whether IRBs "have the resources and are willing to assume additional responsibilities for the conduct of the early phases of clinical investigations." An excerpt from my response (422) will help explain my position on this point.

> Let us consider some of the specific types of data that are submitted with the IND and the expertise required to review them.
> 1. Extensive toxicology studies are required. Review of these data requires experts in toxicology; such expertise is not available to most IRBs.
> 2. Similarly, chemists are not routinely appointed to membership on IRBs. In most universities it should be possible to find suitable consultants in chemistry.
> 3. The expertise required to review preclinical pharmacology data may or may not be available within the institution. Of course, all medical schools have pharmacologists on their faculties. However, with increasing subspecialization within pharmacology, there may or may not be suitable consultants within the institution.
> 4. Review of the proposed clinical protocol and review of any prior work done abroad are required. Medical schools do, in general, have members of the faculty who are competent to perform these tasks. However, the best experts in the institution are likely to be those the sponsor has asked to become investigators in the particular protocol under review. Thus, in general, they would be disqualified from participating in the deliberations of the IRB. Suitable experts are available in most large medical schools to provide advice and criticism at a level compatible with traditional IRB performance. However, the very rigorous review that is necessary will—in many cases—be beyond the capacity of the IRB.
> Thus, the IRB probably will not be capable of providing adequate review to reach a determination that there is sound scientific design. . . .

Giammona and Glantz analyzed randomly selected protocols that had been approved by the IRB at the University of California, San Francisco (260). Of those protocols that resulted in published papers, 52% "included one or more statistical errors that involved results central to the paper's primary thrust. . . . Since virtually all the errors . . . involve elementary statistics covered in introductory statistics textbooks, gaining the knowledge . . . for Committee members to evaluate them properly does not present a difficult problem.'' They recommend that IRBs should require that investigators ''state a hypothesis, outline an adequate experimental design, select their sample size after at least considering statistical power, and propose appropriate statistical analysis for the data to be collected.''

At some institutions, there are separate research review committees charged with reviewing protocols for research design before these protocols are reviewed by the IRB.

Most ethical problems that are related to research design are presented by protocols in which the design is scientifically deficient. However, there is another set of problems connected with some types of research activities even when they are designed in accord with the prevailing standards of their scientific discipline. In these research activities, it is the design itself that is ethically problematic. Examples of such activities are randomized clinical trials and deceptive research strategies; these are discussed in Chapters 8 and 9, respectively. Another example is the Phase I drug study (defined in Chapter 1).

Phase I Drug Studies

The subjects of Phase I drug studies are most commonly normal volunteers. Proponents of this practice justify it on two grounds (58). The first relates to a perception of sick persons as vulnerable and therefore to be protected from harm. Thus, one must first learn the toxic levels of new drugs from studies done on the less vulnerable normal subjects before exposing sick persons to them. Secondly, sick persons commonly have impaired function of various organs that metabolize or excrete drugs and, in addition, commonly take medications that can alter the absorption, metabolism, excretion, and effects, both toxic and therapeutic, of the investigational drug. Thus, interpretation of the results of studies done in normal subjects is relatively uncomplicated. Blackwell has found that a majority of academic and industrial physicians in the United States think that the most suitable subjects for Phase I drug studies are normal volunteers (58).

The chief ethical objection to the use of normal subjects in Phase I drug studies is that it creates an unfavorable balance of harms and benefits for the individual subject. Although the risk of injury to these subjects is very small (Chapter 3), they have been selected so that there is no possibility whatever of any direct health-related benefit to them. Thus, the harm/benefit ratio is always infinite. If the subjects of these studies were patients recruited from the population for which the drug is intended, there would be much more favorable harm/benefit ratios.

The conduct of Phase I drug studies on normal subjects can also be challenged on scientific grounds. For example, Oates argues that studies should first be done on patients for whom the drug is intended to determine whether it is effective and, if so, the doses at which it is effective (548). If the drug is not effective, there is no need for any subject to assume the risks of Phase I drug studies; there is no need to know about the toxicity, metabolism, and bioavailability of the drug. If the drug is effective, knowledge of the doses that produce the therapeutic effect permits more intelligent design of Phase I drug studies; there is no need, for example, to develop data on the disposition and bioavailability of the drug at doses that will not be used clinically.

Azarnoff presents a scientific challenge to the second of the justifications provided by proponents of the current practice (31).

> The metabolism of drugs is so changed by numerous diseases that any transfer of results from healthy volunteers to patients may be dangerously misleading. Drug interactions also modify results and must be evaluated in both normal and patient-subjects.

Oates (548) and Hollister (316) also point out that some pharmacologic effects (both therapeutic and adverse) are highly specific for the disease state and cannot be anticipated from studies done in normal subjects. For example, the effects of many psychotherapeutic drugs in the populations for which they are intended cannot be predicted by observing their effects in persons not having these diseases (316).

Thus, there are important ethical and scientific reasons to question the practice of conducting Phase I drug studies in normal subjects. In its report, *Research Involving Prisoners,* the Commission " . . . suggests that Congress and the FDA consider the advisability of undertaking a study and evaluation to determine whether present requirements for Phase I drug testing in normal volunteers should be modified" (524, p. 12). In most countries other than the United States, the first humans to receive a new drug are patients, not normal volunteers (58).

Even in the United States, not all Phase I drug studies involve normal subjects. In the field of cancer chemotherapy, for example, because most of the drugs employed are so very toxic, Phase I studies usually involve as subjects patients with cancer. According to Lipsett, suitable subjects must have extensive metastatic disease as well as a reasonable life expectancy; they must be expected to live long enough to permit completion of the drug study (451). Further, they must have already tried all alternative "modalities that offer the possibility of success." At his institution, the NIH Clinical Center, prospective subjects are told that the chief purpose of the study is to find an effective dose of the drug. The expected toxic effects are discussed, and the hope of a favorable effect is stated. In some studies such as these, however, the likelihood of a favorable effect may be rather remote (6, 115). Lipsett concedes this point (452).

The President's Commission provides further discussion of the problems presented by Phase I drug studies in patients with cancer (574, pp. 41–43). They emphasize that it is "important that patients who are asked to participate in tests of new anticancer drugs not be misled about the likelihood (or remoteness) of any therapeutic benefit they might derive. . . ."

Competence of the Investigator

The experiment should be conducted only by scientifically qualified persons. The highest degree of skill and care should be required through all stages of the experiment of those who conduct or engage in the experiment (Nuremberg 8).

Biomedical research involving human subjects should be conducted only by scientifically qualified persons and under the supervision of a clinically competent medical person. The responsibility for the human subject must rest with a medically qualified person . . . (Helsinki I.3).

This norm requires that the investigators be competent in at least two respects. They should have adequate scientific training and skill to accomplish the purposes of the research. The purpose of this norm is precisely the same as that requiring good research design; it is responsive primarily to the obligations to produce benefits to society through the development of important knowledge. It is also responsive to the obligation to show respect for research subjects by not wasting their time or frustrating their wishes to participate in meaningful activities. In addition, investigators are expected to be sufficiently competent to care for the subject. The Declaration of Helsinki, as an instrument of the World Medical Association, is addressed only to medical research. Therefore, it places responsibility with " . . . a medically qualified person." The Nuremberg Code, on the other hand, is addressed more generally to research; consequently, it does not call for medical qualification.

Competence to care for the subjects of most clinical research requires that at least one member of the research team be responsible for observing the subject with a view toward early detection of adverse effects of his or her participation or other evidence that the subject should be excluded from the study. The investigator should have the competence to assess the symptoms, signs, and laboratory results of the subject. There should further be the competence to intervene as necessary in the interests of minimizing any harm, e.g., prompt administration of an antidote to a toxic substance.

Holder provides good practical advice on who should be permitted to perform medical acts for research purposes (310). In general, the institution's policy regarding medical practice should govern. Medical students, for example, may draw blood without direct supervision; however, direct supervision may be required for a postdoctoral fellow to perform endoscopy even when the fellow is a licensed physician. Moreover, because "at least in our institution the basic science laboratories are not in patient care areas, we . . . require any procedure in which there is a risk of harm . . . to be performed in the hospital where emergency equipment and adequate personnel are available. . . . This requirement would also be applied in the case of senior clinicians. . . .''

Regulations of DHHS and FDA do not clarify the role of the IRB in determinations of competence of the investigators. The Commission charges the IRB to make such determinations only in its report on those institutionalized as mentally infirm (Chapter 11). In general, IRBs do not engage in formal assessment of the competence of investigators. In most cases, investigators have been appointed to membership in the institution by virtue of demonstrated competence to do research in their fields. Moreover, agencies that provide financial support for research base their funding decisions on considerations of the competence of the investigators as well as the scientific merits of the proposed research.

Form FD–1571 required by the FDA (Section 312.1a) of sponsors of new drugs includes a general statement of "the scientific training and experiences considered appropriate by the sponsor to qualify the investigators as suitable experts to investigate the safety of the drug. . . . " In addition, sponsors are to provide the FDA with the " . . . names and a summary of the training and experience of each investigator. . . . " The sponsor is held accountable for involving only investigators who are " . . . qualified by scientific training and experience. . . . " The sponsor implements this obligation by having investigators in clinical pharmacology (Phase I and Phase II drug testing) and in clinical trials (Phase III drug testing) complete forms FD–1572 and FD–1573, respectively (Section 312.1).

The FDA specifies the following qualifications for investigators (222).

> Phase I studies . . . should ordinarily be performed by investigators skilled in initial evaluation of a variety of compounds for safety and pharmacologic effect.

> Phase II studies should be performed by investigators who are considered experts in the particular disease categories to be treated and/or in evaluation of drug effects on the disease process. Phase III studies may be performed by experts and/or experienced clinicians, depending upon the nature of the studies.

According to the FDA, when Phase I studies are conducted in patients with specific diseases, the qualifications of the investigators should be similar to those specified for Phase II studies.

Williams presents this provocative challenge to IRBs (733):

> The reluctance to evaluate scientific competence is compounded when it is recognized that a serious risk/benefit analysis would also require boards to evaluate the researcher's ethical competence. A researcher who manufacturers or alters data is unlikely to generate . . . benefits. More important but less obvious, an investigator who is morally insensitive—who ignores subjects' privacy or rides roughshod over subjects' autonomy—creates moral "risks" that are serious but difficult to gauge. The inad-

equacy of IRBs' policing of ongoing research necessitates reliance on the moral integrity of the investigator . . . (to keep the) promise to follow his or her protocol. Fortunately this reliance is usually justified. Nonetheless boards seldom, if ever, engage in formal assessment of the virtuousness of investigators. A thorough risk/ benefit analysis would require that they do so.

It is worth noting, however, that many IRBs take the investigator's "virtue" into account in making decisions about protocols. For example, Shannon and Ockene report on their IRB's disapproval of a low risk protocol based, in part, on the fact that the investigator had a poor relationship with the IRB (636). In the same paper, they report their IRB's approval of a high risk protocol; one important factor was that the investigator had "an excellent track record in terms of trustworthiness, exemplified by his willingness to report immediately any problem in research by notifying the appropriate people." Moreover, "within the medical center . . . he was perceived to be skilled and trustworthy."

To say that IRBs do consider the reputations of investigators as they review their protocols is not the same as saying that they should (435). On this point, Wagner concludes that her "IRB does consider the compassion, the credibility, and the conscientiousness of researchers. But to require IRBs to do so would be to acknowledge them as legislators of morality." This, she claims, is not the proper role of IRBs (703). Pattullo observes that IRBs are more likely to respond favorably to good reputations than they are to respond disapprovingly to bad reputations (562). He provides important practical reasons to disregard hearsay evidence about reputations.

Swazey and Scher have edited an excellent book entitled *Social Controls and the Medical Profession* (677). This book represents the results of meetings involving the authors and various advisors over a period of 2 years. Thus, the perspectives of the physicians, lawyers, clergy, sociologists, philosophers, and historians are informed by each other. The book focuses on how society identifies and deals with professional incompetence, particularly among physicians; it further provides perspectives on how these processes might be better accomplished.

Publication of Unethical Research

"Reports of experimentation not in accordance with the principles laid down in this Declaration should not be accepted for publication (Helsinki, I.8)."

Whether or not this principle should be adopted as editorial policy by scientific journals has been a matter of considerable controversy for many years. For example, in 1973, I published an extensive account of my reasons for opposing a similar policy that had been proposed by the Committee on Editorial Policy of the Council of Biology Editors (387). I suggested an alternate policy that in my view would achieve more effectively the same as well as other objectives. As a rebuttal to my proposal, DeBakey published an article that continues to stand as one of the best arguments in favor of the adoption of this principle by editors (178); subsequently, she agreed to a compromise (*infra*).

According to Brackbill and Hellegers (74):

> Most scientists are under great pressure to conduct research and publish it. Publication is the sole route to professional success, to salary increases, to tenure, to promotion. Scientists, therefore, regard the terms and conditions of publication as matters of considerable importance. There is no question that ethical review as a gate to publication is an effective means of maintaining ethical standards in research. It is also the most feasible method.

In their view, editors are most likely to be effective as ethical gatekeepers because they have the opportunity to review research after it has been conducted. Other groups assigned the obligation to review research with the aim of safeguarding the rights and welfare of human subjects, such as IRBs and various granting agencies, review proposals to do research before they are begun. Thus, because they generally perform no monitoring activities, they have no way to know whether the research is actually conducted as described in protocols and grant applications.

One of the earliest commentators on this issue was Beecher, who drew an analogy between the publication of unethically obtained data and the acceptance by a court of unconstitutionally obtained evidence (52). Woodford, speaking for the Council of Biology Editors, opined, "Publication in a reputable journal automatically implies that the editor and his reviewers condone the experimentation" (737). Ingelfinger, speaking on behalf of the *New England Journal of Medicine,* stated, "The *Journal* attempts to observe the policy outlined in an editorial by Woodford: reports of investigations performed unethically are not accepted for publication" (324). Although some other journals have determined to reject manuscripts describing research which, in the view of the editor, has been conducted unethically, many have taken no position on this matter. The Reports of the Commission and federal regulations make no mention of this issue.

In my opinion, the interests of all concerned would be better served if editors proceeded as follows. If in the judgment of the editor an article is scientifically sound but ethically questionable, the decision should be to publish the article along with an editorial in which the ethical questions ·are raised. The author should be notified that this is what is to be done and invited to prepare a rebuttal to be published simultaneously. In the following paragraphs I shall explore some of the expected consequences of this policy, referred to as the acceptance policy, as well as those of the policy required by the Declaration of Helsinki, referred to as the rejection policy.

One inevitable harmful consequence of the rejection policy is that it would deprive the scientific community of valuable new information. The work described in most papers that would be rejected on ethical grounds could be repeated easily under acceptable ethical conditions. One would then be obliged to question the ethics of exposing new subjects to risks in order to obtain information that was already available. Even in the few cases in which it might be impossible to obtain the information ethically, I would make this information, occasionally obtained at very high cost, available to the scientific community.

Moe has accepted this argument even to the extent of justifying the use of data derived from the notorious research conducted by Nazi physicians (508): "A decision to use the data should not be made without regret or without acknowledging the incomprehensible horror that produced them. We cannot imply any approval of the methods. Nor, however, should we let the inhumanity of the experiments blind us to the possibility that some good may be salvaged from the ashes."

The obvious motive of the rejection policy is to discourage the performance of unethical research. The reasoning of its proponents is that if investigators cannot publish the results of unethical research, they will refrain from doing it. If and when all investigators refrain from doing unethical research, it will not be necessary to have policies as to what to do with scientifically sound but ethically questionable manuscripts; there will be none. Meanwhile, denial of publication to such papers will create the false illusion that no unethical research is being done. We would in effect be sweeping these matters under the rug.

Many leading medical journals customarily invite commentary and controversy on the papers they publish. Most commonly, the controversy is concerned with the scientific methods employed or the conclusions reached by the authors. Such material is ordinarily found in the editorial or correspondence sections of these journals. The prevailing view is that all participants in and readers of the material profit by these published exchanges. The same sort of approach to public discussion of controversial ethical issues was recommended by Katz (340); his important book, *Experimentation with Human Beings* (341), published 6 years later, consists largely of a compilation of such published exchanges. Because I believe that this approach—the acceptance policy—has proved more productive than silence, I encourage its increased use in the future.

A minority of investigators are those categorized by Barber et al. as "permissive" (37); they are likely to have the lowest standards of ethics in relation to their research. These individuals are likely to be the relative failures in their profession who perceive themselves as underrewarded. They tend to be mass producers in that they publish a large number of papers and find that they are cited infrequently by their colleagues. Individuals meeting the description of permissive investigators are very likely to yield to the sorts of pressures that can be brought to bear by public exposition and criticism of their questionable ethics.

In the course of implementing the acceptance policy, what might happen at the point at which an author is notified that his or her paper has been accepted, that it is to be published along with an editorial in which the ethical aspects of the work will be challenged, and that he or she is invited to prepare a rebuttal for simultaneous publication? There are several possibilities. Perhaps the author will be able to provide a suitable rebuttal and show that the issue is one on which reasonable individuals might disagree. A good example of this may be found in the published discussion of the ethics involved in the studies conducted at the Willowbrook State School in which mentally retarded children were deliberately infected with hepatitis (cf. Chapter 4).

A second possibility is that the author might be either unable or unwilling to prepare a rebuttal. In this case, the editorial would appear unanswered; all concerned, including the author, could profit by exposition of the ethical deficiency and conduct their future research alerted not to repeat that particular error. Most authors will not wish to be so exposed and probably will take great pains to avoid recurrence of such an experience.

In order to avoid such public exposition, some authors might choose to withdraw their manuscripts. What then? The author might be able to shop around for a permissive journal. One wonders about the responsibility of editors to comment on papers that are unacceptable on both scientific and ethical grounds.

One fundamental question that pervades all of these considerations is that of who actually benefits from the publication of a scientific paper. Most proponents of the rejection policy seem to proceed from the assumption that the principal beneficiary of publication is the author. However, the scientific community and the public are also major beneficiaries of the availability of new information. Through the acceptance policy, it seems possible to protect the interests of society while minimizing the benefit to the author. Specifically, in cases in which the author would rather withdraw a paper than be criticized editorially, the paper could be published without naming either the author or the institution at which the research was performed. Thus, society could get its benefit, access to valuable new information, as well as the additional benefit of exposition of an ethically questionable research maneuver. This approach also appears to offer a direct attack on the permissive investigator in the area of his or her greatest concern.

Members of the Committee on Editorial Policy of the Council of Biology Editors continued their debate on the pros and cons of acceptance and rejection policies for several years after they proposed the latter. After they ceased to function as representatives of the Council of Biology Editors, they published a book in which they recommended a compromise (179, pp. 29–34). They leave it to the discretion of the editor to choose between the acceptance and rejection policies. They further provide recommendations for procedures that the editor may follow in order to form judgments as to whether any particular manuscript describes research that has been conducted unethically.

Brackbill and Hellegers used a questionnaire to survey the editorial policies and practices of major medical journals (74). They found that a majority did not "instruct reviewing editors to judge manuscripts on the basis of ethics as well as substantive material, methodology, and style" or "require authors to submit (evidence of) IRB approval along with their manuscripts." On the other hand, a majority agreed that "the most effective way to eliminate ethical violations in research is for editors to refuse to publish articles based on research in which there are clear ethical violations" and that it is "unethical for an editor to publish unethical research." They also found that a majority of published articles did not mention either informed consent or the use of volunteers. In the same issue of the *Hastings Center Report,* several editors comment on the findings and recommendations of Brackbill and Hellegers; they provide various reasons for not having their editorial boards engage routinely in systematic review of submitted manuscripts for conformance with ethical standards. Yankauer provides examples of how his journal, *American Journal of Public Health,* commonly publishes editorials that address the ethical questions raised by manuscripts published in that journal (743).

It might interest the reader to note, in passing, how many illustrative examples in this book are drawn from these published exchanges in the *American Journal of Public Health.* For further discussions of this issue, see Veatch (695), Holden (296), and Yankauer (744).

Academic Misconduct

About 25 years ago, as a fledgling scientist at NIH, I was accorded a great honor by my laboratory chief; I was put in charge of supervising a post-doctoral fellow. Although he and I carefully discussed our plans for each experiment after detailed review of the results of its predecessors, there were no experiments that I did with my own hands. The paper we published was well-received by our colleagues as a moderately important contribution to our field of interest.

Shortly thereafter, he returned to his own country; I left NIH and started my own laboratory at Yale. My first major project entailed pursuing the physiological ramifications of the same line of pharmacologic research. Nothing worked. After about 1 month of unremitting failure, I decided to repeat the same experiments we had "done together" at NIH. Not only could I not get the same results, I could demonstrate no effects whatever. I was careful to use rats of the same strain (Sprague-Dawley), sex, and size. I tried varying temperatures, lighting conditions, diet, and the like. In short, I tried everything I could think of, but there were no effects whatever of the drugs that had worked so well at NIH.

In the 3 months it took to try to replicate this work, my initial anxiety gave way to desperation that, in turn, became terror.

Apparently, I had become an accomplice to fraud. I knew then, as all scientists knew then, and as the public has learned recently, that fraud is the most heinous sin against the values of science. It is a capital offense. Those who are caught at fraud are excommunicated. After exile from the scientific community, the establishment goes as far as it can to purge the record of traces of the sinner's transient membership; e.g., there are formal retractions of their scientific papers (595, 642).

In those days, the punishment was so swift and certain that most scientists thought that anyone who faked data had to be crazy. Indeed, as with other capital crimes, a plea of temporary insanity resulted occasionally in a therapeutic rather than a punitive response from the scientific community.

"Fraud," in the sense I have been using the term, means the deliberate reporting in the scientific literature or at scientific meetings of "facts" that the reporter knows are unsubstantiated; he or she may not even have done the experiments (forgery). This is the capital offense. The scientific community has not yet reached consensus on the status as felonies or misdemeanors of some of its variants.

"Plagiarism" is perceived as a serious offense when one presents a large body of another's ideas as one's own, as in publishing another's paper and pretending it is one's own. Copying an occasional sentence or paragraph is common and may be treated as serious or minor depending upon who copies what and for what purposes (86).

Trimming data "consists of clipping off little bits here and there from those observations which differ most" from the average with the aim of making the results look better (86, p. 30). It has been alleged that many scientists routinely trim all results that are three or more standard deviations from the mean. I hope that this allegation is untrue. In my view, scientists who do such trimming are obliged to disclose this in their publications. Others may excuse discarding of an unwanted result by saying, "that rat looked funny to me," or "I think I recall having sneezed into that beaker."

"Selective reporting" consists of presenting only those observations that support the point the scientist wishes to establish (86, p. 30). This practice approximates outright forgery in seriousness. When such selected data are subjected to statistical analysis, as they often are, it compounds the mockery of the values of science.

For an excellent and lucid book on fraud in science, its history, its motivations, a comprehensive account of known instances through 1982, and the reactions of the public and its policy makers, see Broad and Wade (86).

Knowing the penalties for fraud, why do some scientists fake data? Knight provides a sensitive analysis of this question (355). With most commentators, he identifies the academic pressures to publish or perish as the leading reason. One must produce in order to get grants, promotions, and recognition from colleagues. This reminds me of the dénouement of the personal anecdote with which I began this discussion.

Recall, please, that I was terrified. I contacted a friend at NIH and asked him to send me 20 Sprague-Dawley rats from the same colony we had used at NIH. Experiments repeated on those rats replicated almost exactly the results I had published with my post-doctoral fellow. This resulted in another publication in which I called attention to the debate among scientists as to whether the rat was a suitable model for predicting pharmacologic and physiologic responses in humans. I had now evidence that the rat was not necessarily even a suitable model for other rats, even those of the same strain (386). This was, to be sure, a minor contribution to the scientific literature. As an assistant professor, however, I was grateful for any legitimate opportunity to enlarge my list of publications.

My only confidant during those awful 3 months was my loyal and proficient technician, who had left NIH to continue to work with me at Yale. Suppose she had been what we now call a whistleblower, with either sincere or malicious motivation. In those days, if she could have convinced the department chairman that I had faked data, I would have been banished suddenly and silently from the world of science. As a physician, I could have gone into medical practice as some data-fakers have (86). These days such matters are handled much more bureaucratically.

By now all universities have or should have established policies and procedures for dealing with instances of academic misconduct. Appendix C of the President's Commission's report, *Implementing Human Research Regulations* (574), presents the policy statements of the Association of American Medical Colleges and of Yale University. Institutional policies must be sensitive to the interests of the accused, the whistleblower, the scientific community, and the public. For a rich discussion of these matters, see *Whistleblowing in Biomedical Research* (573); this volume consists of the proceedings of a workshop sponsored by the President's Commission, the American Association for the Advancement of Science, and Medicine in the Public Interest. The group achieved consensus on four recommendations (573, pp. 205–207).

1. *IRBs should not be expected to perform monitoring, investigative, or adjudicative functions.* Applicable regulations should be clarified as to what is intended (and not

intended) by the charge to IRBs to perform "continuing review" and to report serious and continuing noncompliance. Reasons given in support of this recommendation include the fact that IRBs do not have the time, the resources (staff, money), or the expertise to perform such functions. In addition, adoption of the monitoring role would conflict with the primary role of IRBs: to educate and advise research scientists and to resolve problems in a constructive way. Finally, many, if not most, institutions already have appropriate quality assurance mechanisms in place. . . .

IRBs should be kept informed of all allegations of misconduct in research with human subjects and of investigations, as well as findings, relating to such allegations. Perhaps an IRB member should sit on the institutional quality assurance committee. The IRB might also be consulted as to the seriousness of the misconduct found to have occurred.

2. *Institutions receiving federal research grants and contracts should be required as part of the assurance process to describe to the funding agency their procedures for responding to reports of misconduct.* This should include: (a) A specific office designated to receive and investigate complaints; (b) mechanisms for assuring a prompt investigation; (c) an impartial adjudicator; (d) full opportunity for the complaining parties and the accused to explain their positions, present evidence, call witnesses, and so on; and (e) protection from reprisals for the good-faith complainant and for witnesses. . . .

3. *Institutional administrators, principal investigators, and research personnel should be made aware of their responsibilities to the scientific community and to federal agencies.* . . .

Serious misconduct should be reported to the cognizant federal agency after a formal determination has been made. Administrators and scientists should understand that they have a legal obligation to do so. In fact, knowingly to provide false information to the federal government is a felony. If an institution makes a formal finding that false information has been submitted in a grant application, annual report, or data submitted to a regulatory agency, the institution may incur criminal liability if officials fail to report such a finding. . . .

4. *Federal agencies should respond to reports they receive in a consistent, fair, and timely manner; final determinations that misconduct has occurred should be made known to other federal agencies, state licensing boards, and appropriate professional societies.* . . .

Academic misconduct in research involving federal funding or FDA-regulated test articles must be reported to the appropriate Federal agency. The question is when. Federal officials would like to learn of them early (495, 542). Most academicians prefer to delay making a "federal case" of such matters until the evidence supporting the charge of misconduct is quite secure (729).

For an authoritative survey of NIH's policies and experiences in investigating misconduct, see Miers (495). "NIH relies primarily on awardee institutions to investigate, take actions when warranted, and provide agency staff with the information needed to make reasonable and equitable decisions regarding awards and pending applications."

By contrast, the FDA conducts inspections not only for cause but also routinely. For a careful assessment of the FDA's activities in this regard, see Shapiro and Charrow (638). In my judgment, their data warrant their conclusion: "As long as the problem continues to exist, the FDA should continue or even expand its monitoring program, while ensuring its rigor." This judgment is not inconsistent with my view that all agents involved in safeguarding the rights and welfare of human research subjects should presume that all other agents are trustworthy and that they are performing competently until substantial contrary evidence is brought forward (cf. Chapter 14). Routine audits should be conducted as if investigators are to be trusted. For cause inspections should be conducted differently because they are based upon the availability of substantial contrary evidence.

Chapter 3

Balance of Harms and Benefits

The terms "risk" and "benefit" clearly are not parallel constructions. Risk entails prediction of some future occurrence of harm. Risk may be expressed in terms of probability that a certain harm will occur. The harm or injury itself may be evaluated quantitatively; e.g., it may be described as either a large or small harm. The meaning of such constructions as "small risk" is unclear; it might mean either a small probability of an unspecified amount of harm or an unspecified probability of a small amount of harm.

By contrast, the term "benefit" has no intrinsic connotations of prediction or probability. Benefit denotes something of value that can be supplied upon demand or as one wishes. It is of interest that an antonym of benefit, injury, is the phenomenon the probability of which we are stating when we discuss risk. It should be clear that, when discussing the benefits of research, one is ordinarily discussing the probability of hoped-for benefits.

The common use of such shorthand expressions as "risk-benefit analysis" has unfortunate practical consequences (408). The scholar concerned with the exact meaning of words finds the expression "risk-benefit" offensive because it is dysmorphic; it seems to equate or make parallel things that really are not. Even more perilous is the effect of such constructions on the thought processes of decision-makers. As they are determining the appropriateness of any particular research proposal, they may think that risks can be expressed in the same terms (e.g., units of measurement) as benefits and thus be led to ill-founded decisions.

The use of correct language tends to minimize confusion; risk is parallel to probability of benefit and benefit is parallel to harm.

The normative statements in various codes and regulations require a favorable balance between harm and benefit. Without such a favorable balance there is no justification for beginning the research.

> The degree of risk to be taken should never exceed that determined by the humanitarian importance of the problem to be solved by the experiment (Nuremberg 6).

> Biomedical research involving human subjects cannot legitimately be carried out unless the importance of the objective is in proportion to the inherent risk to the subject (Helsinki I.4).

> Risks to the subjects (must be) reasonable in relation to anticipated benefits, if any, to subjects, and the importance of the knowledge that may reasonably be expected to result (DHHS, Section 46.111a).

There are additional norms in codes and regulations that call for vigilance on the part of those conducting or supervising the research. At any point along the way, the balance of harms and benefits may become unfavorable; under these circumstances the research should be terminated.

> During the course of the experiment the scientist in charge must be prepared to terminate the experiment at any stage, if he has probable cause to believe . . . that a continuation . . . is likely to result in injury, disability or death to the experimental subject (Nuremberg 10).

> The investigator . . . should discontinue the research if in his/her . . . judgment it may, if continued, be harmful to the individual (Helsinki III.3).

> Where appropriate, the research plan makes adequate provision for monitoring the data collected to insure the safety of the subjects (DHHS, Section 46.111a).

> An IRB shall have authority to suspend or terminate approval of research . . . that has been associated with unexpected serious harm to subjects (DHHS, Section 46.113).

The requirement that research be justified on the basis of a favorable balance of harms and benefits is derived primarily from the ethical principle of *beneficence*. In addition, a thorough and accurate compilation of the risks and hoped-for benefits of a research proposal also facilitates responsiveness to the requirements of the principles of *respect for persons* and *justice*. A clear and accurate presentation of risks and benefits is necessary in the negotiations with the subject for informed consent. Similarly, such a compilation of burdens and benefits facilitates discussions of how they might be distributed equitably.

Risks of Research in Perspective

Much of the literature on the ethics of research involving human subjects reflects the widely held and, until recently, unexamined assumption that playing the role of research subject is a highly perilous business. This assumption is shown clearly in the legislative history of the Act that created the Commission (347). Early in the course of its deliberations, because the Commission considered the role of research subject a hazardous occupation, it called upon M. W. Wartofsky, a philosopher, to analyze the distinctions between this and other hazardous occupations (711).

Biomedical researchers have contributed importantly to this incorrect belief (397, 398). For example, they have often stated that accepting the role of subject nearly always entails the assumption of some risk of either physical or psychological harm or, at least, inconvenience. To many members of the public and to many commentators on research involving human subjects who are not themselves researchers, the word "risk" seems to carry the implication that there is a possibility of some dreadful consequence; this is made to seem even more terrifying when it is acknowledged that, in some cases, the very nature of this dreadful consequence cannot be anticipated. And yet, it is so much more common that, when biomedical researchers discuss risk, they mean a possibility that there might be something like a bruise after a venipuncture.

Recently, some empirical data have become available that indicate that, in general, it is not particularly hazardous to be a research subject. For example, Arnold has estimated the risks of physical or psychological harm to subjects in Phase I drug testing (cf. Chapter 2) (26). According to his estimates, the occupational hazards of the role of subject in this type of research are slightly greater than those of being an office secretary, one-seventh those of window washers, and one-ninth those of miners. Zarafonetis et al. found that, in Phase I drug testing in prisoners, a "clinically significant medical event" occurred once every 26.3 years of individual subject exposure (748). In 805 protocols involving 29,162 prisoner subjects over 614,534 days, there were 58 adverse drug reactions, of which none produced death or permanent disability. The only subject who died did so while receiving a placebo. Cardon et al. reported the results of their large scale survey of investigators designed to determine the incidence of injuries to research subjects (116). They found that in "nontherapeutic research," the risk of being disabled either temporarily or permanently was substantially less than that of being similarly harmed in an accident. None of the nearly 100,000 subjects of "nontherapeutic research" died. The risks of being a subject in "therapeutic research" were substantially higher. However, the risk of either disability (temporary or permanent) or of fatality was substantially less than the risk of similar unfortunate outcomes in similar medical settings involving no research.

Even more recently, at the request of the President's Commission for the Study of Ethical Problems in Medicine and Biomedical and Behavioral Research, representatives of three large institutions that had established programs for compensation of subjects for research-related injury summarized their experiences (27, 72, 475). Each reported a very low incidence of injury to research subjects and an extremely low rate of injuries that could be attributed directly to the performance of research. For example, Boström reported that in 157 protocols involving 8,201 subjects, there were only 3 "adverse effects;" of these, 2 were headaches following spinal taps (72). The other was pneumonia in a study on the systemic effects of estrogen and progesterone that included an intravenous adrenocorticotropin infusion. In my view, it is highly probable that the pneumonia was unrelated to the protocol. McCann and Pettit reported that involvement of 306,000 subjects over a period of 8 years resulted in only 13 insurance claims (475). The magnitude of the injuries is reflected in the size of the awards to claimants: 7 awards were for $54 or less, 4 awards were for more than $410, and the largest was for $1,550.

On the basis of all the empirical evidence of which I am aware, it seems proper to conclude that the role of research subject is not particularly hazardous in general. It follows that attempts to portray it as such and arguments for policies designed to restrict research generally because it is hazardous are without warrant. Equally insupportable are arguments that, because research is generally safe, there is no need for any restriction. It is essential to recognize that there are some procedures that are performed in the conduct of some research protocols that present risks of various types of injuries to research subjects. I believe that the most important reason that the record is so good and that there have been so few injuries is that most researchers are keenly aware of the potential for injury and take great care to avoid it.

Mere Inconvenience

In considering the burdens imposed upon the research subject, it is of value to distinguish risk of physical or psychological injury from various phenomena for which more fitting terms are "inconvenience," "discomfort," "embarrassment," and so on; "mere inconvenience" is a general term that may be used to discuss these phenomena (398). Research presenting mere inconvenience is characterized as presenting no greater risk of consequential injury to the subject than that inherent in his or her particular life situation. In determining the risks inherent in a prospective subject's life situation, it is not appropriate to consider risks irrelevant to the design or purpose of the research; for example, in considerations of the risks of biomedical research it is inappropriate to enter into the calculations the risks of being injured in an automobile accident. On this point, I agree with Kopelman (359); although she discusses the "minimal risk" standard of the regulations (*infra*) and not "mere inconvenience," she shows quite clearly the irrelevance of considering risks assumed by individuals in the course of their lives apart from participation in research. If one considered all the risks inherent in each individual's life, it would be necessary to develop different standards for firefighters, gardeners, and persons with agoraphobia. There would, of course, be no standard.

The vast majority of research proposals present a burden that is more correctly described as mere inconvenience than as risk of physical or psychological harm. In general, prospective subjects are asked to give their time (e.g., to reside in a clinical research center, to be observed in a physiology laboratory, or to complete a questionnaire); often there is a request to draw some blood or to collect urine or feces. Although the withdrawal of venous blood may be momentarily painful and be followed by a bruise, no lasting physical harm is done. Removal of some other normal body fluids may be associated with risk of substantial harm, e.g., fluid from the cerebral ventricles or blood from the heart. Ordinarily, such studies can be accomplished using individuals who require removal of heart blood or brain fluid for diagnostic or therapeutic purposes. In these cases, the fact that some of the removed fluid is used for research purposes imposes no additional risk or inconvenience on the subject. Removal of abnormal fluids or tissues can only be done on those who have diseases associated with the development of these abnormal fluids (e.g., pleural effusions) or tissues (e.g., tumors). Again, in such persons, the interests of research and practice usually can be served simultaneously. Removal of normal body tissues may or may not present a risk of physical harm or inconvenience. Thus, removal of a piece of skin by standard biopsy procedures is associated with minor discomfort lasting about 7 to 10 days in suitably selected individuals. Whether the tiny scar that results constitutes an injury or an inconvenience can be determined only by asking the prospective subject. On the other hand, biopsy of the liver is associated with a probability of approximately 0.002–0.005 of complications sufficiently serious to require treatment (e.g., blood transfusion) and of approximately 0.0002 of death. Thus, the performance of a liver biopsy must be considered as presenting risk of physical harm, not mere inconvenience.

In a person who has a tumor for which the administration of methotrexate is indicated, studies on the metabolism of methotrexate present mere inconvenience. The research maneuvers (e.g., drawing of blood or collection of urine) present no risk of physical injury. On the other hand, performing the same studies on an individual for whom this toxic drug is not indicated presents a high probability of serious physical harm. For another example of converting risk of injury to mere inconvenience, see Blackburn et al (57).

Some normal tissues are obtained as by-products of indicated surgery. It is customary surgical practice to remove a margin of normal tissue around the diseased tissue to assure complete removal of a tumor or of an infection. Thus, it is ordinarily possible to secure specimens of most normal tissues without causing even inconvenience to the individual (314).

Classification of Risks and Benefits

In considering the risks and hoped-for benefits of a research protocol, it serves the interests of thoroughness to do so according to a system. The taxonomy of risks and benefits presented here is intended to be sufficiently comprehensive to suggest the dimensions of such systematic considerations; it is a substantially abridged version of two long papers I wrote for the Commission (398, 408).

Risks to Subjects

Risks are classified as physical, psychological, social, and economic. Although some types of risks are generally thought to be linked more or less peculiarly to specific classes of research, on closer examination some of these presuppositions prove incorrect. For example, threats to privacy, most commonly considered in relation to social research, may be presented when left-over surgical specimens are used for research purposes (cf., Chapter 7).

Physical Risks   When the conduct of research presents risks of physical injury, there is commonly little or no difficulty either in identifying or in providing a reasonably adequate description of the risks. This is particularly true when the procedure performed for research purposes has been tested sufficiently to develop adequate data on the nature, probability, and magnitude of the injuries it may cause (*infra*). In some circumstances, however, both identification and description may be problematic. Although I shall draw primarily upon the field of drug development for illustrative examples, there are analogies in most other categories of developmental research.

In the earliest phases of drug development, the nature of some injuries that may be produced by drugs is unknown. Ordinarily, there is some background of experience derived from research on animals that will help predict with varying degrees of confidence what the risks might be to humans. However, it must be understood clearly that one never knows what the adverse (or, for that matter, beneficial) effects of any intervention in humans will be until the intervention has been tested adequately in humans. This statement may be extended further. The effects of any intervention in any particular class of humans cannot be known with certainty until the intervention has been tested adequately in that particular class of humans. For example, some drugs that are acceptably safe for adult males produce severe adverse effects when they are administered to pregnant women or infants (cf., Chapter 10). For another example, it is through adverse reactions to various interventions that we have discovered the very existence of some "classes" of humans. It was through thorough investigation of certain individuals with drug-induced hemolytic anemia that we learned of the existence of glucose 6-phosphate dehydrogenase deficiency. Similarly, through adverse reactions to succinylcholine, we learned of plasma cholinesterase deficiency. These two conditions are genetically determined enzyme deficiency states that have no clinical manifestations unless the afflicted individual is exposed to some drug or other chemical that precipitates reactions peculiar to these individuals. The reactions may be severe or even lethal.

In some cases, physical risks might be totally unknown while an intervention is in its investigational phases, owing to a long period of latency between the intervention and the development of the adverse effect. For example, the fact that some forms of chemotherapy and radiation therapy used in the treatment of cancer could themselves cause other types of cancer was not know until years after their introduction. For another example, it took many years to learn that the administration of diethylstilbestrol (DES) to pregnant women to prevent spontaneous abortion could result in the development of a rare form of cancer of the vagina in their daughters. Similarly, human growth hormone was administered to children for many years before it was implicated in the transmission of Jakob-Creutzfeldt disease.

In the early phases of testing nonvalidated practices, although the nature of their risks may be reasonably well-known, there is relative uncertainty of their probability and magnitude. When they are used in the management of diseases for which there are validated alternatives, there is yet another problem: for the duration of the use of the nonvalidated modality, the patient-subject is deprived of the benefit of the validated alternative. Although an investigational drug may prove superior to its validated alternative drug, this cannot be assumed during the course of its testing. The adverse effects of the investigational drug may prove to be either more or less severe than those of the standard.

In most cases, if the nonvalidated therapy fails, it is possible to revert to standard therapy for the same condition. However, some therapies by their very nature preclude alternatives. For example, in some diseases there is an ever-present possibility of the sudden occurrence of some permanent complication. Thus, during the course of administration of a nonvalidated antihypertensive drug, it may be necessary to withhold known effective therapy. Should the new drug fail to control the blood pressure, the patient is vulnerable to the sudden onset of a serious and irreversible complication, e.g., cerebral hemorrhage (436).

Some other therapies are designed to ablate permanently some tissue or organ by either surgery or radiation. Should the therapy fail, the tissue or organ remains ablated, usually with corresponding loss of function.

Closely related problems are presented by proposals to withhold therapy either for purposes of doing placebo-controlled studies or for the purpose of making physiologic measurements in patients with various diseases without the influence of drugs. Problems presented by proposals to withhold therapy to conduct placebo-controlled clinical trials are discussed in Chapter 8. Houston provides a detailed account of the risks of abrupt discontinuation of antihypertensive therapy for research purposes (318). Weintraub discusses the perils of discontinuing effective therapy in patients with rheumatoid arthritis for purposes of testing new therapies (719); he illustrates that these hazards can be largely mitigated by using an add-on study design in which patient-subjects are permitted to continue taking their routine medications. They are randomly assigned to receive either placebo or the investigational drug in addition to their routine medications.

A subject in a double-blind study may become ill and require the emergency services of a physician who is unfamiliar with the study. It may be impossible for that physician to implement rational treatment without knowledge of the drugs the subject is taking. Mechanisms to assure immediate access in emergencies to the codes for double-blind studies are discussed in the section on minimization of risk.

Some physical risks may be presented to persons other than the subject. For example, inoculation of children with rubella (German measles) vaccine may result in inadvertent transmission of the virus to a pregnant relative, neighbor, or teacher. This, if the pregnancy is in the first trimester, could result in the birth of a baby with serious deformities (300).

Proposals to begin gene therapy in humans have evoked expressions of great concern from scientists, religious leaders, and others; in part, this concern is due to profound uncertainties about the nature, probability, and magnitude of the risks. For authoritative surveys of the problem presented by such proposals, see the President's Commission's report, *Splicing Life* (572), and a more recent article by Grobstein and Flower (275). Plans to proceed with gene therapy with federal support must be reviewed and approved by the Recombinant DNA Advisory Committee (RAC) of the National Institutes of Health (NIH). The RAC's Working Group on Human Gene Therapy has developed guidelines entitled *Points to Consider in the Design and Submission of Human Somatic-Cell Gene Therapy Protocols* (192). These are published from time to time in the *Federal Register* and then revised in response to public comment (383). No doubt they will continue to evolve as the RAC considers specific proposals. Thus far the guidelines are addressed only to proposals to insert genes into somatic cells where, as far as we know, the risks are presented only to the individual who receives the therapy. The development of guidelines for germ line cells will be much more problematic (275, 572); in this case induced defects, if any, could be passed to succeeding generations.

Psychological Risks   In some types of research, threats to the psychological integrity of the subject are obvious and should be anticipated readily. For example, administration of hallucinogens and psychotomimetics may be expected to produce serious behavioral aberrations that may be prolonged or recurrent. For another example, in studies by Bressler et al., traumatic neurosis was induced deliberately by placing normal subjects in an environment of sensory deprivation (341, pp. 365–369). Some of the subjects manifested lingering disturbances after these experiments were terminated. For yet another example the reader is referred to the controversy precipitated by the report of Zimbardo et al. on the causal relationship between deafness and paranoia (445, 753). The subjects, college students, were hypnotized and given a post-hypnotic suggestion that they would be partially deaf. As anticipated, these subjects became paranoid. According to Zimbardo et al., all manifestations of paranoia were removed successfully by hypnotizing the subjects once again following the experiment (753). Lewis contends, however, that "we do not know enough in developmental psychiatry to be confident that an episode of induced paranoia will have no sequelae" (445).

Some other easily anticipated psychological injuries, particularly those associated with the use of deception, are discussed in Chapter 9. Here I shall concentrate on some psychological risks that may be less readily anticipated.

Many investigators prescreen prospective subjects to see if they are "fit" for participation in the research. Some of the effects of this prescreening for psychological studies may be quite traumatic. For example, in one study prospective subjects were informed that the investigators wanted to show a stress-provoking movie to "normal" individuals to determine their physiological reactions to stress. They were further informed that prior to viewing the movie they would have some standard psychological screening tests. After these tests, some were informed that the investigators had decided not to show them the movie, based on the results of the tests. One can imagine the reaction of these prospective subjects who by implication were informed that they had been found psychologically unfit to see a movie (unpublished case).

The phenomenon of rejection based upon failure to pass prescreening tests for normality is not limited to psychological research. Many persons who assume they are healthy learn through such tests that they have chronic diseases. Those who are expecting to purchase life insurance may consider this a harm, while some others may consider this a benefit.

Many nonvalidated therapies have proved, over the years, to produce serious psychological injuries, some of which were reversible and some of which were not. For example, it was not anticipated that the administration of reserpine for purposes of treating hypertension would in a significant number of individuals result in a rather precipitous development of severe, and often agitated, depression; some of these individuals proceeded to commit suicide before their physicians were even aware that they were depressed.

Calling upon a patient to choose between therapeutic innovation and an established therapeutic regimen may provoke severe anxiety. Some persons simply fear venturing into the unknown, particularly when their lives or health are at stake. Some would prefer to abandon the decision-making authority to the physician, and others may abandon the physician-investigator in order to find a doctor "who knows what he or she is doing." Some of those who make a choice may experience subjective guilt (self-blame), particularly if they consider their motivations unworthy. (During the consent negotiations they may or may not have discussed their motivations frankly.) For example, at a time when established therapy for a melanoma on the thigh was amputation of the entire hindquarter, an avid golfer who was invited to participate in a comparative trial of local excision might have a problem. He might be motivated to partake of this relatively untested therapeutic approach largely because he wanted to continue to play golf. On the other hand, he might wonder whether he was behaving responsibly in relation to his family and friends by agreeing to take what might prove to be a greater risk of dying of the disease. A somewhat analogous situation might be seen in a woman with breast cancer who agrees to participate in a clinical trial of lumpectomy at a time when, as she understands it, the standard and accepted practice is radical mastectomy, particularly if her primary and perhaps undisclosed motivation is preservation of her powerful forehand stroke in tennis.

Even if there is relatively little guilt at the outset, there may be severe guilt reactions if the risk actually becomes manifest as harm. Thus, if the individuals mentioned in the preceding paragraph subsequently are informed that they have metastatic tumors, they may develop the belief—which might or might not be accurate—that if only they had chosen to proceed with the established therapy, they would not be in their present predicaments. In extreme cases they may equate their decisions with a choice to commit suicide. In such cases, the guilt reaction becomes manifest only if the cancer recurs. The very same individuals might be quite pleased with themselves if 5 years later there was no evidence of recurrent tumor and they were much less incapacitated or deformed by the experimental surgery than they would have been by standard and accepted therapy.

A physician-investigator may invite a patient with cancer to participate in a clinical trial. Because the patient was informed of the diagnosis by the personal physician, the investigator may assume incorrectly that he or she knows the diagnosis. During the negotiations for informed consent, the investigator may inadvertently deprive the patient of the defense mechanism of denial.

In some cases, the consent negotiations may be even more hazardous than the research procedure. One investigator proposed to approach individuals with suspected or proved acute myocardial infarctions shortly after their admissions to coronary care units with a proposal to perform a complicated but rather harmless research maneuver. It was the judgment of the Institutional Review Board (IRB) that the anxiety that might be produced in patients by subjecting them to the necessity to make such a decision might be substantial. Conceivably, although this cannot be proved, the anxiety might contribute to the development of either further damage to the heart or coronary care unit psychosis (cf. Chapter 5).

The research environment often provokes reactions on the part of subjects or prospective subjects of distrust of the investigator or the institution. During the consent discussions, prospective subjects may perceive, often quite correctly, that the role they are being asked to play (subject) differs in some important way from the role they expected to play (patient). Thus, to some extent, the professional with whom they interact might see the development of new knowledge as being an important goal; this might compete with the goal of proceeding most efficiently to restore the health or well-being of the patient. Some subjects verbalize this as being used as guinea pigs.

In most consent discussions, it should be made clear that the prospective subject is free to refuse to participate in the research as well as free to withdraw from the research at any time. Further, there is an assurance that such refusal or withdrawal will in no way adversely prejudice his or her future relations with the investigator or the institution. Thus, for example, if he or she is a patient, refusal to become a subject will not in any way adversely prejudice this status. One wonders how often a patient questions (perhaps not aloud) whether the physician-investigator is capable of experiencing rejection in terms of refusal to consent to become a subject without retaliating in kind; i.e., rejecting the patient in some way by becoming less responsive to his or her demands or wishes. Thus, a subject who has not been coerced might feel that he or she has or might have been. This, in turn, sows the seeds of a breakdown in trust in the relationship between the professional and the individual.

Similarly, the knowledge that research is being done in an institution (e.g., a university hospital) may create in the community a sense of distrust. In many communities it is general knowledge that if you go to a university hospital you will be used as a guinea pig. Occasionally, people who are asked to become research subjects in a university hospital express surprise. They say, for example, that they didn't know that they would be asked and offered the option to refuse. They just assumed that research would be done on them—perhaps without their knowledge. The peril of this sort of misunderstanding is that some individuals may elect not to go to a hospital when they feel sick, and this decision may be detrimental to their health.

The problem of distrust is discussed further in Chapter 9, particularly in relation to research involving deception.

Some types of research may present threats to the doctor-patient or any other professional-client relationship. Such threats are particularly associated with research designed to examine the nature or quality of the doctor-patient relationship. For example, observers of the relationship might feel impelled, correctly or incorrectly, to inform the patient of the professional's mismanagement (e.g., failure to make a correct diagnosis) or misconduct (failure to disclose alternatives) (408). Researchers may transmit nonverbal expressions of surprise or disapproval upon hearing a patient's response to a question about some aspect of the doctor-patient interaction. A variant of this sort of problem is presented by one common practice of epidemiologists, that of having a pathologist affiliated with the research team confirm diagnoses by examining slides prepared and read months or years earlier by the original surgical pathologist. What if they disagree with the original diagnosis? Who should tell the patient? "It's too bad you had radical surgery. The tumor turned out to be benign after all."

When threats to the doctor-patient or any other professional-client relationship can be anticipated, it is wise to make careful plans for resolution of conflicts before they arise. This is exemplified by a case study in which a medical student proposed to investigate the nature of interactions between nonmedical healers (e.g., acupuncturists and chiropractors) and their middle- and upper-class clients (417). In their review of this protocol, IRB members thought that in the course of these studies, the medical student-investigator might perceive a moral obligation to intervene for what he considered the benefit of a client. For example, if he learned something about a client that had diagnostic implications, he might feel obliged to advise that client to consult a physician. Therefore, the IRB required the investigator to identify the circumstances, if any, under which he might feel compelled to intervene. This resulted in incorporating the following language in the consent forms presented to both the healers and their clients:

> As a scientist, I must minimize my interference with the phenomena that I am studying. Yet, as a medical student I will feel a responsibility for the health of the subjects. Accordingly, in cases where in my judgment a healer's action or inaction will likely raise the possibility of morbidity or mortality of the medically uninformed client, I will intervene—i.e., with the knowledge of the healer I will give the client my opinion.
>
> I will intervene only when all of the following conditions are met: 1) The client is not aware of the medical implications of his or her disorder, i.e., he or she has never seen a doctor for the disorder or has not clearly decided to reject medical advice; 2) orthodox medicine offers a distinctly superior advantage; 3) the treatment

or lack of treatment by the healer is truly dangerous to the client's health. For example . . . in clinically suspicious cases of strep throat, I would mention its dangers and the efficacy of penicillin. I would not, however, comment on a chiropractor's treatment of low back pain or on a faith healer's approach to a client with a terminal or chronic disease for which the client had already rejected orthodox care. . . .

In commenting on this case, Burt observes that it is very difficult for professionals or, for that matter, anyone else to assume the role of neutral observer (96); Swazey illustrates that this role is difficult even for a social scientist who has some experience in the field he or she is observing (675).

Some people believe that some questions asked in the course of conducting survey research may be so stressful to the subject that they might result in psychological injury. McCrae's empirical studies of the effects of stress questionnaires on normal subjects ranging in age from 21 to 91 yielded reassuring results (480). "Merely filling out forms on such topics as stress, mood, and personality does not seem to carry a risk of causing significant distress."

Social Risks   In his analysis of the risks of research, Barber distinguishes risks to the "biological person" from risks to the "social person" (35, 36). What I have described as physical risks consist of threats to the integrity of the biological person (acknowledging that physical impairment may result in social malfunction or injury to the social person).

Barber regards many of the psychological injuries I have described as social injuries, because they have a more direct and immediate impact on the individual's capacity for social interaction. The social person can also be injured through invasions of privacy and breaches of confidentiality (Chapter 7).

In some types of studies, individuals may be labeled inappropriately. During the developmental phases of diagnostic tests or procedures, they may be assigned incorrect diagnoses because the sensitivity and specificity of the diagnostic procedures have not yet been assessed adequately. In other cases, they may be assigned to diagnostic categories that subsequently prove to be illegitimate. In some cases, the result is social stigmatization of the individual as sick or ill when there is no possibility of providing therapy (458, 464). For a discussion of one of the most controversial studies in this field, the putative linkage between the XYY chromosome pattern and "aggressive or sexual psychopathology," see Kopelman (358). Studies in genetics, by falsely indicating that certain disorders are inherited, may stigmatize entire families (592).

In general, investigators owe it to subjects to share with them the results of studies either at the conclusion of the research project or as any particular subject concludes his or her participation in it. Exceptions to this general rule may be justified in several circumstances. When the research is designed to assess the validity of a new diagnostic test, it is appropriate to inform subjects that they will not be informed of the results. In this case, informed consent includes an explanation that results will be withheld because the investigators are unable to interpret the results; until the sensitivity and specificity of the test have been assessed, sharing the results with the subject or even with his or her physician may result in erroneous diagnostic inference.

Reilly has identified some factors that should be considered in determining what preliminary data should be shared with subjects (592). Although his discussion focuses on genetic screening programs, analogous considerations are applicable to determining the necessity for disclosure of other sorts of preliminary data or results (cf., Chapter 8). His criteria are the magnitude of the threat posed to the subject, the accuracy with which the preliminary results predict that the threat will be realized, and the possibility that action can be taken to avoid or ameliorate the potential injury.

At the time of this writing, a particularly poignant case in point is Huntington's disease (458, 612). Huntington's disease is a dominant genetic disorder characterized by incessant writhing and twisting, loss of mental capacity and loss of muscle control. Each child of an individual carrying the gene has a 50% chance of developing the disease if he or she lives long enough. There is no treatment and there is no cure.

At this point it seems very likely that we are on the verge of developing a specific and sensitive test for identification of individuals carrying the gene. Should we tell individuals whether their tests are positive or negative? Those who are informed of negative results can be reassured about their futures and about the safety of procreation. On the other hand, how harmful will it be to individuals to learn that they will certainly develop this dreadful disease? Rosenfeld argues that in almost every case the results of a validated test should be disclosed (612). MacKay offers a sensitive discussion of which results should be disclosed and under what circumstances during various stages of the development of presymptomatic detection tests (458). The President's Commission provides comprehensive and authoritative analysis of problems presented by such developments in the fields of genetic screening and counseling (575).

Economic Risks In protocols designed to evaluate nonvalidated therapies, many more laboratory and other diagnostic tests may be performed than would be necessary in standard medical practice. Occasionally, it may be unclear as to whether these tests are intended primarily to serve the interests of research or of patient care. At times, participation in a protocol may require prolonged hospitalization while standard therapy for the same condition may be accomplished with outpatients. If financial arrangements are not made clear at the outset, subjects may be presented with large and unanticipated hospital bills; some may exhaust their insurance coverage. It is often difficult to understand who should pay for what (720). The Comptroller General of the United States has issued a report on equitable allocation of patient care costs in research conducted or supported by NIH (153). This publication is concerned with NIH policies for determining which costs should be paid by NIH and which should be paid by patient-subjects or their insurance carriers. These policies have been interpreted by the Division of Research Resources of NIH for application to their General Clinical Research Centers Programs (532).

Who should pay for investigational therapies and the tests designed to assess their safety and efficacy? In the case of drugs, it is customary for the industrial sponsor to provide them free of charge during the investigational stages; Food and Drug Administration (FDA) regulations proscribe charging for such drugs. In the case of medical devices, industrial sponsors are permitted by FDA regulations to recover costs of the devices through billing patients. Often a drug that has been approved by the FDA for one indication is evaluated for safety and efficacy for a new indication. In some cases the industrial sponsor may be interested in this development and underwrite the costs of the drug and its evaluation. In other cases, the drug company may have no interest in further studies of the drug. In these cases, if other sources of funding are unavailable, it may be necessary to present bills for the drug to the patient. In some cases, insurance carriers may label such therapy experimental and therefore refuse to pay for it (130, 448). Lind has analyzed the ethical and scientific problems presented by such situations (448). For example, some patients may have to be excluded because they cannot afford to participate in the clinical investigation. This raises questions of fairness in distribution of both the burdens and the benefits of research. This may also introduce a selection bias that may undermine the generalizability of the results.

Some subjects may suffer loss of income when they take time off from jobs to participate in research; they may also have to pay baby-sitter fees, transportation costs, and so on. Some may become uninsurable as a consequence of diagnoses made by screening tests for eligibility for entry into protocols.

Some research is done without the awareness of third parties who might be asked subsequently to share the economic burdens imposed by failure of an investigational drug to accomplish the hoped-for benefit. For example, in studies of post-coital contraceptives, an unaware third party (coital partner) might be called upon after the fact to share the burdens (economic and decision-making) in regard to abortion or carrying a potentially damaged fetus to term (306).

Risks to Society

Physical and Psychological Risks The risks presented by research to society may also be classified as physical, psychological, social, and economic. Though some types of research may present serious risks of physical injury to society, clinical research ordinarily does not, unless one counts such things as the risk of spreading infections to persons other than the subject. Basic biological research may create the potential for widespread, grave physical injuries. For example, consider the possibility of developing an alien and particularly virulent strain of microorganism. An appreciation of the magnitude of this potential problem may be gained by considering the introduction of measles into previously unexposed populations by sixteenth century explorers. Dramatic fictional descriptions are presented in Michael Crichton's *Andromeda Strain* and Leo Szilard's *My Trial as a War Criminal*. Recent developments in the recombinant DNA field suggest that the potential development of alien viruses and antibiotic-resistant bacteria are no longer the exclusive domain of science fiction.

Premature or otherwise inappropriate dissemination of either the findings or the opinions of researchers may present psychological risks. For example, phobias may be developed on a rather grand scale when the public is informed that low cholesterol diets may cause cancer, that small breasts are associated with low I.Q., that citizens of certain cities are likely to be violent owing to the low lithium content of their local water supply, and so on. On the other hand, false hopes may be raised by premature or otherwise inappropriate dissemination of either the findings or opinions of researchers. Consider, for example, how many different surgical "cures" for coronary artery disease have been discovered and abandoned in the past 30 years. In addition to these activities of legitimate researchers, one might also consider the consequences of public announcement of such miracle cures as Krebiozin for cancer, copper bracelets for arthritis, and rainbow pills for obesity.

Social Risks Studies designed to compare certain social, ethnic, racial, religious, or political groups may develop findings that in the view of some members of a group might have pejorative implications. Others in the same group may view the same results as beneficial to the group. For example, when research revealed that the incidence of suicide was much higher in female physicians than in male physicians, some female physicians found this to be pejorative and supportive of the male chauvinist position. Others welcomed the information as beneficial, supporting their efforts to develop constructive affirmative action plans. Similarly, some individuals perceive plans to offer abortion and sterilization services in publicly funded health care delivery systems as attempted genocide. Others in the same group perceive regulations to proscribe such activities as depriving them as a class of their right to benefit from technology available to others to aid them in what they consider rational family planning.

Some research either by its very nature of by the sometimes callous manner in which it is performed may shock the sensibilities of society or some subsets thereof. Historically, public sensibilities as they were perceived about 30 years ago caused investigators who were then developing oral contraceptives to conduct the early clinical trials in Puerto Rico (714). One doubts that this would be necessary today. In its time, the Kinsey Report offended the sensibilities of many citizens. At the time the first Masters and Johnson work was being publicized, almost no one would have considered the Kinsey Report offensive. Similarly, the second Masters and Johnson Report shocked public sensibilities at a time when the first report probably would have shocked relatively few.

In some cases relatively little issue might be taken with the actual research; rather the apparently callous manner in which it was performed is offensive. Thus, graphic descriptions of decapitation of dead abortuses have shocked public sensibilities (221); possibly the very same research done with greater sensitivity would have attracted little or no attention.

In the course of conducting certain sorts of research, proposals may be made that challenge prevailing assumptions about the nature of the human person. The question of personhood, i.e., who or what is a person as a bearer of rights, obligations, and dignity has been raised in debates over research involving the human fetus (Chapter 13) and in relation to the dead or dying individual (Chapter 4). Such challenges are perceived by many as extremely threatening. Similarly threatening to some are attempts to explain human behavior in biological terms. Consider, for example, recent controversy over whether premenstrual syndrome is a legitimate defense against a charge of murder (382). Connell foresees the possibility that biologic determinism might displace from our culture such fundamental notions as free will and individual responsibility (156).

According to Department of Health and Human Services (DHHS) regulations (Section 46.111a(2)), "The IRB should not consider possible long-range effects of applying knowledge gained in the research (for example, the possible effects of the research on public policy) as among those research risks that fall within the purview of its responsibility." For an explanation of the Commission's reasons for recommending this addition to the regulations, see Gray (272). In short, predictions of the effects of the results of research on policy are highly speculative, essentially political, and in no way peculiar to research involving human subjects.

Regulations notwithstanding Ceci et al. provide empiric evidence that IRBs express their political preferences without necessarily being clear that this is what they are doing (125). Hypothetical protocols were submitted for IRB review that were identical in their proposed treatment of human subjects but differed in sociopolitical sensitivity. The sensitive study was designed to document discrimination or nondiscrimination according to race or sex in corporate hiring practices. Sensitive protocols were twice as likely as the others to be rejected. For those with ethical problems (e.g., deceptive strategies), this was generally

cited as the warrant for nonapproval. For the others, most commonly method-
ological problems were identified (e.g., poor control groups). Nonsensitive protocols without ethical problems were approved 95% of the time; this contrasts
with a 40 to 45% approval rate for comparable sensitive studies. Content analysis
of the narratives accompanying the decisions revealed that the primary reason
for rejection was usually the anticipated political impact of the study. These
findings are in harmony with my personal observations of IRBs dealing with
real protocols.

Economic Risks   The actual cost of doing research must be considered society's
risk capital to develop information assuming that it will benefit society. Thus,
one criterion for determining the appropriateness of any proposed activity should
be whether the information we might develop is worth that much money. For
an interesting and provocative analysis of how society responds to such questions,
see Calabresi (100).

Calculations of the economic risk borne by society should take into account
the cost of taking care of individuals who are damaged by research. The topic
of compensation for research-related injury is discussed in Chapter 6.

Scientists often claim that as we are weighing the risks or costs of doing
research, we should equally consider the risks or costs of not doing research.
These costs are essentially a deprivation of society of the benefits of doing
research. Commonly, to support this argument, one points to the great savings,
expressed either in economic terms or in terms of human suffering, brought
about by the development of immunizations (small pox, rabies, and polio provide
dramatic examples), general anesthesia, antibiotics, heart surgery, and amniocentesis.

Others point to the great savings that could be achieved by demonstrating the
lack of validity of certain widely used, very expensive, and possibly ineffective
therapies; a recent workshop sponsored by the Arthritis Foundation and the
National Multiple Sclerosis Society concluded that the savings could be so great
that such third-party payors as the Health Care Financing Administration (HCFA),
Blue Cross/Blue Shield, and other private insurance carriers would be wise to
finance such research activities (130). Meanwhile, HCFA has developed a policy
of reimbursing hospitals for patient care under the Medicare program according
to their Diagnosis Related Groups (DRGs), which permit no recovery of costs
of prolongation of hospitalization for research purposes; Rabkin is among those
who fear that this policy will have highly detrimental effects on the conduct of
clinical research (580). The therapeutic orphan phenomenon has been identified
correctly as a harm of not doing a specific type of research (cf. Chapter 10).
For a particularly scholarly analysis of costs of the not doing research, see Comroe
and Dripps (154).

Benefits

An extensive discussion of the benefits of research is beyond the scope of this book. Those who review research proposals will almost always find that the hoped-for benefits to both society and the subjects tend to be well described; their probability and magnitude both tend to be overestimated. This does not reflect dishonesty. Rather, investigators, like all humans, require some degree of optimism for motivation. For a discursive survey of the benefits of research both to society and to subjects see Levine (408).

Judgments about the probability and magnitude of hoped-for benefits require technical expertise. Many other judgments about benefits are not technical. For example, how can one find that the hoped-for benefits to society outweigh the risks to the individual subject? What sorts of benefits to subjects merit inclusion in considerations of whether there is a favorable balance of harms and benefits? What sorts of benefits are appropriate to offer to prospective subjects?

With regard to benefits to subjects, I shall assume that it is always appropriate to weigh and to offer direct health-related benefits if their probability and magnitude are stated correctly. Let us instead focus on some sorts of direct benefits to subjects about which there are controversies as to whether they ought to be weighed or offered. These are economic, psychosocial, and kinship benefits.

For a discussion of cash payments and other material inducements to subjects see Chapters 4 and 12.

Psychosocial Benefits Patients who know they have terminal illnesses and patients who are depressed will often respond favorably to the notion that investigators are not only interested in them but also are attempting to devise something that might offer relief or, perhaps, cure. Some patients with cancer in whom all validated modes of therapy have been tried without success become optimistic when the prospect of trying an investigational drug is offered. The patient may be relieved to learn that he or she need not give up hope because there is yet another possibility for relief and that he or she is not about to be abandoned by health professionals. Many individuals who are depressed or anxious or both will experience relief as they assume the role of subject; in the relatively sheltered research environment, they are largely divested of the burdens of some sorts of decision-making.

Individuals who are concerned about their sense of worth may welcome the opportunity to appear valuable to themselves as well as to others; doing something that they consider altruistic enhances their sense of personal worth. Among examples of such individuals are some elderly persons and some prisoners (who might, incidentally, hope that their altruistic tendencies will be appreciated by those who make parole decisions) (cf. Chapter 12).

In some social groups, playing the role of subject may bring an individual considerable prestige. Some may be flattered to be the subject or object of attention of so many important people. This is particularly true of individuals who become eligible for the role by virtue of having a rare disease. Others gain what they consider substantial prestige or satisfy their tendency to exhibitionism through participation in research that attracts great publicity, e.g., Walter Reed's studies on yellow fever.

Many bored person find that participation in research is a welcome diversion. Although this is discussed most commonly in relation to prisoners and those institutionalized as mentally infirm, patients hospitalized in acute general medical services, after the busy first day or two, commonly state that they prefer being used as research subjects or teaching material to the ennui of daytime television.

Kinship Benefits   Some persons experience as a personal benefit the belief that their actions will produce direct benefits to others. Thus they may be willing to assume the burdens of the role of research subject in order to better the lives of others. In general their motivation to do this is higher when they either are related to or have a sense of kinship with the prospective beneficiaries. To the extent that the individual is motivated by a sense of kinship with increasingly large groups of humans, the largest group being the entire human species, the motivation increasingly approximates altruism or charity.

Perhaps the closest sense of kinship one can feel is with one's self. Some persons may become subjects of basic research on diseases with which they are afflicted hoping that they will contribute to the development of knowledge about their disease. This knowledge, in turn, might lead to the development of improved therapy for their disease; thus, they may hope for a direct health-related benefit in the future, particularly if they have a chronic disease. Alternatively, they might feel a sense of kinship with others with the same disease, hoping that in the future some direct benefit might accrue to them.

Within families, persons are often motivated by the prospect of kinship benefits; this is particularly relevant to research in the fields of genetics and transplantation. Persons are generally more likely to offer bone marrow to a relative than to a stranger. The father or mother of a child with phenylketonuria might be willing to participate in research designed to perfect techniques for detecting heterozygous carriers although it will bring no direct benefit to them or their child; they already know they are carriers. However, if a better method for carrier detection is discovered, it would be likely to provide direct health benefit to another relative who is phenotypically normal but who might be a carrier.

A sense of kinship might be based on racial or ethnic factors. Thus, some Jews might be motivated to serve as normal controls for research designed to explore the pathogenesis or therapy of Tay-Sachs disease; blacks might volunteer from similar roles in research related to sickle cell anemia (320).

Women who are about to have abortions may feel a sense of kinship with other pregnant women who expect to continue their pregnancies to term; among such women in the future might be the woman who is now planning an abortion. Thus, she might be motivated to participate in research made possible by virtue of the fact that she has planned to have an abortion, research designed to develop knowledge that might be of benefit to pregnant women who expect to carry their pregnancies to term (cf. Chapter 13).

It is customary to appeal to kinship interests and altruism when prospective subjects are capable of consenting for themselves. Kinship interests are commonly weighed by those who are considering whether one can justify research involving vulnerable subjects or those incapable of consent. For example, arguments for justification of research on the dying person or dying fetus have been grounded in presumptions of their ''interest'' in the welfare of others like them (cf. Chapters 4 and 13). Some arguments supporting the involvement of children in research have even gone so far as to construe a limited moral obligation to such acts of charity (cf. Chapter 10.)

Description of Harms and Benefits

Investigators are obligated to prepare descriptions of risks and hoped-for benefits for at least two types of readers; these are IRB members and subjects. Each of these readers will determine whether they find the balance of harms and benefits favorable. An adequate description of the harms and benefits facilitates these determinations. Ideally, for each harm or benefit there should be an identification of its nature and an estimate of its probability and magnitude.

An adequate description of the *magnitude* of either a harm or a benefit should include as complete a statement as possible of its *expected duration*. Thus, for example if a possible harm is paralysis of one leg, how long is this paralysis expected to last? Certainly, the magnitude of the harm will be considered greater if the paralysis were ordinarily expected to last 1 year than if it were expected to last 1 hour. If the harm is ordinarily expected to be irreversible, i.e., expected to continue unabated for the duration of the subject's life, this represents the greatest possible magnitude of that particular harm.

Some potentially irreversible harms, if detected early, may either be avoided entirely or reduced in magnitude by discontinuation of the potentially harmful research procedure. Minimization of a developing or nascent harm may also be accomplished by therapeutic intervention, e.g., timely administration of an antidote to a poison. In such circumstances, adequate descriptions include 1) a list of procedures that might be employed for timely detection of the developing harm, 2) a clear statement of criteria that will be used to determine when to terminate the research procedure or administer the antidote, and 3) an assessment of the probability and magnitude of success that can be reasonably expected of the monitoring procedures and corrective interventions.

Similarly, the magnitude of a hoped-for benefit should be analyzed in terms of its expected duration. For example, if an investigational modality provides the hoped-for benefit, what provisions have been made to assure the subject's continuing access to this modality? The beneficial modality might be a drug or a device which proves effective (beneficial) in a particular subject but whose sponsor (e.g., industry) decides to discontinue producing it because it has not been found beneficial to a sufficient number of individuals to make its further development worth the sponsor's investment (448, 610). Alternatively, it might be an "experimental" health delivery system developed under public funding in a community lacking the economic resources for its continuation at the termination of the period of public funding.

Similarly, the *probability* of the occurrence of both harms and benefits should ordinarily be elaborated. In consideration of harm, is there any means by which individuals who are most susceptible to harm might be identified? If so, will these means be used and will those individuals either be excluded from the research or informed that they are especially vulnerable? For example, in planning research designed to test the effects of strenuous exercise in normal humans, one would ordinarily plan to perform various screening tests to identify individuals with coronary artery disease in order to exclude them.

In consideration of benefit, is there any means by which individuals who are most likely to be benefited might be identified? If so, will these means be used to assist in recruiting research subjects who are most likely to be benefited? A necessary consequence of using such means is the exclusion of those who are relatively less likely to receive benefit. Thus, in the development of a therapeutic innovation, particularly one designed to alleviate a serious disorder or one whose administration or implementation entails consequential risk, it is generally most appropriate to select subjects in whom standard modalities have been tried without success.

In general, in the early stages of development of diagnostic and therapeutic modalities, it is appropriate to include in the description and in the weighing deliberations the fact that there may be harms the nature of which remain to be discovered.

Those who describe harms and benefits should be encouraged to provide quantitative estimates of their probability and magnitude. Ideally, such estimates should be based upon empirical data developed in well designed studies. Unfortunately, such data are difficult to find. Harvey and Levine reviewed the literature on the risks of injury that could be attributed to 20 invasive procedures commonly employed in the conduct of research involving human subjects (282, 283). For each procedure, they found a wide range of estimates of the probability of each type of injury. The probability of injury seemed to be related to whether the estimates were based upon retrospective or prospective studies, as well as to the experience of the person performing the procedure and various attributes of the population upon whom the procedure was performed. Almost all available data are derived from experience in the practice of medicine where the risk of injury is greater than it is in the conduct of research. In research it is generally easier to exclude subjects who are vulnerable to injury.

What I have said about description of risks and benefits should be viewed as an ideal standard. Keeping this standard in mind is helpful in reminding us of what we ought to be trying to do. We must acknowledge, however, that only rarely can we reach this ideal standard; often, for good reason, we fall far short of this ideal.

Consider, for example, the risks of radiation exposure. In the early days of radiology, no one had any idea that there were any risks. Researchers and radiologists exposed themselves insouciantly to massive doses; many of them died of radiation-induced cancers. People of my generation will recall the routine use of fluoroscopy in shoe stores to evaluate the fit of new shoes. What fun it was to watch the bones of our toes wriggling in our new shoes!

Now we know a good deal about the adverse effects of high doses of radiation (10, 472). We know little or nothing, however, about the effects of low doses, although professionals in the field have negotiated agreements on how to predict them based upon extrapolations of what we know about high dose exposure (10, 165). What is a reasonable amount of radiation exposure for research purposes? Should maximum permissible doses per year be defined as they have been for laboratory workers (472)?

What should one disclose to prospective research subjects? If the risk of all persons of developing a fatal cancer is 1700 per 10,000 and the risk of inducing a fatal cancer from one rem of radiation exposure is 1 per 10,000, is the fact that such exposure increases the risk to 1701 per 10,000 material for purposes of disclosure (10)?

How should we explain these risks to prospective subjects? Is it fair to use language of the sort recommended by Massé and Miller (472):

> The amount of radiation exposure you will receive from this procedure, if averaged over the entire body, would be equivalent to the amount of naturally-occurring background radiation all people in this region receive in each one-year period. The concentrated dose to the primary region is less than one-half the allowable whole-body radiation dose workers (for example, x-ray technologists) may receive each year of their working lifetimes.

Is it fair to relate the exposure to the amount of excess radiation one would receive by flying from New York to Los Angeles or by spending a week in Denver or Brazil (10)?

I have no answers to these questions. For further discussion of these problems, the reader is referred to articles by Alazraki (10), Veatch (699), Crawford-Brown (165), and Massé and Miller (472).

One aspect of the risk of radiation exposure that many research subjects find hard to understand is its additivity. Here is an example of a statement in a consent form designed to warn research subjects about the hazards of multiple radiation exposures:

> Please let us know if you have ever before been exposed to radiation as a patient, as a research subject or at work. This will enable us to check to be sure that your cumulative radiation exposure does not exceed accepted safety limits. If you participate in this study and then wish to participate in future studies that involve the use of x-rays or radioisotopes, you should discuss the safety limits of radiation exposure with the investigator who is performing that study.

One may encounter similar problems in estimating and explaining the risks of exposure to other forms of energy. Anbar discusses the problems associated with mechanical energy (including ultrasound), thermal energy (as, for example, produced by microwave exposure), electrical energy, and electromagnetic energy (as, for example, that associated with magnetic resonance imaging), as well as the effects of exposure to various materials such as contrast agents used in diagnostic x-ray angiography, metallic pins in orthopedics, and elastomers (organic artificial polymers used, for example in hemodialyzing membranes) (16).

Minimization of Risks

Ethical codes and regulations require not only that risks be justified by being in a favorable relationship to hoped-for benefits but also that they be minimized.

> The experiment should be so conducted as to avoid all unnecessary physical and mental suffering and injury (Nuremberg 4).

> The IRB shall determine that . . . risks to subjects are minimized (i) by using procedures which are consistent with sound research design and which do not unnecessarily expose subjects to risk, and (ii) whenever appropriate, by using procedures already being performed on the subjects for diagnostic or treatment purposes (DHHS, Section 46.111a).

In this chapter, many procedures are discussed that are oriented toward the minimization of risk. For example, the monitoring requirement mentioned in the introduction to this chapter is designed to serve this end. Prescreening tests may be done to identify prospective subjects who ought to be excluded because they are vulnerable to injury. The opportunistic use of procedures that are to be done for diagnostic or therapeutic purposes, as called for in Part ii of the regulation cited above, has been discussed in the section on mere inconvenience. Additional procedures are discussed in many other chapters.

Perhaps the most important maneuver that can be performed in the interest of minimization of risk is to include in protocols an exhaustive list of adequately described risks. Upon examining such lists, investigators and IRB members commonly develop resourceful approaches to enhancing the safety of subjects. For an example, let us consider a policy developed by the IRB at Yale Medical School when it became aware of the special hazard presented to subjects in double-blind studies who may require treatment in emergencies. This policy is designed to assure physicians who need to know what drug the patient is taking immediate access to the codes for these studies.

Each subject is provided with a wallet card or an identification bracelet that has on it the information that he or she is the subject of a particular double-blind study, the name and telephone number of the investigator, the subject's code number in the study, and the message that in an emergency the code may be broken by contacting the Hospital Pharmacy Department (telephone number provided) at any time of the day or night. Codes for all studies are kept in the Drug Information Service of the Hospital Pharmacy Department, which is staffed by professionals 24 hours daily. If necessary, these professionals can also provide expert advice to physicians who may be unfamiliar with the drug the patient is taking.

Maximization of Benefit

In pursuit of the obligatory goal of increasing the favorableness of the balance between harms and benefits, the mirror image of the requirement to minimize harm is the requirement to maximize benefit. As noted in Chapter 1, the proscription against inflicting harm is generally regarded as a stronger ethical obligation than that of promoting good. Correspondingly, although the ethical codes and regulations forcefully interdict causing death and injury, obligations to promote good tend to be implicit in the norms calling for good scientific design, competent investigators, and favorable balances of harms and benefits. The Nuremberg Code focuses exclusively on benefits to society in its references to experiments that "will yield fruitful results for the good of society" and "the humanitarian importance of the problem to be solved." The Declaration of Helsinki is concerned with benefits both to society and to subjects. One of its direct admonitions to enhance benefits to subjects is discussed in Chapter 1 as an illustration of the infelicitous consequences of the spurious concept of therapeutic research. The other requirement to promote direct health-related benefit seems to be motivated at least in part by the obligation to avoid harm. "In any medical study, every patient—including those of a control group, if any—should be assured of the best proven diagnostic and therapeutic method (principle II.3)." As noted earlier in this chapter, this principle is violated frequently. However, withholding known effective therapy for diseases that, if untreated, may produce death or disability is generally not condoned. The principle is an important component in the justification of the randomized clinical trial (Chapter 8). DHHS and FDA regulations do not embody this principle; rather they require that prospective subjects be informed of any alternative therapies that might be advantageous to them (Chapter 5).

Devices that may be used to maximize benefit are discussed in several sections in this chapter. For example, prescreening tests may be designed for the selection of those subjects who are most likely to derive benefit. Arrangements may be made to assure subjects continuing access to investigational drugs after completion of a clinical trial. Suggestions for maximizing benefit to subjects are also provided in several other chapters.

Justification of Risk

Justification of risk is never an isolated event. To say that the imposition of risk in the interest of research is justified presupposes that the plan to do research is also in accord with requirements of all relevant ethical norms and procedures.

Maneuvers employed with the intent and reasonable probability of providing direct benefit for the individual subject, including all diagnostic and therapeutic maneuvers whether validated or nonvalidated, are justified differently from nonbeneficial procedures. The risk is justified by the expectation of benefit for the particular subject, a strictly personal calculation precisely as in the practice of medicine. One additional criterion is that the relationship of anticipated benefit to the risk presented by the modality must be at least as advantageous to the subject as that presented by any available alternative, unless, of course, the individual has considered and refused to accept a superior alternative. The Commission first explicated this justification in its recommendations on *Research Involving Children;* in Chapter 10 there is further discussion of this justification and a demonstration of how the Commission recommended different justifications for the risks of nonbeneficial procedures performed in the interests of research, particularly when the subjects are either incapable of consent or, for other reasons, highly vulnerable.

It is more problematic to justify the risk presented by maneuvers performed for research purposes, i.e., to contribute to the development of generalizable knowledge. The benefits one hopes for in this case will be for society rather than to the individual subject. How does one find that the balance of hoped-for benefits is in a favorable relation to the risks to the individual subject?

As we approach an answer to this question, two points must be kept in mind. First, for the present purposes it will not be possible to construct a satisfactory algorithm for determining favorableness. Modern approaches to decision analysis, which rely on the development of algorithms, may be quite helpful in reaching judgments about the favorableness of particular proposals. Cost-effectiveness and cost-benefit analyses may afford helpful insights into the favorableness of developing specific new therapies by forcing participants in such policy decisions to express their assessments of costs, harms, values, disvalues, benefits, and utilities in concrete terms (98). Similarly, they may be helpful in reaching judgments about the management of individual patients (745). But for our purposes "favorable" is intended as a dispositional attribute like "elastic" or "responsible." Attempts to pin down its meaning in absolute terms will defeat its purpose. "Favorable" is a term used to suggest to reasonable persons that there is something about the balance of harms and benefits that other reasonable persons are likely to find felicitous. Secondly, judgments about the balance are expected of at least three classes of agents or agencies. These are 1) investigators, in order to justify their proposing to do the research; 2) IRB members, in order to decide whether to approve it; and 3) subjects, in order to decide whether to consent to or refuse participation in the research.

Let us now consider risk justification in protocols in which prospective subjects will be only those who are in all respects capable of negotiating informed consent. Should the IRB assume a paternalistic stance by constraining the investigator from offering what it considers unfavorable harm-benefit ratios to subjects? An obligation to such paternalism was explicit in the Department of Health, Education, and Welfare (DHEW) regulation that required the IRB to determine that:

> The risks to the subjects are so outweighed by the sum of the benefit to the subject and the importance of the knowledge to be gained as to warrant a decision to allow the subject to accept these risks (Section 46.102, rescinded).

By contrast, the Commission decided more on the side of autonomy (527, pp. 24–25):
> . . . (I)f the prospective subjects are normal adults, the primary responsibility of the IRB should be to assure that sufficient information will be disclosed in the informed consent process, provided the research does not present an extreme case of unreasonable risks.

In accord with the Commission's recommendation, DHHS revised its statement in the regulations to the one quoted earlier in this chapter calling upon the IRB to determine that the risks to subjects are reasonable in relation to anticipated benefits.

> No experiment should be conducted where there is *a priori* reason to believe that death or disabling injury will occur; except, perhaps, in those experiments where the experimental physicians also serve as subjects (Nuremberg 5).

This principle is grounded in the premise that no person as rational as a scientist would ever deliberately take such risks unless the research objectives were extremely important. However, history reveals that while some self-experimenters are quite properly placed in the heroic tradition of science (e.g., members of Walter Reed's Yellow Fever Commission), many have taken extreme risks for relatively unimportant goals and, further, their results were often invalid owing to a lack of proper controls (326). Since Nuremberg, there has been a growing recognition that the occasionally overpowering motivations of investigators may lead them to erroneous or, at least, quite different value judgments from those of lay subjects. The Declaration of Helsinki does not mention self-experimentation and the NIH Code for Self-experimentation requires the same standards of "group consideration" of proposed research, whether it is to be done on oneself or on normal volunteers (53, pp. 304–305). In general, however, an expression of willingness on the part of investigators to participate as subjects lends credibility to their claims that the benefits to be expected merit the taking of the risks (53, p. 233; 160).

I shall not attempt here to resolve the eternal tensions between paternalism and autonomy. With DHHS I shall leave it to reasonable persons to determine on a protocol-by-protocol basis whether risks are reasonable in relation to hoped-for benefits. In my casuistry, when considering particular proposals to involve autonomous adults as research subjects, if I must err, I am inclined to err on the side of autonomy. This is because I consider overprotection to be a form of disrespect for persons (377). For further discussion of this issue, see Wikler (731).

Justification of the imposition of risks in the interests of conducting research is much more problematic when the subjects cannot protect themselves through negotiating informed consent. Until recently, the codes and regulations did not deal with these problems except to the extent that they called for the consent of the "legally authorized representative" (proxy consent). As we shall see in Chapters 10 through 13, standards for justification of risk in the special populations are now much more sophisticated.

Threshold Standards

Both DHHS and FDA have established risk threshold criteria. When the risk burdens presented by research proposals exceed these thresholds, the regulations call for special procedural or substantive protections, particularly if the subjects are drawn from one of the special populations. On the other hand, when the subjects are to be consenting adults and the risks do not exceed the threshold, the IRB may waive the requirement for certain routine protective maneuvers.

The major threshold established by DHHS (Section 46.102) and adopted by the FDA is minimal risk.

"Minimal risk" means that the risks of harm anticipated in the proposed research are not greater, considering probability and magnitude, than those ordinarily encountered in daily life or during the performance of routine physical or psychological examinations or tests.

This definition is derived from definitions of minimal risk developed by the Commission for research involving children and those institutionalized as mentally infirm (Chapters 10 and 11). In my view, minimal risk is too low a threshold to serve the purposes DHHS and FDA prescribe for it when the research subjects are to be autonomous adults; e.g., justification of expedited review, waiver of requirements for consent forms, and so on, as discussed in most subsequent chapters. I have argued, without success, that a more suitable threshold is mere inconvenience (398). Kopelman has published a detailed account of the various problems presented by the regulations' definition of minimal risk (359).

The somewhat higher threshold of minor increments above minimal risk and its purposes are discussed in Chapters 10 and 11.

The FDA has defined another threshold as "significant risk device" (Section 812.3m). The definition is very long, consisting of four components of which the fourth is: "Otherwise presents a potential for serious risk to the health, safety, or welfare of the subject." Because all these words (except otherwise) appear in each of the other three components, the operative definition is: significant risk device means an investigational device that . . . presents a potential for serious risk to the health, safety, or welfare of the subject.

FDA regulations require that sponsors may not begin testing an investigational device without having first secured an Investigational Device Exemption (IDE) if it is a significant risk device; if it is not, no IDE is required (Section 812.20). The determination of whether any article is a significant risk device is the duty of the IRB (Section 812.65a). The FDA retains the authority to overturn the decision of an IRB that an article is not a significant risk device.

The multiple problems presented to IRBs by this regulatory requirement have been reviewed by Holder (301). Among other things, IRBs are not designed to be competent to make such judgments in many cases and IRB members do not understand the definition of significant risk device. I am unable to assist in fathoming this meaning.

Chapter 4

Selection of Subjects

In its report on Institutional Review Boards (IRBs) (527), the Commission recommended that the IRB shall determine that: ". . . selection of subjects is equitable. . ." (Recommendation 4B). The brief commentary on this recommendation elaborates, "The proposed involvement of hospitalized patients, other institutionalized persons, or disproportionate numbers of racial or ethnic minorities or persons of low socioeconomic status should be justified." The reaction of the Department of Health and Human Services (DHHS) is reflected in its substantially identical Section 46.111a(3).

This requirement is derived from the principle of *justice*, which requires equitable distribution of both the burdens and the benefits of research. Until very recently codes of ethics and regulations were relatively silent on this matter; however, the preamble to the Nuremberg Code reflected a concern with issues of social justice. It pointed out that the ". . . crimes against humanity . . ."

were particularly egregious in that they were perpetrated on ". . . non-German nationals, both prisoners of war and civilians, including Jews and 'asocial' persons. . . ." Implicit in this statement is the perception that, because these subjects were not considered persons in the full sense of the word, they were not accorded the respect due to fully enfranchised persons. As a consequence, Principle I of the Nuremberg Code established the high standards for consent discussed in Chapter 5. When a totally honest offer is made to fully autonomous persons, they are presumed capable of defending their own interests and of selecting themselves as research subjects. Because the Nuremberg Code does not entertain the possibility of involving less than fully autonomous subjects, no requirements for their selection are provided.

Federal regulations and the Declaration of Helsinki reflect the understanding that at times it is necessary to involve the legally incompetent as research subjects. In these documents, the issue is addressed by calling for the consent of the legal guardian or legally authorized representative. Thus, until recently, consent, either by oneself or by a legally authorized representative, was the only device specified in codes and regulations for the distribution of the burdens and benefits of research.

The concept that we should not rely exclusively on consent as a criterion for selection of subjects was brought sharply into focus in 1969 by Jonas (329). Jonas argued for a "descending order of permissibility" for the recruitment ("conscription") of subjects for research. His criteria for selection related directly to the prospective subject's capacity to understand the goals of the research and to participate in it as a partner with the investigator. Accordingly, he proposed that the most suitable subjects would be researchers themselves, because they had the greatest capacity to give a truly informed consent. He also argued that very ill or dying subjects should not be used in research even if they gave consent unless the research related directly to their own illnesses. Underlying this argument is a perception of very ill or dying subjects as peculiarly vulnerable to pressures which make their consent insufficiently free or informed. In this way, the strict concern for consent was supplemented by Jonas with a concern for the situation of the subject and the ways in which the situation might render the subject vulnerable; vulnerable subjects were afforded extra protection against selection even if they wished to participate.

Recognition that extra protection is required for those who are vulnerable by virtue of being sick is expressed in the 1975 revision of the Declaration of Helsinki. However, the requirement established in the Declaration is precisely the opposite of what Jonas proposed; sick persons are to be recruited as subjects of nontherapeutic research only when it is unrelated to their illness. As discussed in Chapter 1, this is one of the unfortunate unintended consequences of classifying research as either therapeutic or nontherapeutic.

In Germany, in 1931, federal law was promulgated that specified the obligations of physicians and the rights of patients and subjects in medical research; Sass has published these regulations in German along with his translation of them into English (625). On matters relating to autonomy and minimizing injury, they are very similar to the Nuremberg and Helsinki principles. They go much further than Nuremberg in their specification of requirements for special protections for such vulnerable populations as children and dying persons. Although these were binding law in Germany from 1931—before the establishment of Third Reich—to 1945, they were obviously ignored by the Nazis who were tried and convicted of crimes against humanity at the Nuremberg tribunals. Sass provides, in addition to this valuable translation, a very thoughtful analysis of the principles that informed their writing and of what the consequences might have been of adopting these principles rather than those set forth in the Nuremberg Code and the Declaration of Helsinki. In his view, which I share, excessive bureaucratization of research ethics has an undesirable tendency to decrease the physician-investigator's sense of personal responsibility.

Public Concerns with Justice

Until recently, our ethical codes and regulations have dealt explicitly with only two of the three fundamental ethical principles: *respect for persons* and *beneficence*; yet considerations of *justice* have, for many years, been implicit in expressions of public concern over the ethical propriety of research involving human subjects. In general, the cases that have evoked the greatest public outcry have violated or have seemed to violate the requirements of all three fundamental ethical principles; these research activities seem to have imperiled the life or health of vulnerable or disadvantaged persons without their informed consent. The common reaction to these activities—"That's not fair!" reflects the concept of *justice* as fairness (cf. Chapter 1).

Let us now consider briefly four of the cases that have been the object of the most extensive commentary in literature addressed to both lay persons and professionals.

The Tuskegee Syphilis Study

This study was designed to determine the natural history of untreated latent syphilis (80). The subjects were over 400 black men with syphilis and approximately 200 men without syphilis who served as controls. The studies began in 1932, when standard treatment for syphilis involved injections with various drugs containing the heavy metals arsenic and bismuth. The preponderance of medical opinion at the time was that such treatment reduced the mortality and morbidity of syphilis; however, it was suspected that some complications commonly attributed to syphilis were caused by the treatment.

The men were recruited without informed consent. In fact, they were misinformed to the extent of being told that some of the procedures done in the interests of research (e.g., spinal taps) were actually "special free treatment."

By 1936 it became apparent that many more infected men than controls had developed complications; 10 years later a report on the study indicated that the death rate among those with syphilis was about twice as high as it was among the controls.

In the 1940s, when penicillin, known to be safe and effective for the treatment of syphilis, became available, the men were not informed of this. Although there was clear evidence that syphilis shortened the life expectancy of these men by "about 20 percent," the study was not interrupted and antibiotic therapy was not offered. The study continued until the first accounts of it appeared in the national press in 1972, shortly before the Department of Health, Education, and Welfare (DHEW) formed the Tuskegee Syphilis Study Ad Hoc Advisory Panel (688).

The Willowbrook Studies

These studies were designed first to contribute to understanding of the natural history of infectious hepatitis and subsequently to test the effects of gamma globulin in preventing or ameliorating the disease (341, pp. 633 and 1007–1010; 710); they were conducted at the Willowbrook State School, a New York State institution for "mentally defective persons." The subjects, all children, were deliberately infected with the hepatitis virus; early subjects were fed extracts of stool from infected individuals and later subjects received injections of more purified virus preparations. The investigators defended the deliberate infection of these children by pointing out that the vast majority of children admitted to Willowbrook acquired the infection spontaneously and, perhaps, it would be better for them to be infected under carefully controlled conditions in a program designed to provide the best available therapy for hepatitis.

An additional criticism was leveled against the recruitment policies. During the course of these studies, Willowbrook closed its doors to new inmates owing to overcrowded conditions. However, the hepatitis program, because it occupied its own space in the institution, was able to continue to admit new patients as each new study group began. Thus, in some cases parents found that they were unable to admit their child to this institution unless they agreed to his or her participation in these studies.

The controversy over the ethical propriety of this study has continued over the years (618). It is of interest that a follow-up report was published in the *New England Journal of Medicine* along with an editorial by Ingelfinger (324) stating,

> . . . reports of investigations performed unethically are not accepted for publication. Thus, appearance of another Willowbrook report in this issue indicates that the study, on balance, is not rated as unethical.

The Jewish Chronic Disease Hospital Study

These studies involved the injection of live cancer cells into patients who were hospitalized with various chronic debilitating diseases (341, pp. 9–65). Their purpose was to develop information on the nature of the human transplant rejection process. Previous studies had indicated that healthy persons reject cancer cell implants promptly. Patients with widespread cancer also reject such homografts; however, rejection in these patients is delayed substantially as compared with healthy subjects. These studies were designed to see if delay in the rejection is due to the presence of cancer or if it is a more general manifestation of debility. The investigators hypothesized that the delayed rejection was related to impaired immunity due to cancer, not to debility in general, and thus predicted that there would be rapid rejection characteristic of healthy persons.

Consent was said to have been negotiated orally but not documented. In the view of the investigators, documentation was unwarranted because it was customary to employ much more dangerous medical procedures without the use of consent forms. Further, the patients were not told that they were going to receive cancer cells because in the view of the investigators this would frighten them unnecessarily. The investigators defended this view on the basis that they had good cause to predict that the cancer cells were going to be rejected.

The San Antonio Contraceptive Study

This study was designed to determine which of the alleged side effects of an oral contraceptive were due to the drug and which "reflect the symptomatology of every day life" (341, p. 791; 692). The subjects of this study were mostly impoverished Mexican-American women who had previously had multiple pregnancies. They came to the clinic seeking contraceptive assistance. The study was a randomized, placebo-controlled, double-blind clinical trial. In addition, there was a crossover design, in which one-half the patients received placebo first and the other one-half received the active contraceptive; midway through the study, the medications were switched. None of the women were told that a placebo was to be used as part of the study. However, all women were advised to use a vaginal cream as a contraceptive for the duration of the study. Of the 76 subjects, 11 became pregnant during the course of the study, 10 while taking placebo and 1 while receiving the active contraceptive.

Other Studies

In 1966, Henry Beecher published in the *New England Journal of Medicine* his classic exposé of examples of "unethical or questionably ethical procedures" he found in published reports of research involving human subjects (51). Each of these studies was characterized by a high ratio of risk to benefit. In addition, almost all of these studies involved as subjects vulnerable or disadvantaged persons. Little was said about informed consent because the published reports available to Beecher did not customarily discuss such matters.

Vulnerable and Less Advantaged Persons

There are various criteria for identifying individuals as vulnerable or less advantaged in ways that are relevant to their suitability for selection as subjects. In this section there is a systematic survey of these criteria, along with several examples of each suggested vulnerability-producing attribute. In general, we identify as vulnerable those who are relatively (or absolutely) incapable of protecting their own interests. More formally, they have insufficient power, prowess, intelligence, resources, strength, or other needed attributes to protect their own interests through negotiations for informed consent.

Most discussions of the ethics of research involving human subjects depict the norms calling for informed consent and those requiring equitable selection of subjects as distinct. They reflect, respectively, the obligations arising from the two fundamental ethical principles of *respect for persons* and *justice*. However, because we define as vulnerable those persons who are relatively or absolutely incapable of protecting their own interests through negotiations for informed consent, in the practical arena of IRB review, there is often an interplay between considerations of informed consent and selection of subjects. To the extent that the subject population can be made less vulnerable in the sense of becoming more capable of protecting their own interests, fewer procedures are needed to assure the validity of their consent. In many sections of this chapter, there are illustrations of the trade-offs between requirements arising from these two sets of norms. In general, procedures designed to mitigate the problems associated with involvement of vulnerable persons as research subjects focus on improving the quality of consent; they are designed either to increase the capacity of vulnerable persons to consent or to exclude vulnerable persons and select as subjects individuals who are more capable of consent.

It should be understood that each person, when measured against the highest standards of capability, is relatively vulnerable. We are all dependent upon someone or something and susceptible to temptation by what we consider very large sums of money. Thus, it is necessary to introduce a note of caution into these considerations. It is easy to identify too many persons as vulnerable and to apply procedures designed to protect the interests of vulnerable persons too extensively. Some judgment is required. In each case it is worthwhile to reflect on the implications of labeling persons as vulnerable. Are we being disrespectful of persons by repudiating their authority to live according to their considered judgments? Are we inappropriately stigmatizing groups of people as being unable to take care of themselves? Are we contributing to the trivialization of the process of protecting the rights and welfare of human research subjects by applying cumbersome and unwarranted procedural protections? On the matter of trivialization, see Ingelfinger (325), Cowan (160), Holder and Levine (314), and Chapter 14.

Each of the categories of vulnerable subjects should be viewed as consisting of a spectrum ranging from slight to absolute vulnerability, dependency, impoverishment, and so on. Moreover, in most categories the vulnerabilities may be transient, prolonged, permanent, or intermittent. These factors should be taken into account in deciding on the need for the various procedural protections discussed in this and succeeding sections.

Uncomprehending Subjects

Inclusion of uncomprehending persons may present two types of problems. Consent may be invalidated by lack of comprehension. Also, in research activities requiring cooperation, the subjects may not understand how they are to cooperate and, thus, either defeat the purposes of the study or contribute to their own harm.

Persons having prolonged or permanent incapacity to comprehend include the mentally retarded, the uneducated, and those with chronic organic brain syndrome (senility). Difficulties with comprehension also arise when investigators and subjects do not share a common language.

Persons with various psychological disturbances may be incapable of comprehension transiently, for prolonged periods, permanently, or intermittently. In some persons this incapacity may be absolute. Some very withdrawn schizophrenics might comprehend but give no evidence that they do. Some intelligent individuals, during obsessional states, may be so preoccupied with their obsessions that they have no energy or interest available to comprehend anything else.

Persons who are inebriated may have incapacities to comprehend ranging from barely perceptible to absolute. In some cases the inebriated person may insist upon overestimating his or her capabilities. Inebriation is ordinarily intermittent.

Most often, persons who can reasonably assume that they will be inebriated at some future time are capable of comprehending plans to involve them in research during future inebriated states. In some cases, for example, inebriation will be induced by administration of some drug for therapeutic purposes, e.g., a narcotic or barbiturate. Under some circumstances, it might be appropriate for a person to consent in anticipation of inebriation to being treated in a certain way while inebriated, even though while inebriated he or she might object to being treated in that way.

Unconscious persons are, of course, absolutely uncomprehending. Unconsciousness may be intermittent and reasonably predictable, as in individuals with grand mal epilepsy. Such individuals are ordinarily capable of valid consent in anticipation of their next unconscious period. Similarly, a plan may be developed between a physician and a patient to produce unconsciousness, e.g., through general anesthesia, at some future time. This individual can consent in advance to various research procedures that might be done during the period of unconsciousness.

An interesting case in point is provided by the early stages of development of the total artificial heart. The IRB decided to select as subjects only those patients who in the course of open heart surgery could not be "weaned off the heart-lung machine," persons who were facing certain death (203). Thus, the investigators could not predict who the recipients would be. The IRB required that in order to be considered eligible to receive the device, a patient must have consented preoperatively. Thus, many patients were to consent to an intervention that would never be done on them. In a somewhat similar case involving a left ventricular assist device (LVAD), an IRB required consent only from those patients who seemed relatively likely to require its use (298); for others who developed an unanticipated need for the LVAD, consent from family members was authorized.

Binding commitments to be treated in a specified way even though at the time one may protest being treated in that way are variously called "Ulysses contracts," "Odysseus pacts" or "Odysseus transfers," and "voluntary commitment contracts" (197). The names for these contracts are derived from the story in *The Odyssey* in which Ulysses instructed his men to bind him to the mast of his ship before sailing past the Sirens and, further, to refuse his requests to be released while he was under the irresistible influence of the Sirens' song. These contracts are designed to show respect for one's rational will and to see to it that this will is followed subsequently when one becomes irrational. They were designed originally to make plans for voluntary commitment of individuals who are prone to become psychotic intermittently. It has been proposed that such contracts could also be used to authorize research or investigational therapies during episodes of irrationality (734). Although reliance on such contracts is highly controversial, I think their use can be justified in some circumstances in which the prospective subject, while rational, is unquestionably capable of negotiating informed consent and in which no more than minimal risk is presented by procedures designed to serve the interests of research—e.g., drawing blood from inebriated persons. However, plans to use such contracts to authorize research procedures that present more than minimal risk or to authorize non-validated practices should, in general, be supplemented by the procedural protections recommended by the Commission for those institutionalized as mentally infirm (Chapter 11). At least, one should take care when selecting the person who will play the role of Circé, the lady who in the original story advised Ulysses to make his literally binding agreement.

These contracts are somewhat related to the research living will (cf. Chapter 11).

Sick Subjects

Persons who have assumed the role of patient or sick person may be vulnerable in several respects. Assumption of the sick role tends to place limits on a person's autonomy. To illustrate the barriers presented to the achievement of autonomy by having assumed the role of sick person, let us examine Talcott Parsons' view of this role in relation to the physician (341, p. 203). Parsons discusses the roles of patient and physician in the context of his definition of health and the overall role of medicine in society. Health is defined in terms of a given individual's capacity to perform effectively the roles and tasks for which he or she has been socialized, and the concept of a person's health is defined with respect to that person's participation in the social system.

Illness is also defined within the context of the social system and, therefore, is judged as being indicative of a disturbance of the capacity to perform roles and tasks effectively. Parsons identifies four aspects of the institutionalized expectation system relative to the sick role.

> 1. (There is an) . . . exemption from normal social role responsibilities, which . . . is relative to the nature and severity of the illness. This exemption requires legitimation . . . and the physician often serves as a court of appeal as well as a direct legitimatizing agent . . . being sick enough to avoid obligations cannot only be a right of the sick person but an obligation upon him. . . .
>
> 2. The sick person cannot be expected by "pulling himself together" to get well by an act of decision or will. In this sense also he is exempted from responsibility—he is in a condition that must "be taken care of." . . . the process of recovery may be spontaneous but while the illness lasts he can't "help it." This element in the definition . . . is crucial as a bridge to the acceptance of "help."

3. The state of being ill is itself undesirable with its obligation to want to "get well." The first two elements of legitimation of the sick role thus are conditional in a highly important sense. It is a relative legitimation as long as he is in this unfortunate state which both he and alter [authority] hope he can get out of as expeditiously as possible.

4. (There is an obligation upon the sick person) . . . to seek technically competent help, mainly, in the most usual sense, that of a physician and to *cooperate* with him in the process of trying to get well. It is here, of course, that the role of the sick person as patient becomes articulated with that of the physician in a complementary role structure. (Emphasis in the original.)

Parsons describes in considerable detail the practical consequences of this definition of the sick role (555). The second component of the sick role may be perceived by the sick person as a type of personal gain. At least temporarily, he or she is relieved of various duties and obligations which he or she might consider onerous. However, this social definition of illness imposes upon the responsible sick person the obligation to seek and cooperate with competent help (ordinarily health professionals); if he or she fails in these responsibilities, his or her sick role will eventually come to be seen as illegitimate (irresponsible).

Individuals with short term illnesses from which they might be expected to recover completely without the aid of health professionals need not be considered especially vulnerable. Similarly, persons with illnesses from which they might be able to recover completely but only with the aid of standard (noninvestigational) technically competent help need not be considered particularly vulnerable, assuming that such help will never be withheld as punishment for refusal to become a research subject.

Persons having prolonged chronic illnesses that are refractory to standard therapies should be considered seriously vulnerable. This becomes increasingly important when such people perceive themselves as desperate and willing to take any risk for even a remote possibility of relief. In this category are some infertile persons or couples who desperately want a child; some persons with chronic, painful, and disabling disorders such as rheumatoid arthritis; some obese persons who cannot lose weight following standard procedures; and some with severe depression or obsessive-compulsive disorders. Depressed persons and others who question their self-worth may be especially vulnerable to inappropriate inducement by offering benefits other than those directly related to their health; they are especially susceptible to appeals to their altruism.

It must also be recognized that persons who are or who perceive themselves to be seriously ill suffer. In his important article on this subject, Eric Cassell observes that a distinction must be made between suffering and physical distress (120).

Suffering is experienced by persons, not merely by bodies, and has its source in challenges that threaten the intactness of the person as a complex social and psychological entity. Suffering can include physical pain but is by no means limited to it. . . . Even in the best of settings and with the best physicians, it is not uncommon for suffering to occur not only during the course of a disease but also as a result of its treatment.

I shall not do violence to the concept of suffering by attempting here to present it concisely. Suffice it to say that suffering persons are, in an important sense, "not themselves." Their judgments may differ drastically from those the same individuals would make when not suffering. In another article, Cassell argues convincingly that the seriously ill person is in a state of damaged autonomy and that the function of the physician is to preserve or restore autonomy (119).

In some circumstances, persons may be offered the choice between the sick role and the criminal role. The special problems associated with justifying research on such persons are discussed in Chapter 12.

Dying Subjects No topic in bioethics exemplifies the Tower of Babel phenomenon more vividly than contemporary discussions of death and dying (439). So many definitions have been given for such terms as "terminal" and "imminent" that the reader often can not discern the author's meaning unless the author specifies which definition he or she is using. Although standard dictionaries are quite clear in defining death as a final irrevocable event, consider some phrases gleaned from the recent medical literature (546). "The patient was brought back to life." ". . . the first case of resuscitation of sudden death from myocardial infarction. . ." One author claims that "death itself in its early stages should be regarded only as an illness of penultimate severity." A recent editorial in the *New England Journal of Medicine* was entitled "Prevention of Recurrent Sudden Death." Perhaps discussion of death and dying evokes so much anxiety that we obscure our meanings deliberately (431, 442).

In this book, the term "death" refers to a final, irrevocable event. "Dead" persons are those who will never again be alive. The term "brain-dead" refers to those who are pronounced dead according to criteria appropriate for the jurisdiction in which they are so pronounced; in general, this term will be used to describe dead individuals whose vital functions (e.g., respiration, circulation, and nutrition) are being continued with the aid of such instrumentalities as ventilators and intravenous feedings. Words such as "terminal," "imminent," and "dying" will be used as dispositional terms (Chapter 3) to suggest that the death of the person is anticipated soon; as illustrated so vividly in the case of Karen Quinlan, predictions about the timing of death for an individual may be very inaccurate.

For a philosophical analysis of the duties owed the dead and dying, see Preus (577). Youngner et al. provide sound practical advice for dealing with the psychosocial problems presented by the retrieval of organs for transplant purposes from dead (including brain-dead) persons (746). In my view their advice is relevant to the involvement of the dead and dying as research subjects.

Among those who are in the sick role, perhaps the most vulnerable are those who believe, correctly or incorrectly, that their own death in imminent. Katz (341, pp. 1054–1068) has assembled several case studies involving the use of the dying as research subjects and as organ donors.

Robertson has argued that the use of brain-dead subjects can be justified both legally (if performed in accord with the consent requirements of the Uniform Anatomical Gift Act) and ethically (607). He further argues that because a brain-dead individual does not conform to the DHHS definition of subject, it is not clear that the IRB has jurisdiction over proposals to do such research. In reflecting on Robertson's arguments, Smith concurs with his conclusions (660). However, he calls attention to the likelihood that such activities may offend public sensibilities. The President's Commission concludes its consideration of this issue (574, p. 41), "Given the kinds of research that might conceivably be conducted on the deceased, we suggest that consideration be given to requiring IRB review of certain kinds of such research to determine . . . it is consistent with commonly held convictions about respect for the dead." For other reports of research activities involving brain-dead subjects, see Maugh (474), Fost (234), Carson et al. (117), and the President's Commission (574, pp. 39–41). Abramson et al. have proposed that patients may be involved in randomized clinical trials of cardiopulmonary resuscitation without consent (3); this proposal is discussed in Chapter 8.

Within the last few years, there has been a good deal of discussion of presumed consent to organ donation by dead and brain-dead persons (105). According to this concept, unless an individual has made known his or her wishes to the contrary, one may presume consent to the artificial maintenance of vital functions while awaiting the most propitious time for such harvesting. Arguments used to support presumed consent in this context are as follows: The demand for organs for transplantation exceeds the supply. The results of public opinion polls indicate that a much larger percentage of the population is willing to donate organs after death than have signed organ donor cards. Even when such cards have been signed, physicians and other hospital personnel are reluctant to bring up the subject of transplantation; they simply disregard the organ donor cards in order to protect the family from unnecessary anguish during this period of grief and vulnerability. Further justification is found in the fact that the harvesting of organs imposes so little by way of burden on the donors; a brain-dead person is incapable of experiencing discomfort. It seems to me that it would be most difficult to justify presumed consent for the involvement of brain-dead persons as subjects in research. In any event, presumed consent to either organ donation or research is at the time of this writing not authorized by law in any state.

Caplan argues that physicians ought to be required by law to request permission for organ donation in appropriate circumstances (105, 109). By August 1985, New York and Oregon had passed laws requiring routine requests for organ donation (Hastings Center Report, August 1985, p. 36); similar legislation was pending in California.

Gaylin has proposed the establishment of a new class of subjects which he would call the "neomort" (256). A neomort is an individual who has been declared dead by virtue of current criteria for establishing brain death whose other bodily functions are maintained more or less indefinitely with the aid of various devices. Presuming their lack of sentience, such beings could be used for a variety of biological research procedures that for various reasons one might not wish to perform on sentient persons. Additionally, organs and tissues might be harvested as necessary for transplant purposes.

Dependent Subjects

While most persons are dependent upon some other persons or institutions for most of their lives, we shall be most concerned with dependent relationships that present either of two potentials. The first is that by virtue of the relationship, the dependent individual is administratively more available to the investigator to be selected as a subject than are other individuals not having the same dependent status. The second is that in which the dependent individual might fear that he or she might forfeit the desired dependent status by virtue of refusing to participate in research.

Administrative Availability Some types of institutions tend to accumulate relatively large populations of individuals who are suitable as subjects for various types of research; these include hospitals and other health care facilities, schools, welfare agencies, places of employment, and so on. Many investigators capitalize on this ready availability of suitable prospective subjects. By establishing their research units in or near such institutions, they minimize the expense and inconvenience of recruiting subjects. This has the effect, or at least seems to have the effect, of imposing an unfair share of the burdens of research participation on these dependent populations.

In Chapter 12 there is a discussion of a classical example of capitalizing on the administrative availability of institutionalized persons; this is the once widespread use of prisoners as subjects in Phase I drug studies. Phase I drug studies require subjects who are healthy as well as reliably available for frequent examinations, timed urine collections, and so on. Such individuals are more abundantly available in prisons than in any other place. Most other healthy populations are not so readily regimented.

Many research projects use as subjects individuals who are administratively available to the investigator. For examples, see all four of the cases presented earlier in this chapter in the section on public concerns with justice. Three of the four cases drew on patients in hospitals or clinics in very close proximity to the investigators. In the other case, the Tuskegee Syphilis Study, it is clear that the investigators thought that there was no other locale in which they were likely to find a higher number of suitable subjects (80).

That investigators seek subjects in close proximity to their laboratories should come as no surprise. However, the fact that most researchers are located in universities presents some problems. First, it is commonly alleged that patients in university hospitals bear an unfair share of the burdens of research participation. Secondly, patients in university hospitals are, at times, subjected to tests that are not relevant to their conditions. Some examples were provided by Beecher (51). 1) A controlled, double-blind study of the hematologic toxicity of chloramphenicol, well-known as a cause of aplastic anemia, was conducted on patients chosen at random. These patients had no disease for which chloramphenicol administration is indicated (Example 5). 2) A new technique for left heart catheterization, transbronchial catheterization, was first performed on patients both with and without heart disease. Bronchoscopy was performed on these patients; it is not clear whether this procedure was medically indicated. During bronchoscopy, a "special needle" was inserted through the bronchus into the left atrium of the heart. The hazards of this procedure were ". . . at the beginning quite unknown" (Example 19).

The inner city location of many university hospitals is problematic in that it increases the likelihood of disproportionate use of racial and ethnic minorities as well as impoverished people as research subjects.

In the university environment, students are recruited frequently as normal volunteers for biomedical research. Social and behavioral scientists commonly use students in their own classes as research subjects. Such practices are a matter of continuing and spirited controversy. The University of Massachusetts Medical Center has a policy restricting participation of medical students to research presenting no more than minimal risk and no more than "minimal interruption of one's daily routine" (634). Publication of this policy by Shannon in 1979 elicited charges of "paternalism" and "covert elitism" from Nolan, then a Yale medical student (544).

More recently, Christakis, a Harvard medical student, criticized his institution's policy, which is even more highly protective of medical students than that of the University of Massachusetts (135). He argues cogently that their double standard is unwarranted. Angoff skillfully carries his argument even further, claiming that medical students, particularly those considering careers in research, have an obligation to serve as research subjects (20); ". . . how better to know what it is like to be a subject than to be one!'' She further argues that ''. . . treating medical students differently by applying different standards to their participation in research than to the participation of the general public has the flavor of elitism. One may wonder why it is acceptable to ask the masses to accept risk in the name of science but not the very people whose futures are linked to the successful perpetuation of biomedical research.'' Further, she defends payment of the medical student subjects; among other reasons: ''Earning money from participation in research is at least as reputable a way as a variety of others available to students, such as selling their blood, tending bar, or baby-sitting for a faculty member's children.''

Yale University School of Medicine has no policy on the involvement of medical students as research subjects. However, its position is reflected in its disposition of a protocol involving multiple catheterizations, infusions of drugs, and exposures to moderate doses of radiation (433). The IRB decided that there was a very low likelihood that any subject would sustain a serious injury. However, it further decided that most ''normal volunteers'' not only would have great difficulty understanding the massive amount of information to be presented about the risks of the procedures, they would also have very little appreciation of the benefits the investigators were pursuing. Therefore, it required that subject selection be limited to those who could reasonably be expected to have the desired amount of understanding. These included ''Physicians, nurses, third or fourth year medical students, technicians who have been employed or who are employed in an animal physiology laboratory which performs techniques similar to those . . .'' to be used in the research protocol. In response to the investigators' arguments, the IRB gave permission to ''. . . extend the population eligible . . . to include all Yale students in health professional schools including medicine, nursing, and physicians' associates. However, if they do not meet the specifications in the protocol as approved originally, they must be invited into the catheterization laboratory in order to observe this procedure or a very similar procedure being performed . . .'' before being invited to participate as subjects. In general, in order to avoid exploitation of particular faculty-student relationships, student-volunteers should be recruited only by posted advertisements, not by direct invitation to particular students.

A special problem is presented by the common practice of academic psychologists to use as research subjects students in their own classes. Often their participation in research is mandatory; in some cases participation in research is a device through which students may improve their course grades. These practices have been analyzed by Gamble, who concludes that it is questionable as to whether students can consent freely under such conditions (255). He further suggests that the exchange of extra credits for student participation is an improper use of the academic grading system. Students do not demonstrate the degree of mastery of psychology suggested by grades inflated in this fashion. Gamble concludes that students should be offered cash payments rather than course credits because this "removes a benefit unique to students and allows them to judge the advantages of research participation in the same manner as paid subjects." Cohen endorses Gamble's position in principle but calls attention to the fact that many investigators do not have sufficient funding to comply with it (145). He presents the policy at the State University of New York at Albany, which emphasizes the provision of realistic alternatives. The American Psychological Association offers a sensitive discussion of this problem and how to deal with it (14, pp. 47–48).

Another environment in which large numbers of investigators are concentrated is the pharmaceutical industry. An increasing amount of their drug studies is done on their employees. Some have developed special guidelines for the use of employees as research subjects (491).

Threatened Relationships   Prospective subjects commonly are in relationships of dependency with institutions or individuals that either conduct research or are closely affiliated with those who do. Under these conditions, prospective subjects may fear that they will place their relationships in jeopardy if they refuse to cooperate with investigators. The types of relationships that are of concern here are the same as those that create the condition of administrative availability, e.g., physician-patient, teacher-student, employer-employee, and so on. The problems of threatened relationships are discussed more completely in Chapter 12; devices for dealing with these problems are discussed in Chapter 5.

Impoverished Subjects   It is very difficult to develop a definition of impoverishment that is at once sufficiently inclusive and exclusive (469). For the present purposes, impoverishment is defined as a condition in which persons consider it necessary to assume extraordinary risk or inconvenience in order to secure money or other economic benefits that will enable them to purchase what they consider the necessities of life. Their willingness to assume extraordinary burdens is based upon their belief that they are unable to secure sufficient amounts of money by ordinary means.

Virtually all modern ethical codes and regulations proscribe undue inducement on grounds that it invalidates informed consent. However, none of them specify what might be considered a due inducement. Some of the problems involved in offering cash payments and other material inducements to impoverished subjects are discussed in Chapters 5 and 12. For an illuminating analysis of the problems in distinguishing due and undue inducements, see Macklin (459). The problems are further exposed in the subsequent debate between Newton (537) and Macklin (460).

The Commission concluded that protection from undue influence could best be accomplished by "Limiting remuneration to payment for the time and inconvenience of participation and compensation for any injury resulting from participation . . ." (527, p. 25). While I can think of no better resolution of this issue, it does present some problems (398, p. 42). For example, this represents a departure from our society's custom of paying persons high salaries to assume large risks. If this formulation is appropriate for research, why is it not also appropriate for test pilots and deep sea divers?

Provision of cash payments for research subjects will inevitably undermine the goal of distributing the burdens of research participation more uniformly throughout the population. If the rate of pay is low, subjects are most likely to be recruited from the ranks of the unemployed. On the other hand, if salaries are set high enough to attract corporate executives, they will almost certainly be perceived as irresistible inducements to impoverished persons.

It is my impression that most IRBs, as they consider appropriate levels of cash payments, view the role of the research subject as an ordinary job generally requiring relatively unskilled laborers. Wages are determined by customary market factors in that they are established at sufficiently high rates to attract adequate numbers of suitable subjects. Thus, the burdens of research participation are distributed as they are for other types of employment. One notes without serious dismay that corporate executives are not competing for employment as office secretaries, meter readers, jurors, or research subjects. This seems appropriate to me for the vast majority of research projects in which persons capable of informed consent volunteer to assume such burdens as minimal risk and mere inconvenience. However, I would avoid application of this policy to research presenting more than minor increments above minimal risk (cf. Chapters 10 and 11) until adequate systems of compensation for research-induced injury have been established.

Subjects of research designed to test the safety or efficacy of nonvalidated practices are not usually paid. This reflects a general presumption that access to the nonvalidated practice modality is, of itself, sufficient reward. However, in this context, there are commonly subtle material inducements that are much more problematic than cash payments (Chapter 3). These include offers of free medical care, hospitalization, medications, and so on. In the face of modern hospitalization costs, inadequately insured individuals may easily be considered impoverished by the definition used in this section. At times, such impoverished individuals may be faced with the necessity of assuming large financial burdens to purchase needed medical care. For those who are subjects of research, the costs of the needed medical care may be largely underwritten by the agency sponsoring the research. Thus, the "choice" between being a patient or a patient-subject may be based primarily on such financial considerations. Although this problem is not susceptible to easy resolution, the degree of financial constraint on the subject's freedom to choose can be minimized through application of this policy guideline. In calculating the costs of health services to an individual who will simultaneously play the roles of subject and patient, it should be calculated how much medical care would have cost had he or she not agreed to play the role of subject. This amount should be paid by the patient. The additional costs incurred as a consequence of an agreement to play the role of subject should be paid by the investigator or the sponsoring agency.

Departures from this guideline occur commonly; as discussed in Chapter 3, the investigator's authority to act according to this guideline may be limited by Food and Drug Administration (FDA) regulations; by policy decisions made by granting agencies, insurance companies, and other third-party payors; and by decisions made by industrial sponsors of therapeutic modalities.

Holder has reported on an interesting case in which a researcher wished to interview persons who were applying for the first time for unemployment benefits; he wished to pay prospective subjects $4.00 per hour to participate in his protocol (309). According to state law, subjects who agreed to participate could lose their unemployment benefits because they had a job. On the other hand, if they refused to participate, they could lose unemployment benefits because they had refused to accept a job for which they were eligible. In this case it was necessary for the investigator to negotiate with the State Unemployment Office an agreement that they would not interpret a refusal to participate as a refusal of employment. Further, although they would lose some payments from the state during the week in which they participated in the study, their period of eligibility to receive unemployment benefits would be extended proportionately.

Minority Groups

Members of various minority groups (as determined by race, sex, age, ethnicity, and so on) are commonly portrayed as vulnerable. Because members of such groups may be the objects of discriminatory societal customs, an inordinantly high percentage of them may be vulnerable owing to impoverishment, sickness, dependency, and so on; such persons should be treated accordingly. However, it is not appropriate to treat entire minority groups as if they were homogeneous with respect to vulnerability. Among other things, it adds unnecessarily to the burden of stereotypes already borne by such groups.

Elderly Persons   The establishment in 1974 of the National Institute on Aging (NIA) signaled as a national priority a need to do research addressed to the special medical and social problems of elderly persons. NIA's founding director, Robert Butler, coined the term "ageism." In his important book, *Why Survive?*,

Butler defines "ageism" as a "deep and profound prejudice against the elderly" (97, pp. 11–12). Its manifestations include "stereotypes and myths, outright disdain and dislike, or simply subtle avoidance of contact; discriminatory practices in housing, employment and services of all kinds; epithets, cartoons and jokes." Thomas Cole's insightful essay on the evolution of America's attitudes toward the elderly helps very much to put ageism and modern efforts to combat it in their proper perspective (149). As he observes:

> Apart from its class bias and its empirical deficiencies, the attack on ageism perpetuates the existential evasiveness of its Victorian forebears. The currently fashionable positive mythology of old age shows no more tolerance or respect for the intractable vicissitudes of aging than the old negative mythology. Whereas health and self-control were previously virtues reserved for the young and middle-aged, they are now demanded of the old as well. Unable to infuse decay, dependency, and death with moral and spiritual significance, our culture dreams of abolishing biological aging.

Cole urges our attention to the fact that aging is more than a "problem" requiring a scientific "solution." "Unless we grapple more openly with the profound failure of meaning that currently surrounds the end of life, our most enlightened view of old age will amount to perpetual middle age."

In considering the recruitment of elderly persons as research subjects, it is worth keeping in mind that, although some are feeble, fragile, or forgetful, most are not. Of the 1.5 million Americans who are over 65 years old, 5 to 10% show signs of senile dementia and a similar percentage live in institutions for a variety of reasons (586); this means that over 90% of elderly people do not have the criteria for defining vulnerability. Many elderly persons are relatively risk-aversive and make choices designed to maximize their comfort and avoid disrupting their habitual routines (586). Thus, they are more inclined than others to refuse participation in research involving minor discomforts such as venipunctures (587) or interviews (672). Some seem to fear the implications of signing consent forms, and some are inexperienced in interpreting "quasi-legal documents" (754).

Taub and Baker report that there are age-related deficits in both comprehension and memory in reading "meaningful prose materials" (679). They find that they can improve both by giving repeated multiple choice tests on prose materials such as consent forms and giving immediate feedback on the results. With time, the scores on these tests improved for most people. Taub and Baker recommend that those prospective subjects who do not eventually get a passing score be rejected as research subjects.

Stanley et al. studied the choices made by elderly medical patients about participation in hypothetical research projects (668). They conclude, "Thus, as a group, geriatric patients may have some impairment in their competency to give informed consent for research; however, this impairment does not appear to have a significant impact regarding the quality of their decisions. . . . They generally seem to make equally reasonable decisions.''

Ratzan (587) supports the need for research involving elderly subjects. In his view their general tendency to cautiousness and risk aversion stands in the way of conducting such research. He argues that to counteract these tendencies we should translate risk information into ''commonly acceptable terms encountered in every day life.'' Thus, ''it might help a 75-year-old subject to know that a 0.1% incidence of serious complications from an invasive procedure was equivalent to a 1 in 1000 risk which is roughly equivalent to one-fifth his chance of dying that year as a one pack per day smoker.'' The sorts of translations he recommends are analogous to those presented in Chapter 3 in relation to expressing the risks of exposure to radiation. Ratzan further recommends that a person be identified as a ''risk advisor.'' Such a person would need and, as an IRB member, would soon acquire expertise in understanding and interpreting risk. The risk advisor's responsibility would be to assist elderly subjects in assessing risk. They would be ''directly answerable to the group at risk, and given standing by their other interests in risk negotiations.''

There is no doubt that many persons including some elderly persons require assistance in interpreting risk. The need for involvement of third parties in the consent process is discussed in Chapter 5. In general, I resist the mandatory intrusion of such persons in the negotiations between investigator and subject. Some reassurance may be found in the observation that elderly persons often insist upon consultation with their sons or daughters before signing consent forms (754).

Several authors have lamented the fact that the Commission did not identify elderly persons as a special population. Ratzan, for example, argued for the development by the National Institutes of Health (NIH) of guidelines similar to those proposed by the Commission for those institutionalized as mentally infirm (586). Parenthetically, Ratzan subsequently retreated from this position (587). Others claim that the elderly population is highly heterogeneous and, for that reason, the development of uniform guidelines is unwarranted (551). NIA considered and rejected a plan to develop guidelines for the conduct of research involving elderly subjects (531); among other reasons, they wished to avoid adding unnecessarily to the burden of stereotypes already borne by elderly persons.

For a fine overview of the state of the art in clinical research in geriatric populations, see the summary of the 1984 NIA symposium on that topic (754). Among other things, this symposium takes note of the marked heterogeneity of the elderly population. This creates problems not only in subject protection but also in research design.

Justification of the Use of Vulnerable Subjects

The use of vulnerable persons as research subjects is not forbidden by any ethical code or regulation. The Commission did not recommend any such proscriptions; rather, it called for justification of any plan to involve vulnerable persons as research subjects. In general, as the degree of risk and the degree of vulnerability increases, justification becomes correspondingly more difficult. Justification usually involves a demonstration that less vulnerable subjects would not be suitable, e.g., the condition to be studied does not exist in less vulnerable persons. In this section I shall review various procedures that may be employed to mitigate the ethical problems presented by plans to recruit vulnerable persons as research subjects.

Reducing Vulnerability

Commonly, it is possible to take steps designed to reduce the vulnerability of research subjects. Examples of such procedures were presented in the preceding section in the discussion of each type of vulnerability, e.g., negotiating consent with those who are intermittently uncomprehending during their lucid periods and assuring dependent persons of their freedom to refuse to participate in research or to withdraw at any time without adversely prejudicing their dependent relationships. Further examples are provided in each of the chapters concerned with persons having limited capacities to consent, e.g., the involvement of consent auditors in plans to involve those institutionalized as mentally infirm in some types of research, the requirement that payments to prisoners be on a par with those for other jobs in the prison, and the requirement that procedures used in some types of research involving children be commensurate with those encountered in the standard medical care for their illnesses.

Levine and Lebacqz have examined the widespread practice of conducting randomized clinical trials in Veterans Administration (VA) hospitals (441). Veterans are highly dependent upon the VA hospital system. Although private patients may choose another physician or, in most cases, another hospital setting if they prefer not to participate in a randomized clinical trial, veterans wishing to exercise such an option will often be obliged to assume large financial burdens. In order to receive the benefits due them as veterans, they must use certain specified facilities; their options for receiving medical care thus are usually more limited. To this extent, they must be considered more vulnerable than other patients with similar conditions receiving private care. Several steps may be taken to alleviate the disadvantaged position of patients in the VA hospital as they consider participation in a randomized clinical trial. For example, they might be provided the option of personal care or participation in the randomized clinical trial, thus increasing their options to be more similar to those enjoyed by private patients. Alternatively, veterans refusing participation might be referred for personal care to a private hospital at the expense of the VA.

Involving Less Vulnerable Subjects

Plans to involve vulnerable persons as research subjects are generally more easily justified if it can be shown that less vulnerable persons are willing to assume the burdens of participation in the same project. Such involvement would be responsive to any charges that vulnerable subjects were being exploited because less vulnerable persons would not agree to participate. In Chapter 12, there is a discussion of the Commission's recommendations to this effect on research involving prisoners; Levine and Lebacqz have discussed the application of similar procedures to the conduct of randomized clinical trials in VA Hospitals (cf. Chapter 8).

At times it is essential to recruit vulnerable subjects because no other persons have the conditions to be studied. Under these circumstances, one should generally select the least vulnerable representatives of these populations. The Commission made detailed recommendations to this effect in their reports on research involving children and those institutionalized as mentally infirm (see Chapters 10 and 11).

Some protocols entail the use of a large number of complicated or unfamiliar procedures. In such cases, ordinary persons who are not usually considered vulnerable may have difficulty achieving sufficient mastery of the information about risks, benefits, and purposes to give an enlightened consent. Under these circumstances, it may be appropriate to limit the subject population to individuals who are more sophisticated in relevant respects. This is exemplified by the requirement to use only health professionals, students, and laboratory technicians in a complex cardiac physiology protocol (*supra*). For similar reasons, children may be involved in protocols in which more than minimal risk is presented by procedures that do not hold out the prospect of direct benefit to the children only when the procedures are "reasonably commensurate" with those inherent in their actual or expected medical, psychological, or social situations (Chapter 10).

Distribution of Benefits

In general, research done on vulnerable subjects should be relevant to the condition that causes the subjects' vulnerability. The Commission made this point most consistently in each of its reports on persons having limited capacities to consent (see Chapters 10 through 13). Jonas also makes this point most forcefully in his consideration of proposals to do research involving sick persons (329). Such procedures create more favorable balances of harms and benefits. At the very least, the benefits one hopes to secure would be returned to the vulnerable population whose members bore the burdens. One of the chief objections to the conduct of Phase I drug studies on prisoners is the belief that prisoners, in general, are not among the first to reap the benefits of new developments in drug therapy (Chapter 12). Similarly, the subjects in the Tuskegee studies did not receive the benefits of advances in the treatment of syphilis even during the conduct of the study.

Ideally, one should attempt to offer the benefits of the results of research directly to those who served as subjects. An example of this was provided by the way in which the first trials of the Salk polio vaccine were conducted (253, p. 147). The first use of this vaccine was in a double-blind, randomized, placebo-controlled trial involving thousands of children. The vaccine was not available outside the trial. Parents of the children were told that if the vaccine proved safe and effective, the control group and the families of all participants would be the first to be offered subsequently developed supplies of the vaccine.

It is customary in the United States to introduce new therapeutic modalities in the context of clinical trials designed to test their safety and efficacy. As we have seen in Chapter 1, the Commission recommended that this custom be continued and extended. Thus, it is commonly the case that a patient cannot gain access to a form of therapy without agreeing to participate in a clinical trial. My argument that this is quite appropriate is based upon viewing new therapeutic modalities as benefits the distribution of which is either conducted or encouraged by society (398, pp. 46–47). The choice to receive such a benefit imposes upon the individual the *reciprocal obligation* of assuming the inconvenience of tests necessary to demonstrate its safety and efficacy. In cases in which the modality is scarce, refusal to participate in such tests may justify exclusion of that individual from receiving that modality. However, a distinction must be made between depriving a person of a single therapeutic modality on the one hand and, on the other, of an entire dependent relationship. It is usually not appropriate to terminate a professional-client relationship because the client refuses to be a research subject (cf. Chapter 5). Moreover, it must be emphasized that the obligation is to assume the burden of inconvenience, not substantial risk of physical or psychological harm.

It must be emphasized that many therapeutic innovations are perceived by their intended beneficiaries as benefits, indeed very precious benefits. A good case in point is the totally implantable artificial heart. While journalists were portraying the plight of its early recipients as terribly burdensome, candidates for implantation, whose only alternative was imminent death, found themselves in fierce competition for this scarce resource (434). The Utah program has developed criteria for selection of early recipients for this device (55, 739). In general these criteria are patterned after those of the Stanford Heart Transplant Program (137). The guiding principle for selection of such recipients is "to each according to medically defined need." The Working Group on Mechanical Circulatory Support of the National Heart, Lung, and Blood Institutes refers to "medically defined need" as "suitability criteria" (740, pp. 26–27) reflecting its position that eligibility for implantation should not reflect exclusively the judgment of physicians. Although most of the suitability criteria are noncontroversial, there are some adopted by both Stanford and Utah the interpretation of which may be problematic—e.g., psychosocial factors such as a history of poor compliance with medical advice or an inadequate social support system (434). Great care must be taken not to conflate legitimate psychosocial exclusion criteria with "social worth" criteria.

The Working Group calls attention to another important consideration for selection of subjects (740, p. 23):

> In the early stages of developing mechanical circulatory support systems there is a legitimate interest in efficiently evaluating the new device. It is both customary and ethically acceptable to supplement suitability criteria with various additional grounds contributing to effective collection of such data. Patient-subjects should be selected on the basis of their having the biological, psychological, and social attributes necessary to accomplish efficiently the evaluation of the device. Also, it would impede development of the device if some of its early recipients could not live through a period of evaluation because of concomitant diseases or inability to comply with medical instructions.

The Commission recommended in each of its reports on persons having limited capacities to consent that their involvement in research designed to test the safety and efficacy of innovative practices was justified if there was no other way to receive the modality being tested and the procedures designed to develop generalizable knowledge presented no more than minimal risk.

Community Consultation

It is widely believed that persons having even modest degrees of vulnerability are at a serious disadvantage in negotiations with investigators for informed consent to participate in research. They may be afraid to appear stupid by asking questions, afraid to risk relationships of dependency through apparently uncooperative behavior, afraid to seem to question the wisdom of the more learned physician, and so on. Barber suggests that this intimidation is maximized under conditions in which investigators approach prospective subjects one at a time to negotiate for informed consent (36, pp. 72–74): "The isolation of individuals one from another in the experimental situation is one source of the great power of experimenters.'' Similarly, Spiro claims that a physician having a close relationship with a patient can ordinarily persuade that patient to consent to nearly anything (665).

Community consultation is a device that may be used to reduce the degree of intimidation (398, p. 18). It consists of assembling groups of prospective subjects for the purpose of discussing plans to conduct research. In such assemblies, prospective subjects provide support for each other's efforts to secure what they consider adequate explanations. They should learn that their questions and concerns are dealt with responsively and respectfully by investigators.

Community consultation may also serve another very important function. Often it is not clear to investigators and IRB members what weight should be assigned to the various burdens and benefits of participation in a particular research project. In fact, groups of professionals may be unable to determine with any confidence whether any particular feature of a research project should be considered a burden. For example, it is commonly argued that, in the conduct of a randomized clinical trial, the deprivation of the good of personal care is an important burden (cf. Chapter 8). Identification of something as a burden is a value judgment, and it is always possible that the value judgments of the patients might differ from those of investigators and IRB members. At an assembly of prospective subjects, investigators can learn whether these patients consider the deprivation of personal care a burden and, if so, what weight they would assign to it.

Assemblies of prospective subjects were first convened by investigators largely in the interests of efficiency; this is a most efficient way to provide information of the sort necessary to recruit subjects and to present the elements of informed consent to large numbers of them. Such meetings have been conducted in drug companies to recruit employees as subjects as well as in schools, and prisons. In addition, investigators commonly meet with organizations devoted to dealing with the problems of persons having particular diseases; many members of these organizations either have or are related to persons who have the disease. Thus, there is often a high degree of sophistication about the disease among members of the organization. In recent years it has become increasingly apparent that meetings with assemblies of prospective subjects serve much more than the purposes of efficiency in recruitment; they also accomplish the goals of reducing individual intimidation and learning about the value judgments of prospective subjects.

Lottery Systems

In consideration of just distribution of scarce resources, one ordinarily attempts to devise rational criteria for determining who among seemingly similar candidates should be considered most eligible to receive the resources. For example, in determining who should be most eligible to receive medical therapy for "health crises," Outka argues that the guiding principle should be: "To each according to his essential need" (552). Essential needs are determined largely by medical criteria. How serious is the illness? What is the prognosis with or without the therapy? After the use of such rational criteria has been exhausted, there may still be more individuals eligible to receive the benefit than the supply can accommodate. Under these circumstances, distribution can be accomplished most fairly by resorting to either lottery or queuing (first come, first served) (552, 582, p. 252ff).

By analogy, several commentators have proposed that when the pool of suitable prospective research subjects is larger than necessary to accomplish the objectives of the research, the fairest device for the selection of subjects would be a lottery. Queuing seems unsuitable for most research projects, with the possible exception of well publicized clinical trials. While the theoretical basis for the use of lotteries in selection of subjects seems sound, such systems are not used often (cf. Chapter 8). Specific proposals for the development of lottery systems have been published by Capron (111), Fried (253, pp. 64ff), and Levine (398. pp. 51–57).

Biological and Social Criteria

Research subjects should have those attributes that will permit adequate testing of the research hypothesis. In most biomedical and in some behavioral research, these attributes can and should be stated precisely in biological terms. In some behavioral and in most social research, the attributes can and should be stated in social terms. An adequate statement of biological or social attributes that establishes eligibility for participation in a project includes criteria for exclusion as well as inclusion.

Selection of subjects based upon biological or social attributes serves purposes unrelated to other selection criteria discussed in this chapter. The purposes are more closely related to the norms calling for good research design and a favorable balance of harms and benefits. Much damage can be done by including in the subject population individuals who are not legitimately part of that population. The problem may be exemplified by considering the consequences of inappropriate selection of subjects for studies designed to develop diagnostic tests or therapeutic modalities. Owing to improper selection of subjects, the efficacy of a diagnostic test or therapy may be either overestimated or underestimated. A means to diagnose or treat a condition may be utilized inappropriately by others who read the results of the research, who may be unaware that they are applying these modalities to individuals differing in important respects from those who were subjects of the research. For an exhaustive discussion of the problems deriving from inappropriate or inadequate definition of the biological attributes of research subjects, see Feinstein (215).

The development of inclusion and exclusion criteria based upon biological and, at times, social attributes also serves the interests of producing more favorable balances of harms and benefits by identifying for purposes of exclusion those individuals who are most susceptible to injuries. This process is discussed in detail in Chapter 3.

In many studies it is necessary to include in the subject population individuals who are commonly called "normal controls" or "healthy volunteers." It should be recognized that there are no such persons. Normality and health are states of being that cannot be proved scientifically (399). Thus, such individuals should be described in scientific publications and research protocols as being free of certain specific attributes of nonhealth or non-normality.

Chapter 5

Informed Consent

For of this you may be very sure, that if one of those empirical physicians, who practice medicine without science, were to come upon the gentleman physician talking to his gentleman patient, and using the language almost of philosophy—beginning at the beginning of the disease, and discoursing about the whole nature of the body, he would burst into a hearty laugh—he would say what most of those who are called doctors always have at their tongue's end: foolish fellow, he would say, you are not healing the sick man, but you are educating him; and he does not want to be made a doctor, but to get well. (Plato, *Dialogues. Laws*, B. IX. C. 4)

Thus began the critical commentary on the doctrine of informed consent. Subsequently, a massive literature on this topic has accumulated with a conspicuous crescendo during the past 40 years. I do not intend to provide a comprehensive survey. In 1979, William Woodward announced the availability of a bibliography of over 4000 references to publications on informed consent (738); by now I suppose the number has at least doubled.

The empirical literature on informed consent is, in general, characterized by many flaws; some of the more common ones are discussed in the Section on Comprehension. Specific errors are noted in passing in relation to specific points. These problems notwithstanding, it is necessary to call attention to these studies that provide some knowledge of how informed consent is negotiated and what it accomplishes.

This chapter begins with an attempt to define informed consent and to identify its purposes and components. The definition for which I strive is sufficiently comprehensive to cover most contingencies that might arise in negotiating and documenting informed consent to various types of clinical research. After constructing this comprehensive model, I discuss some situations in which some or all of the processes and procedures are either unnecessary or detrimental to the interests of all concerned. Although everything in the comprehensive model is relevant to some clinical research, I have never seen a protocol for which even one-half of these considerations were relevant.

Ethical and Legal Bases

According to the President's Commission, "Although the informed consent doctrine has substantial foundations in law, it is essentially an ethical imperative" (571, p. 2). The President's Commission refers repeatedly to "ethically valid consent" and in this way reflects a perspective differing from that of federal regulations, which refer to "legally effective informed consent." In *Making Health Care Decisions*, the President's Commission provides an authoritative overview of the law and ethics of informed consent focusing on the patient-practitioner relationship (571); although much of this commentary is relevant to the subject-investigator relationship, I shall comment on some important distinctions later in this chapter. The requirement for informed consent is designed to uphold the ethical principle of *respect for persons* (377, 378). It is through informed consent that we make operational our duty to respect the rights of others to be self-determining, i.e., to be left alone or to make free choices. We are not to touch others or to enter their private spaces without permission. As stated by Justice Cardozo, "Every human being of adult years and sound mind has a right to determine what will be done with his own body . . ." (341, p. 526).

Respect for persons also requires that we recognize that some individuals are not autonomous. To the extent that their autonomy is limited, we show respect by protecting them from harm. Devices used to show respect for those with limited autonomy are discussed in Chapters 10 through 13 on special populations.

The legal grounding for the requirement for informed consent to research is based on the outcome of litigation of disputes arising almost exclusively in the context of medical practice (253, pp. 18–25). There is virtually no case law on the basis of which legal standards for consent to research, as distinguished from practice, can be defined; there is one Canadian case, *Halushka v. University of Saskatchewan* (341, pp. 569–572).

The law defines, in general, the circumstances under which a patient, or by analogy a subject, may recover monetary damages for having been wronged or harmed as a consequence of a physician's failure to negotiate consent adequately. Traditionally, failure to negotiate adequate consent was treated as a battery action. In accord with the view that *respect for persons* requires us to leave them alone, the law of battery makes it wrong *a priori* to touch, treat, or do research upon a person without consent. Whether or not harm befalls the person is irrelevant; it is the "unconsented-to touching" that is wrong.

The modern trend in malpractice litigation is to treat cases based upon failure to obtain proper consent as negligence rather than battery actions. The negligence doctrine combines elements of patient benefit and self-determination. To bring a negligence action, a patient (subject) must prove that the physician had a *duty* toward the patient, that the duty was *breached*, that *damage* occurred to the patient, and that damage was *caused* by the breach. In contrast to battery actions, negligence actions remove as a basis for the requirement for consent the simple notion that unconsented-to touching is a wrong; rather, such touching is wrong (actionable) only if it is negligent and results in harm. Otherwise, the patient (subject) cannot recover damages.

Under both battery and negligence doctrines, consent is invalid if any information is withheld that might be considered material to the decision to give consent.

For extensive surveys of the legal aspects of informed consent as it relates more particularly to research, see Annas et al. (23) and Katz and Capron (344). More recently, Katz has provided a lucid historical perspective on the legal doctrine of informed consent (343). I also wish to call attention to a brief but valuable editorial that deals with the roles of metaphysics, ethical principles, laws, and judgments in thinking about informed consent (331).

Definition of Informed Consent

Principle I of the Nuremberg Code provides the definition of consent from which definitions contained in all subsequent codes and regulations are derivative (emphasis supplied).

> The *voluntary* consent of the human subject is absolutely essential.
>
> This means that the person involved should have *legal capacity* to give consent; should be so situated as to be able to exercise *free power of choice*, without the intervention of any element of force, fraud, deceit, duress, over-reaching or other ulterior form of constraint or coercion; and should have sufficient *knowledge* and *comprehension* of the elements of the subject matter involved as to enable him to make an understanding and enlightened decision.

Thus the consent of the subject in order to be recognized as valid must have four essential attributes. It must be competent (legally), voluntary, informed, and comprehending (or understanding).

It is through informed consent that the investigator and the subject enter into a relationship, defining mutual expectations and their limits. This relationship differs from ordinary commercial transactions in which each party is responsible for informing himself or herself of the terms and implications of any of their agreements. Professionals who intervene in the lives of others are held to higher standards. They are obligated to inform the lay person of the consequences of their mutual agreements.

According to Katz (341, p. 521) and Holder (305), the relationship between investigator and subject is a fiduciary relationship. The fiduciary is bound by law "to the duty of constant and unqualified loyalty" (305). Although she argues that the investigator-subject relationship is a fiduciary relationship, Holder acknowledges some differences. For example, federal regulations require that under certain circumstances confidential information must be shared with agents of the federal government (Chapter 7). As Holder observes, "The fiduciary . . . has no authority to disclose any information from confidential records without explicit alteration of normal fiduciary requirements." Further, as discussed in Chapter 6, many institutions offer free medical therapy for research-induced injury; in the traditional fiduciary relationship, no remedy is required for non-negligent harm. Other ways in which public policy acknowledges and individual authors recommend distinctions between the investigator-subject relationship and fiduciary relationships (as exemplified in the physician-patient relationship) are discussed subsequently in this chapter and in Chapter 8. These distinctions notwithstanding, I must agree with Holder's prediction that "courts would hold that a physician establishes a fiduciary relationship with anyone with whom he or she interacts professionally regardless of the reason for the interaction. . . .

One could not, in effect, contract with the subject by saying 'I am a physician, but you agree that because I am not acting as a physician, I owe you no fiduciary obligation.' "

The process of creating the condition of informed consent commonly is seen as having two components. The first component is that of informing; this is the transmission of information from the investigator to the prospective subject. The second component is that of consenting, signified by that persons' declaration that, having assimilated the information, he or she is willing to assume the role of subject. Moreover, it may be assumed that these two components are accomplished sequentially; i.e., information transmission is followed by consent or refusal. Katz, who envisions the process ideally as an invitation to the prospective subject to join the investigator as a partner in a collaborative venture, portrays a more complex interaction (342).

> Informed consent would entail, if it is truly seen as an invitation, asking for consent, seeking authorization to proceed, and not making a demand under the guise of a symbolic egalitarian gesture. It would necessitate sharing knowledge and admitting ignorance, answering questions and identifying unanswerable questions, appreciating doubts and respecting fears . . . It requires that the interaction between investigator and subject become a partnership, giving the subject the right to determine what should be done for and with him, and forcing the investigator to be explicit in what he wants to do and why. Thus the controversy over the subject's capacity and incapacity to understand, on which the debate about informed consent has focused, is a displacement from the real issue, which is the dread of an open and searching dialogue between the investigator and his subject. This displacement is caused by the unacknowledged anxiety over making the invitation in the first place.

As individuals communicate with one another toward the goal of informed consent, each of the component processes may occur in any order. The most appropriate single word for these communications is "negotiation;" these individuals are negotiating informed consent. Many documents use different words; investigators often are admonished or advised to secure or to obtain informed consent. Such words do not capture the full dimensions of the desired interactions; viz, an interaction involving dialogue, encounter, and so on. The negotiations will be presented as if they had four separate component parts: 1) informing, 2) assessment of the prospective consentor's comprehension, 3) assessment of the prospective consentor's autonomy, and 4) consent. In the real world, negotiations for informed consent are virtually never conducted as four separate component processes.

Informing

This section identifies the elements of information that should be communicated to the prospective consentor. In each case I shall note whether the same or a similar element is included in DHHS Regulations. The reader may assume, unless otherwise indicated, that FDA regulations call for identical elements of information.

1. Invitation   The negotiations should begin with a clear invitation (not a request or demand) to the individual to become a research subject. The implications of playing the role of subject, rather than any alternative role available to the person, should be made clear. Most importantly, when one agrees to play the role of subject, one ordinarily agrees to become, at least to some extent, a means to the ends of another.

If the protocol involves innovative therapy, the physician-investigator may be held liable for failure to negotiate informed consent merely by virtue of having failed to explain that the procedure used represented a departure from customary practice (297, p. 310).

2. Statement of Overall Purpose   DHHS requires: "A statement that the study involves research, (and) an explanation of the purposes of the research. . ." (Section 46.116a). There should be a clear statement of the overall purpose of the research. When appropriate, one should state that there is not only an immediate purpose but also a larger ultimate purpose. For example, the immediate purpose might be to develop a more sophisticated understanding of normal kidney function. If the immediate purpose is achieved, one hopes this information might contribute to our ability to identify and treat persons with diseases of the kidney.

One of the most important consequences of stating the purpose of the research is that it alerts prospective subjects to decline participation in research the goals of which they do not share (262, 287). For example, some individuals might not wish to contribute to the general fund of information that would enhance our capacities for genetic engineering. Some others may decline participation in research that might identify their racial or ethnic group as having certain qualities, e.g., as having lower intelligence than the general population.

This element of information may partially duplicate Element 7. In some situations, it might be appropriate not to reveal the true purpose of the research (Element 14).

3. Basis for Selection   Prospective subjects should be told why they have been selected. Ordinarily it is because of some specific disease or other life situation that they or their relatives might have. In other cases, the investigator may presume that they do not have that disease or condition, as when they are asked to serve as controls. Occasionally, subjects will exclude themselves because they recognize that they do not have the attributes the investigator thinks they have. Such self-exclusions are in the interests of all concerned.

Prescreening, which is the performance of diagnostic or other testing solely for purposes of identifying individuals who are eligible for inclusion in protocols, requires consent negotiations and often documentation according to standards established for research. The negotiations should include all relevant elements of information. In addition, the consequences of being found eligible for participation in the protocol should be explained. Prospective subjects should receive a fair explanation of the protocol in which they will be invited to participate in the event they "pass" the tests. They should also be informed of the consequences of "failing," as described in Chapter 3.

4. Explanation of Procedures DHHS requires: "A statement (of) . . . the expected duration of the subject's participation, a description of the procedures to be followed, and identification of any procedures which are experimental" (Section 46.116a). One should be meticulous in identifying those procedures that are done solely in the interests of research. In situations in which persons are invited to play dual roles—patient and subject—it is important that they be informed which procedures will be recommended to serve their personal interests even if they decline participation in the research. This fosters their ability to assess the burdens of the role of subject, e.g., the inconveniences, risks of injury, and, at times, economic burdens. In some protocols, procedures that would be performed in the routine practice of medicine will be done more frequently; the incremental burden should be calculated and expressed clearly.

The description of each procedure or interaction should be designed to fit the interests of the prospective subject. Thus, if one proposes to draw blood from a vein for purposes of assaying some chemical, the prospective subject might be expected to be more interested in knowing how much blood will be drawn, how often it might be repeated, where to go to have the procedure done, and what practical consequences to him or her there might be of the results of the assay than in the details of the assay technique. Further, he or she might be interested in who (if not the individual negotiating the consent) might draw the blood and what their experience and qualifications are. The language necessary to convey the meaning of each of these bits of information will vary enormously depending upon the experience of the prospective subject with previous blood drawing. For example, it might be possible to relate the amount of blood to be drawn in terms of what fraction it is of the amount removed when one donates a pint of blood or in relation to the amount the individual has had drawn for various diagnostic tests in the past.

In addition to describing each of the research procedures, particularly as they affect the prospective subject, it is generally advisable to anticipate that the prospective subject will want to know: 1) With whom will I interact? 2) Where will the research be done? 3) When will the research be done? 4) How often will the various procedures be performed? 5) How much of my time will be involved?

1. Most prospective subjects will be reassured to learn that the individual negotiating with them for informed consent will play a key role in the actual conduct of research. However, many protocols require interactions with a larger number and variety of professionals and their assistants. In general, it is better to advise prospective subjects of the numbers and types of individuals with whom they will interact rather than to surprise them during the course of the research. For example, some prospective subjects may have strong biases against having physical examinations by individuals of the opposite sex or by students.

2. For various reasons prospective subjects will be interested in where the research is to be conducted. Some might feel reassured to learn that a questionnaire will be administered in their own homes; others might regard this as an unwelcome intrusion. In some cases, e.g., in some hospitals, the research unit might be more or less attractive to prospective subjects than the alternative facilities available to those who choose the role of patient. A statement of where the research is to be done will also allow the prospective subject to assess the amount of inconvenience there might be in traveling to and from that location.

3. An explanation of when the research is to be done will allow prospective subjects to determine whether there are any essential time conflicts with their own schedules. Some sorts of research are dependent upon repeating observations at precisely timed intervals. If this is discussed frankly at the outset, it may be possible to negotiate a mutually satisfactory schedule; alternatively, it may be found that a subject must drop out during the course of the research owing to a prior commitment.

4. A precise statement as to how often various procedures will be performed will also assist the subjects in assessing the totality of their personal commitment of time and other inconvenience. In some research, it is necessary to have various follow-up procedures done at intervals as long as a year or more. If prospective subjects know they will not be available that much later, they can advise the investigator that full participation will be impossible.

5. In explaining how long the research will take, there should be an estimate of not only how much time each component of the research may reasonably be expected to occupy, but also the total duration of the research.

In complicated research activities, it is occasionally of value to invite the prospective subject to visit the site of the proposed research (e.g., the metabolic research unit, the office of the investigator, the physiology laboratory) to see the personnel, facilities, and apparatus that will be involved. In complicated activities in which persons are invited to play the role of patient and subject simultaneously, special care should be given to explaining the differences between the role of patient and that of patient-subject. The information discussed above should be elaborated as follows. "If you agree to participate in this research you will be interacting with (specify) additional types of individuals; it will take (specify) additional time; procedures that might have been done *n* times will be repeated *n* plus *x* times; the location will change in a specific way . . . " and so on.

5. Description of Discomforts and Risks DHHS requires: "A description of any reasonably foreseeable risks or discomforts to the subject" (Section 46.116a). For a survey of the risks, discomforts, and inconveniences of research and what constitutes an adequate description of them, see Chapter 3. Here we shall focus on how to determine which of them ought to be disclosed.

There are distinct perils to the process of informed consent presented by either overdisclosure or underdisclosure. Some of the perils of overdisclosure were documented by Epstein and Lasagna (206). They presented consent forms of various lengths and thoroughness to prospective subjects of a drug study. They found that the more detail was included, the more likely were the prospective subjects to be either confused or intimidated. In their study they found a high incidence of refusal to take an experimental drug based upon its apparent danger. At the conclusion of the study, they informed the individuals who refused that the drug they were describing—the drug they refused to accept money for taking—was aspirin; many of those who refused were regular users of aspirin. Almost all of them reported that although they had declined participation in the study, they intended to continue to use aspirin as they had before.

In the medical practice context, a study was done to determine the influence of full disclosure on the willingness of patients to consent to angiography (11). The consent forms published with the report of the study seem to contain an adequate description of the risks of the procedure, a small probability of serious complications which might include death. In understanding the implications of this study, it is important to know that, because angiography is a diagnostic procedure, its performance may or may not result in any information upon which further therapy might be recommended. Of 232 patients asked to consent, all but 2% did. Responses to the questionnaire indicated that the majority of patients were pleased to have the information conveyed in the consent form. The author concludes that he is convinced of the value of disclosure to the extent contained on the form.

Underinforming, on the other hand, may be perilous to all participants in the research process. Inadequately informed subjects may make incorrect choices. Harmed subjects who had not been informed of the possibility of that particular harm might sue the investigator.

With regard to determining which risks should be disclosed, the standards provided in ethical codes and regulations are not particularly instructive. DHHS's standard is "reasonably foreseeable" and Nuremberg's is "reasonably to be expected." The Declaration of Helsinki simply states that "each potential subject must be adequately informed of the . . . potential hazards of the study and the discomfort it may entail." However, Principle I.7 requires that, "Doctors should abstain from engaging in research projects . . . unless they are satisfied that the hazards involved are believed to be predictable."

Adherence to this principle would, of course, preclude virtually all research designed to develop new therapies. As noted in Chapter 3, during the development of new therapies, there is risk of injury the very nature of which remains to be determined; the possibility of unforeseeable risks must be disclosed (DHHS, Section 46.116b) (*infra*).

The legal criterion for disclosure in the context of medical practice is "material risk," i.e., any fact that is material to the patient's decision must be disclosed (297, p. 233). The determination of which risks are material in that they must be disclosed may be accomplished according to three different standards or tests (173, p. 25). Until recently, the prevailing standard was that of the reasonable physician; the determination of whether any particular risk or other fact should be disclosed was made on the basis of whether it was customary to do so in the community of practicing physicians.

The standard that is applied most commonly is the "reasonable person" or "prudent patient" test. In the case of *Canterbury v. Spence* (104), the court held that the disclosure required was determined by the "patient's right of self-decision," a right that can be "effectively exercised only if the patient possesses enough information to enable an intelligent choice. . . . A risk is thus material when a reasonable person, in what the physician knows or should know to be the patient's position, would be likely to attach significance to the risks or cluster of risks in deciding whether or not to forego the proposed therapy."

Some courts have adopted the rule that a risk is material if the particular patient making the choice or decision considers it material. Of the three standards, this rule, which some authors call the "idiosyncratic person" standard, is most responsive to the requirements of *respect for persons*. However, as the *Cobbs v. Grant* court noted in its articulation of the reasonable person standard (141):

> Since at time time of trial the uncommunicated hazard has materialized, it would be surprising if the patient-plaintiff did not claim that had he been informed of the dangers he would have declined treatment. Subjectively he may believe so with the 20–20 vision of hindsight, but we doubt that justice will be served by placing the physician in jeopardy of the patient's bitterness and disillusionment.

Each of these three legal standards was designed to guide determinations as to whether a patient should have disclosed a particular fact; the determinations they intend to guide are to be made when the patient-plaintiff has already been injured.

The approach to planning negotiations for consent to research should be different. In my view, the minimum amount of information that should be imparted by the investigator to each and every prospective subject should be determined by the reasonable person standard. Then, in the course of the consent negotiations, the investigator should attempt to learn from each prospective subject what *more* he or she would like to know.

Whether or not a prospective subject should be permitted to choose less than the minimum amount of information determined by the reasonable person standards is a matter of some controversy. Freedman contends that it is most respectful of autonomous persons to respect their relatively uninformed decisions (246, 247). Veatch, who agrees in principle, argues that anyone refusing to accept as much information as would be expected of a reasonable person should not be permitted to be a research subject (696). My approach offers each at least the amount of disclosure determined by the reasonable person standard. Prospective subjects are free to be more or less interested in any particular bit of information. While they may reject a fact by either ignoring or forgetting it, they may not do so without having been made aware, however transiently, of its existence.

How does one determine what the reasonable person or prudent prospective subject might wish to know? In the context of research, the Institutional Review Board (IRB) determines the minimum standards for disclosure of risk. However, the capability of the IRB to perform this function has been challenged by some commentators such as Veatch (693), who suggests that since the IRB is dominated by scientists it does not truly reflect the need to know of the reasonable layperson.

Fost has proposed that a surrogate system might be used (232). The surrogates would be selected from a population that matches as closely as possible that from which prospective subjects might be drawn in all respects but one. They should be aware that, for some reason, they are not eligible to become subjects, although they are asked to pretend that they are. The surrogate system is designed to inform the negotiator for informed consent of the range and diversity of factors of material interest to prospective subjects. Fost emphasizes that he is *not* proposing this system as a *necessary* adjunct to the approval of research proposals.

DHHS requires that, when appropriate, there should be: "A statement that the particular treatment or procedure may involve risks to the subject (or to the embryo or fetus if the subject is or may become pregnant) which are currently unforeseeable" (Section 46.116b). The risks to which this requirement refers are those discussed in Chapter 3 as risks the nature of which remain to be determined.

In some cases prospective subjects are called upon to assume responsibility for minimizing the chance of harm by performing certain functions during the course of the research. For example, when a woman of child-bearing age participates in a research activity in which there is risk to the fetus, the nature of the risk being either known or unknown, she should be advised that if she wishes to be a subject she should avoid becoming pregnant. Her plans for avoiding conception should be reviewed during the consent negotiations. At times, if her plans seem inadequate, it will be necessary either to exclude her from the research or to ask her to agree to more certain plans for contraception. She should further be advised that if, during the course of the research, she deviates from the plans discussed at the outset, she should advise the investigator immediately.

In cases in which there is a high probability of serious injury, pain, or discomfort, stress inoculation should be considered. For a good recent review of the literature on this subject, see Janis (327).

> Stress inoculation involves giving people realistic warnings, recommendations, and reassurances to prepare them to cope with impending dangers or losses. At present, stress inoculation procedures range in intensiveness from a single 10-minute preparatory communication to an elaborate training program with graded exposure to danger stimuli accompanied by guided practice in coping skills, which might require 15 hours or more of training. . . . The process is called stress inoculation because it may be analogous to what happens when people are inoculated to produce antibodies that will prevent a disease.

Stress inoculation has been used with success to prepare people for surgery and childbirth and in the treatment of various emotional and physical disorders.

6. In Case of Injury   For research activities involving more than minimal risk, DHHS and Food and Drug Administration (FDA) regulations both require a statement on the availability of medical therapy as well as compensation for disability. This element of information is discussed in Chapter 6.

7. Description of Benefits   DHHS requires: ''A description of any benefits to the subjects or to others which may reasonably be expected from the research'' (Section 46.116a). For a survey of the benefits of research and what constitutes an adequate description of them, see Chapter 3. Often it will be appropriate to incorporate the description of benefits in the statement of overall purpose (Element 2). Decisions as to which benefits to describe and how to describe them are generally negotiated between the IRB and the investigator. There is usually little controversy over the acceptability of accurate descriptions of benefits to society or, if they are intended, direct health benefits of nonvalidated practice modalities. It must be emphasized that the benefits are hoped for, not something that can be guaranteed. With nonvalidated therapies, it should be made clear that, while a major purpose is to attempt to bring direct benefit to the subject, an additional purpose is to try to develop a systematic body of new knowledge. Thus, the prospective subject is not the only intended beneficiary of the activities.

Proposals to include descriptions of psychosocial and kinship benefits are often controversial (Chapter 3). Descriptions of economic arrangements are incorporated in Element 10.

It is appropriate to describe the possibility of fortuitous direct health-related benefits? This question, too, often provokes argument. When it can be justified, an acceptable statement may take the following form.

> The purpose of these tests is to develop knowledge that may be useful in developing improved therapies for your disease. Thus, we hope to provide benefits in the future for persons like you. It is possible, though very unlikely, that the results of one of the tests may aid your physician in planning your treatment; if so. . . .

8. Disclosure of Alternatives   DHHS requires: "A disclosure of appropriate alternative procedures or courses of treatment, if any, that might be advantageous to the subject" (Section 46.116a). The obligation to disclose alternatives arises when a component of the protocol is designed to enhance the well-being of the subject, as in the testing of nonvalidated therapies. There should be at least a statement of whether there are any alternatives and, if so, a general statement of their nature. For example, "There are other drugs available for the treatment of your condition and, in some cases, some physicians recommend surgery." More elaborate statements are in order when a choice between therapies may depend on the prospective subject's personal values or when physicians may disagree as to whether there are superior alternatives to the proffered practice modality.

> While we hope to demonstrate that this procedure is just as likely to cure your disease as the much more extensive surgery recommended by the majority of surgeons, we cannot be certain that it is until the study is over. The main disadvantage of the standard surgery is that it causes impotence in almost all cases; the procedure we are testing does not cause this problem. In the event this procedure fails to cure your disease, we shall not be able to offer the standard surgery because your disease will have advanced too far for the standard surgery to work.

For further discussion of alternatives, see Chapters 3 and 8.

The work of McNeil and her colleagues provides some important insights into disclosures of alternatives and the consequences of such disclosures. In an early study they found that some patients with operable lung cancer chose radiation therapy over primary surgical therapy even though the latter offered a higher probability of long term survival (483). Radiation therapy is more attractive to some, apparently because there is no perioperative mortality. This choice is rather straightforward because it involves almost exclusively quantity and not quality of life considerations.

In subsequent studies, McNeil et al. asked 37 healthy volunteers to imagine that they had cancer of the larynx in order to determine how much longevity they would exchange for voice preservation (481). The volunteers were 37 middle and upper management executives and 12 firefighters, each group averaging age 40. Using principles of expected utility theory to develop a method for sharpening decisions, they found that to avoid artificial speech, the firefighters would trade off about 6% of their full life expectancy. Although executives would trade off more—an average of 17%—the difference was nonsignificant. "Although most subjects were willing to accept some decrease in long-term survival to maintain normal speech, virtually none would ever accept any decrease below 5 years."

Subjects were informed that with surgery 60% could expect to survive 3 years and with radiation, 30 to 40%. The subjects reacted as follows: if radiation offered only a 30% chance of survival for 3 years, practically none would decline surgery. By contrast, if the chance of survival for 3 years were 40%, 19% of the subjects would choose radiation therapy alone and 24% would choose radiation followed by delayed laryngectomy if necessary. The authors conclude "that treatment choices should be made on the basis of patients' attitudes toward the quality as well as the quantity of survival."

In a third set of studies involving a much larger number of subjects, McNeil et al. demonstrate that disclosure of alternatives may be a very complicated matter (482). In these studies, 238 ambulatory patients with different chronic medical conditions, 491 graduate students, and 424 physicians were asked to imagine that they had lung cancer; further, they were asked to choose between radiation and surgical therapies after they had received various sorts of information comparing the therapies. In all three populations, surgery was chosen over radiation when information was presented in terms of probability of living rather than in terms of probability of dying. This means that subjects would choose rather consistently a 90% chance of living over a 10% chance of dying even though it is the same thing. Surgery was chosen over radiation when the information was presented in terms of life expectancy rather than in terms of cumulative probabilities (what percent will die per unit of time). In this particular study, this is not surprising owing to the relatively high probability of perioperative death with surgery, which was disclosed in the cumulative probability condition but not in the longevity condition, in which outcomes were expressed in terms of average years of life expectancy. When the treatments were identified as either surgery or radiation, this had a very powerful effect that apparently took precedence over other considerations. According to the authors, this "indicates that people relied more on preexisting beliefs regarding the treatments than on the statistical data presented to them."

In short, different presentations of what might appear to be the same information about alternatives can yield very different choices. As noted by the authors, the preferences expressed by the physician-subjects were subject to the same biases as were those of the students and patients. The authors suggest that these biases "are likely to play an important part in the advice (physicians) give to patients."

I find the results of these studies intimidating. The authors suggest that these effects might be offset by varying the types of data presented to prospective patients or subjects.

> If a patient prefers surgery . . . whether the data are presented as cumulative probabilities or as life expectancy and whether the probabilities are presented in terms of mortality or in terms of survival, the preference may be assumed to be reasonably certain. If, on the other hand, a change of presentation leads to a reversal of preference, then additional data, discussions, or analyses are probably needed.

9. Confidentiality Assurances   DHHS requires: "A statement describing the extent, if any, to which confidentiality of records identifying the subject will be maintained" (Section 46.116a). This element of information is discussed in Chapter 7.

10. Financial Considerations   DHHS requires: "When appropriate . . . a statement of . . . any additional costs to the subject that may result from participation in the research" (Section 46.116b). There should be a careful discussion of all the financial implications of agreeing to the role of subject. Subjects should be told who is expected to pay for what. For further discussion of this matter, see Chapter 3.

Although not mentioned in regulations, it is equally important to explain the economic advantages and material inducements. Economic advantages may include access to improved facilities; free physical examinations, treatments, or food; or subsidized transportation or baby sitters. Cash payments should be detailed meticulously, including bonuses for completion of a series of tests or partial payments for early termination (414). For example:

> 1. Participation in this project involves reporting to the clinic every Wednesday morning for six consecutive weeks. You will be paid $10.00 for each of the six weekly sessions. If you participate satisfactorily in all weekly sessions, you will receive a bonus of $30.00; thus, the total payment for complete participation in this project will be $90.00.

> 2. If your participation in this protocol is terminated by the investigator in the interests of your safety, you will receive the $30.00 bonus in addition to $10.00 for each session that you have completed. On the other hand, if you choose to withdraw from this study without providing a suitable reason, you will be paid only $10.00 for each session you have completed. An example of a suitable reason would be the development of unpleasant side effects.

3. You will be paid $10.00 for participation in the initial screening procedures. If the results of these procedures indicate that you are eligible for inclusion in the rest of the protocol, your further participation will be invited. In that case, you will be offered an additional $45.00.

There is now a public debate on whether patients or subjects are entitled to patent or royalty rights to marketable products developed from their cells or tissues. A hearing on this matter was held on October 19, 1985, by the Subcommittee on Investigations and Oversight, Committee on Science and Technology, U.S. House of Representatives (438). Congressional attention was attracted to this issue by two cases in which patients or subjects claimed such rights to monoclonal antibody-producing hybridomas derived from their cells. One of these cases was settled by, among other things, assigning to the patient's family an exclusive license for use of the marketable product in Asia (620). This case is highly peculiar in several respects, including the fact that the patient's son is a molecular biologist who collaborated in the development of the hybridoma from his mother's cells. Litigation on the second case has not yet been completed (166).

Murray has written an excellent philosophical analysis of property rights in human tissues (520). He argues that cells and tissues "contributed" by patients and subjects should be seen as "gifts." He further argues that such contributors will feel wronged if their gifts are exploited to create substantial commercial gains for researchers and their academic or industrial sponsors. Accordingly, "biotechnology entrepreneurs should be strongly encouraged to adopt policies that assure that individuals and corporations do not unduly prosper from the gifts of individual's tissues and organs."

I agree with the intent of Congress expressed in the Organ Procurement and Transplantation Act, in which it proscribes "sale for valuable consideration of human organs. . . ." People should not be encouraged to sell either themselves or parts of themselves. Although the act excludes explicitly "replenishable tissues such as blood or sperm," for various reasons I do not think that persons should be encouraged to sell *any* parts of themselves. It may be somewhat hypocritical or evasive, but when we give people money in exchange for "replenishable tissues" we usually call it a payment for services rather than a purchase of tissue. Moreover, the amounts of money that change hands in such transactions are trivial when compared with the amounts at stake in some types of research that could yield marketable products from human materials.

In my view, which is elaborated in a recent publication (438), it would be nearly impossible to identify which subjects, if any, should be entitled to a share in the profits from the sale of such products. Among other reasons, most efforts to develop valuable products fail. Researchers may take cells from hundreds of people before they succeed. Are all of these subjects entitled to negotiate for a share in the proceeds? Or, instead, is it more like a lottery in which there will be but one winner?

It is a commonplace that all scientists stand on the shoulders of their predecessors. They all draw on the work of those who have earlier discovered, validated, and published. Such is also the case for these research subjects. Should we have discussed potential market value with all of the many men and women who donated their cells so that researchers could learn how to make hybridomas and monoclonal antibodies in the first place? If so, we are talking about a very large number of people.

Until public policy in this field has been settled, I recommend that in appropriate cases—cases in which it appears likely that a marketable product will be developed from human material—information to the following effect should be added to consent forms. "In the event this research project results in the development of any marketable products, you will have no right to share in any profits from its sale and no obligation to share in any losses."

On this point I disagree with Murray, who argues that such language would inappropriately raise expectations of commercial profitability and perhaps decrease the numbers of people willing to donate their tissues to science and education. Moreover, "we might find an alteration in the public's attitudes towards science and scientists such that they would no longer be seen as engaged in a socially valuable enterprise the market value of which is only a pale reflection of its worth to society" (520). He and I agree that the problem is likely to arise only in rare cases.

11. Offer to Answer Questions   DHHS requires: "an explanation of whom to contact for answers to pertinent questions about the research and research subjects' rights, and whom to contact in the event of a research-related injury . . ." (Section 46.116a). Devices for enhancing subjects' opportunities to have their questions answered are discussed later in this chapter.

12. Offer of Consultation   When appropriate, there should be a suggestion that the prospective subject might wish to discuss the proposed research with another. When the proposed research entails a consequential amount of risk, discomfort, or inconvenience to the prospective subject or when there are difficult choices between reasonable alternative therapies, consultation with a trusted advisor should be suggested, particularly if there are factors limiting the prospective subject's autonomy or capacity for comprehension.

Most commonly, the trusted advisor will be the prospective subject's personal physician when he or she has no involvement in the research. When the prospective subject has no personal physician or when the personal physician is involved in the conduct of the research, it might be appropriate to offer the services of another physician. In other cases, depending upon the nature of the research, the prospective subject might wish to consult a trusted minister, lawyer, some other appropriate professional advisor, or a friend who need not be a professional.

Prospective subjects should be informed that they are free to choose whether to consult an advisor and, if they do, whether the advisor should participate in the negotiations for consent. If the involvement of a third party is mandatory, as it will be in some cases, the prospective subject should be so informed (*infra*). If participation could have economic or other consequences for someone other than the prospective subject, there should be advice to consult with such persons (see Chapter 3).

13. Noncoercive Disclaimer   DHHS requires: "A statement that participation is voluntary, refusal to participate will involve no penalty or loss of benefits to which the subject is otherwise entitled, and the subject may discontinue participation at any time without penalty or loss of benefits to which the subject is otherwise entitled" (Section 46.166a). Such assurances are especially important when there is a relationship between the investigator or any colleague of the investigator and the prospective subject which has any potential for coercion; such relationships include physician-patient, employer-employee, faculty-student, and so on. Patients, students or employees, or persons who are applying to a professional or to an institution for one of those roles, must be assured that they will not deprived of their rights to enjoy all of the usual expectations of such roles as punishment for having refused (cf. Chapter 3).

In some situations it may be impossible, or at least very difficult, to make these promises. For some diseases, nonvalidated therapies are all that is available for their definitive treatment. For example, the only definitive treatment for some children with growth defects is synthetic human growth hormone (HGH); this is available only to those who agree to participate in research designed to test its safety and efficacy. HGH, which is in very short supply, is supplied only to investigators who agree to administer it according to its sponsor's research protocol. Thus, the offer to a dwarf's parents of freedom to refuse without prejudice would be rather vapid. Standard medical care consists of watching the child fail to grow over the years.

A somewhat analogous situation is presented when patients have malignant tumors or other inevitably lethal diseases for which all standard modes of therapy have been tried without success. The only definitive approach to therapy might be a nonvalidated therapy. However, in this situation patients who refuse to consent may at least be assured that they will continue to receive all the supportive and palliative therapy at the physician's disposal.

Often, when the therapeutic innovation involves a manipulation of the health delivery system or the introduction of new types of health professionals, a particular institution can offer no alternative (392). For example, methadone maintenance programs may be obliged to state that as a precondition of participation in the program, the patient (subject) is obliged also to participate in some forms of research. Similarly, experimental clinics have been established to assist smokers in abandoning this habit; if the smoker is unwilling to participate in the research, there may be no alternative facility or service offered by the institution.

A rather novel example was presented by a proposal to videotape the interactions of certain individuals on a psychiatric research ward. The individuals whom the investigators wished to study comprised approximately one-third of the ward's total population. However, the videotapes would also record the activities of the other two-thirds of the research ward population as well as the staff and visitors. Thus, in conducting negotiations for informed consent with the two-thirds who were not the intended subjects of the videotape research, it was necessary to inform them that their continued participation in their own research projects would also involve videotaping of their activities. Those who objected to this did, in fact, have their relationships to the institution prejudiced. The only way they could escape was to discontinue their participation as research subjects in studies in which they wanted to be involved (303).

In projects involving questionnaires or interviews, it is customary to advise subjects of their freedom to refuse to answer any particular question that they may find embarrassing or offensive.

In some projects it may seem ludicrous to offer an opportunity to withdraw. For example, it seems pointless in such evanescent interactions as the drawing of a single blood sample or in cases in which the research procedures will be done while the patient is under anesthesia. Even in such studies, it is customary to remind subjects that even though they have consented, they may change their minds before the procedure is actually done.

Limits to Liberty. All ethical codes and regulations require that subjects should always be at liberty to withdraw without prejudice; none suggest any limits to this freedom. This requirement derives from the assumption that the subject is always doing something for the good of others; such supererogatory acts are generally not considered obligatory.

Levine and Lebacqz have surveyed several alternative views of the role of research subject (441). Several authors have identified participation in research, when such participation involves little or no risk, as a moral obligation; this is the argument used by McCormick and Pence to justify the involvement of children as research subjects (Chapter 10). Ackerman contends that parents have a "duty" to guide the child in becoming the "right kind of person." If one agrees that service to the community as a subject of certain types of research should be considered an obligation, one might extrapolate from this position to contend that under some circumstances the subject might not have the freedom to withdraw. On the other hand, one could equally argue that individuals having such a duty should be free to choose when they will exercise it.

But what if they express a free choice to exercise this duty now? Once they have begun to serve in the role of subject, can they be construed to have made a commitment? Some authors, including Ramsey (582), Jonas (329), and Katz (341), characterize research as a "joint venture" between the subject and the investigator; ideally, they may be considered "coadventurers." In this view, the agreement to participate in research is a form of promise. Ethically, promises may be broken only for certain justifying reasons, not simply at the whim of the promisor.

Newton argues that the completely unfettered freedom to withdraw required by ethical codes and regulations "is an anomaly in ethics, since it appears to be . . . in direct conflict with (one's) ordinary duty to keep (one's) promises" (540). She recommends that future formulations of regulations should recognize that the relationship between the investigator and subject is "binding on both sides, hence to be taken very seriously on both sides, and go on to specify the circumstances that shall be taken to negate or cancel that mutual commitment."

There is yet another line of reasoning that yields the conclusion that participation in one type of research may be seen as an obligation (398). The subject ordinarily chooses to receive an innovative therapy because of the good (benefit) he or she expects to derive. In some circumstances, the subject may be viewed as having assumed a "reciprocal obligation" of bearing the inconvenience of tests necessary to prove its safety and efficacy. This argument relies on the premise that the innovative therapy is a good that is provided by society; hence the reciprocal obligation to serve society by participating in research designed for its validation. It must be emphasized that this obligation is to assume a burden characterized as "mere inconvenience;" it is not extended to create an obligation to assume risk of physical or psychological harm.

Relying on these and other arguments, Caplan concludes that persons have a limited moral obligation to volunteer to serve as subjects in biomedical research (108).

Whether or not investigators should be allowed to impose limits on the freedom of subjects to withdraw from participation in research is an issue that requires further ethical analysis. Limits to this freedom are implicit in the Commission's recommendations for research involving the special populations. For example, in *Research Involving Children*, Recommendation 7b states: "A child's objection to participation in research would be binding unless the intervention holds out a prospect of direct benefit that is important to the health or well-being of the child and is available only in the context of the research.'' In the commentary under this recommendation, the Commission recognizes the validity of making some therapies available only in the context of protocols designed for their validation.

What other limits we might impose on the subject's freedom to withdraw remain to be determined. I do not foresee, however, the development of a satisfactory argument in defense of punishing withdrawal through deprivation of benefits to which the subject is otherwise entitled.

DHHS requires: "When appropriate . . . the following . . . information shall also be provided . . . The consequences of a subject's decision to withdraw from the research and procedures for orderly termination of participation by the subject . . ." (Section 46.116b). Abrupt withdrawal from various forms of drug therapy may present serious hazards to the subject's health. In such cases, the subjects should be informed of the perils, and plans must be made in advance for the orderly conversion to some other form of therapy if such is available. If there is no alternative therapy, the implications of this should be made clear.

Shortly after Barney Clark became the first recipient of a totally implantable artificial heart, it was reported that he had been given a key with which he could turn it off if he decided to withdraw from the "research" (581). Although he probably was never actually handed a key, much was made of this "fact" in publications addressed to both lay and professional audiences. In response to this, the Working Group on Mechanical Circulatory Support of the National Heart, Lung and Blood Institute recommended (740, p. 23):

> In some cases, it may become appropriate to consider turning off the (artificial heart) even though the device is functioning well. At times the condition of the patient will be such that continued use of the device serves no useful purpose. Decision-making in such circumstances will be similar to that of choosing to disconnect a patient from a ventilator. At times continued use of an (artificial heart) would result in prolongation of a life that some people might find useful, but the patient may choose to discontinue use of the device, knowing that this act will result promptly and certainly in his or her death. Such situations cannot be resolved ethically simply by handing a key to the device to the patient and saying, "You are free to do as you choose." While all persons have the legal and ethical right to discontinue therapy, this situation may create considerable disturbance. Such situations therefore call for careful and searching discussions involving the patient, health professionals, and as appropriate, members of the patient's family.

In some cases it may be necessary to advise subjects that the act of withdrawal will result in the cancellation of various commitments made by the investigator to the subject. For example, it may be necessary for the investigator to cancel a commitment to provide continuing access to an investigational therapy to subjects who withdraw even though the investigator has promised to make it available to those who complete the study after the study has ended. It may be necessary for some physician-investigators to terminate their professional relationships with subjects at the conclusion of the study. Ordinarily, such subjects may be referred back to their personal physicians. Termination of a pre-existing physician-patient relationship as punishment for premature withdrawal should usually be regarded as an impermissible "loss of benefits to which the subject is otherwise entitled."

In some cases the only reason that a person has a relationship with a physician-investigator or with an institution is that he or she has agreed to play the role of subject. When the involvement in research ends, prematurely or otherwise, it may be necessary to terminate the relationship. If the subjects' health condition requires it, the institution and the physician-investigator have the obligation to assist the exsubject in securing suitable health care. The financial consequences of premature withdrawal should be clarified (see Element 10).

In some protocols, various efforts are made to reestablish contact with subjects who miss appointments. Commonly, the investigators attempt to contact them by telephone or mail. Prospective subjects should be made aware of such plans; some may consider these attempts unwelcome intrusions or threats to their privacy. At times such plans may be rather elaborate. For example, one program enlisted the aid of an investigative services firm in locating subjects with whom they had lost contact for 12 months or more. After attempts to locate the subject through "local resources and Social Security" failed, they provided the investigative services firm with the subject's name, last known address and telephone number, place of employment and work phone, and names and addresses of persons indicated by the subjects as those likely to know his whereabouts (354).

During the consent negotiations, the investigator might wish to ask the subject whether there is any *a priori* reason that the subject feels that he or she might have to withdraw. If there is, the investigator may advise the subject to decline participation. It seems reasonable to explain to the prospective subject that premature withdrawal is wasteful and that it should be avoided unless there are important reasons.

The problem of dropouts from research is a serious one. Most properly reported randomized clinical trials, case-control studies, and longitudinal epidemiological studies report their dropout rates and assess the impact of these rates on the validity of the results. The IRB at the University of Pittsburgh undertook a general survey of research, largely biomedical, in its institution (444). They found that 9% of all subjects dropped out. The fact that these were from less than one-half the research protocols suggests that in some studies the dropout rate may have exceeded 20%; this number is high enough to undermine the validity of many types of studies.

DHHS requires: "When appropriate . . . the following . . . information shall also be provided . . . Anticipated circumstances under which the subject's participation may be terminated by the investigator without regard to the subject's consent . . . '' (Section 46.116b). It may be necessary to discontinue involvement of a particular subject because the results of monitoring procedures indicate that continuing involvement might result in injury to that subject. Inadequate cooperation on the part of the subject may also justify termination. A general statement of the grounds for terminating the entire project should also be made. For example, in a randomized clinical trial, one of the therapies being compared may prove to be superior (cf. Chapter 8).

14. Consent to Incomplete Disclosure   In some studies it is necessary to inform prospective subjects that some information is being withheld deliberately. Most commonly, the element of information that must be withheld is the disclosure of the purpose either of the entire study or of some of the maneuvers that will be performed in the course of the study; this is necessary when disclosure of the purpose would vitiate the validity of the results. In such cases, all other elements of information that are material to the subject's decision must be disclosed. There should also be an offer to disclose the purpose at the conclusion of the study.

Nondisclosure of the purpose without the subject's permission is a form of deception; it is, in Nuremberg's term, "fraud." As such, it invalidates consent and shows disrespect for persons. However, if subjects agree to the withholding of this information, there is no fraud. Notifying them of the plan to reveal the purpose after they have completed their participation provides them with the means to make a free choice. To justify such nondisclosure, there must be grounds to presume that upon later learning the purpose, almost all reasonable persons will approve their having been used for this purpose. There is further discussion of incomplete disclosure subsequently in this chapter and in Chapter 9.

15. Other Elements   DHHS requires: "When appropriate . . . the following . . . information shall also be provided . . . The approximate number of subjects involved in the study" (Section 46.116b). Holder argues and I agree that this information generally should be considered immaterial to the prospective subject's decision (301). The "when appropriate" standard is likely to be interpreted differently by various IRBs.

In its commentary, the Commission stated (527, p. 26):

> While Recommendation 4F contains a list of topics about which it can generally be presumed that subjects would want to be informed, it should be recognized that no such list is wholly adequate for this purpose. Thus, there may be research in which it is not reasonable to expect the subjects would want to be informed of some item on the list (e.g., who is funding the research). More frequently, it can be expected that research will involve an element that is not on the list but about which it can be expected that subjects would want to be informed. Such information should, of course, be communicated to subjects.

DHHS Regulations omit two elements of information that the Commission recommended should be disclosed if the IRB considers them necessary: "Who is conducting the study, who is funding it. . . ." In the Commission's view, some persons might not wish to invest their time and energy in advancing the interest of, for instance, the drug industry or the Department of Defense.

16. Continuing Disclosure   DHHS requires: "When appropriate . . . the following . . . information shall also be provided. . . . A statement that significant new findings developed during the course of the research which may relate to the subject's willingness to continue participation will be provided to the subject . . . " (Section 46.116b). Debates over whether and when to disclose preliminary data relating to harms and benefits are discussed in Chapters 3 and 8. Decisions about the necessity for disclosure of other types of information can usually be made in an analogous fashion.

Commonly, investigators offer to tell subjects the results of the study at its conclusion. They may offer to tell the general results of the study as well as any particular results having to do with the individual subject. When there is a good cause not to share results with subjects, this should be explicated during the initial consent discussions. One good reason is to avoid the "inflicted insight" (cf. Chapter 9). Some data require sophisticated technical knowledge for their interpretation and for determination of their relevance to the particular patient-subject; in general, these should be provided to the personal physician, who may then interpret them for or with the patient (cf. Element 9). The results of some innovative diagnostic tests should not be disclosed until the tests are validated; their disclosure might lead to erroneous diagnostic impressions, occasionally with seriously harmful consequences (cf. Chapter 3).

Should a subject be informed that he or she has been injured in a way that nobody would ever learn unless the investigator so informed the subject? Angoff has reported a case in which an IRB considered this question (18). In this case, a catheter placed in the subject's heart for research purposes perforated the anterior wall of the right ventricle; this was discovered in the course of heart surgery immediately following completion of the research procedure. It was agreed by the IRB that the perforation was, itself, a benign event; there was no cause to predict that the subject would suffer any harmful consequences. The IRB further acknowledged that in the course of routine medical practice, the fact that there had been such a perforation usually would not be revealed to the subject. This is particularly true when considering the context in which many "perforations" were made in the patient's heart for therapeutic purposes. The IRB reasoned, however, that the fact that the perforation occurred while pursuing research interests made a difference. Among other reasons, "there was the possibility this subject might . . . again be invited to have cardiac catheterization for research purposes. The fact that he had previously sustained a perforation could be material to his decision . . ." in the future.

"One member of the IRB felt that this consideration was outweighed by the more likely possibility that the man might need cardiac catheterization for clinical indications. . . . Knowledge of the perforation could result in a severe neurosis that might hamper his decisionmaking ability and thus affect his future health."

The IRB asked the subject's attending cardiologist to inform him of the perforation. However, if in the judgment of the cardiologist, such disclosure would be highly detrimental to the subject's health interests, the information could be withheld, but this fact should be documented for the medical record and for the IRB. In this case, the cardiologist informed the patient without apparent adverse consequences.

Comprehension

"The information that is given to the subject or the representative shall be in language understandable to the subject or representative" (DHHS, Section 46.116). Informed consent is not valid unless the consentor comprehends the information upon which consent is based. In the context of medical practice, our legal system is evolving the concept that the professional has the responsibility not only for disclosing what will be done but also for seeing to it that the patient understands the information (341, p.521). Informed consent has been found not to exist because patients have not understood such words as mastectomy (341, p. 651), laminectomy (297, p. 236), and so on. In *Reyes v. Wyeth Laboratories*, the fact that the plaintiff had "a seventh grade education, but her primary language is Spanish" was taken to imply that she may have "lacked the linguistic ability to understand" the significance of a form she had signed (598). Parenthetically, in the latter case, this was not a pivotal point because the form lacked the information that the court considered important. Such failures in communication ordinarily can be obviated by explaining technical procedures in lay terms and by the use of interpreters and translators as necessary.

In some cases, although the patient might have an adequate command of the English language, the physician may not. Morris cites several trial court cases in which hospitals were found liable for malpractice because, owing to inadequate command of the English language, a foreign medical graduate had either taken an inadequate history or failed to provide instructions to patients (514).

Investigators who are uncertain about the prospective subject's comprehension may ask them some questions about the proposed research. Miller and Willner propose that in some research projects, particularly those in which there is a large amount of complex information presented to the prospective subject, this process might be formalized in what they call the "two-part consent form."

(502). The first part is a standard consent form. After the information has been presented to prospective subjects and after they have had their questions answered, they are presented with the second part, which consists of a brief quiz on the essential elements of information. Typical questions include, "How much time will you be spending in the hospital if you agree to participate in this research? How much time would you spend in the hospital if you do not participate in this research?" Responses may be retained as evidence that prospective subjects have a clear grasp not only of what they are consenting to, but of the consequences of their consent.

As noted in Chapter 4, such tests have been used repeatedly with salutary results, improving comprehension in elderly subjects.

There have been many empirical studies of comprehension by subjects of information disclosed to them in the course of negotiations for informed consent to medical practice and research; for a good review of these, see Meisel and Roth (488). According to these authors, the literature reveals little or nothing about whether informed consent accomplishes its purposes. Most studies have serious flaws. Many studies are based upon examination of consent forms rather than what physicians or investigators actually say to patients and subjects. In many studies, patients or subjects (sometimes normal volunteers) are asked to deal with hypothetical situations; this does not tell us much about how real patients or subjects might perform when presented with real decisions. The results of many studies do not have generalizable validity because there was either a very small number of subjects or a large number of subjects having diverse illnesses and social attributes. In some studies, the investigator of the consent process is also the physician or investigator whose consent negotiations are being evaluated; for this reason, interpretation of the results may be biased. Stanley has also identified many problems in the design and execution of these studies (669). She suggests that comprehension might not be nearly so poor as the results of these studies seem to indicate (cf. Chapter 12).

In most empirical studies, subjects' comprehension is measured by submitting them to oral or written quizzes quite some time after they have consummated their agreements to participate in research. The fact that a subject has forgotten some information does not mean that it was not understood and applied at the time of the decision. (34, 488). Most responsible persons as they are preparing to make a decision become as informed as they feel they need to be at the time. Then, after making the decision, they forget many details, clearing their minds, as it were, to prepare for the next decision. How many readers can recall the relative merits, as compared with alternatives to which they gave serious consideration, of the schools they chose to attend, the houses and automobiles they purchased, and so on? I suggest that at the time you made these choices you had what you considered good reasons. A lot more was at stake in these decisions than in the average decision to play the role of subject.

Woodward has shown that careful teaching of prospective subjects results in high levels of comprehension (738). His project was designed to study the pathogenesis of cholera in normal volunteers recruited through advertisements in local newspapers. Each volunteer was given a lengthy explanation of the disease by at least two investigators. They were each informed that they would not be accepted into the study unless they achieved a passing grade of 60% on a multiple-choice examination on this information. Parenthetically, this seemed to provide some motivation to learn the material.

Woodward administered the examination to academic physicians, almost all of whom were specialists in tropical infectious diseases and more than one-half of whom had had direct professional experience with cholera. He also administered the examination to medical students and housestaff (interns and residents in internal medicine). The volunteers achieved higher scores than any other group.

Marini et al. provide evidence that subjects are capable of a high degree of understanding and information retention (467). However, their prisoner-subjects had a tendency to forget some information that investigators considered most important (e.g., adverse effects of drugs), while remembering almost perfectly those matters that were of greatest interest to them (e.g., how much money they would be paid and where they had to go to get paid).

Brady argues that our usual approach to informing is not designed to produce meaningful comprehension (78). Our usual approach, says Brady, represents an attempt to establish "echoic control of a verbal response." Subjects are expected to "echo" the words told to them. "Those of us who 'learned' the Lord's Prayer and the Star Spangled Banner by rote need hardly be reminded of how little 'comprehension' may be involved." He suggests that for complicated research activities, subjects gain a far superior understanding if they are brought into the research environment and allowed to experience its routines and procedures.

In Chapter 4 there is a survey of those classes of persons that may have difficulty consenting owing to impaired capacities for comprehension. In Chapter 11, there is a discussion of standards for determining whether individuals are incompetent to consent by virtue of their inability to comprehend.

Autonomy

"The person involved . . . should be so situated as to be able to exercise free power of choice without the intervention of any element of force, fraud, deceit, duress, over-reaching or other ulterior form of constraint or coercion" (Nuremberg 1). In the biomedical ethics literature, the term "autonomy" is commonly used in a very broad sense. All persons having limited capacities to consent are regarded as having reduced autonomy. Writers who focus on freedom most commonly do so in discussion of institutionalized persons such as prisoners and those institutionalized as mentally infirm. In the context of medical practice and clinical research there is also great concern about the imbalance of power between the physician or investigator and the patient or subject (cf. Chapter 4).

There is considerable debate as to whether a physician who is involved in a physician-patient relationship can negotiate fairly for informed consent with the patient to become a subject. Spiro, for example, asserts that a physician having a close relationship with a patient can usually persuade that patient to do almost anything (665). Unlike most commentators, because Spiro emphasizes the importance of the closeness of the relationship, he feels the problem is greater in private practice than it is with ward or clinic patients.

Cassel considers this problem in the specific context of conducting research on patients with senile dementia of the Alzheimer's type (SDAT), a most vulnerable population of patients (118, pp. 103–104).

> It needs to be mentioned here, however, that a trusting relationship with any physician is an immensely valuable thing for a patient with SDAT and therefore an absolute prohibition against the physician playing both those roles (researcher and physician) may finally not be in the patient's best interest. An optimal clinical setting for an SDAT patient includes consistent relationships with a relatively small number of caregivers. Therefore, the moral character of investigators is a crucial aspect of the ethical conduct of research. Devising a test of ethical competence for investigators may be as relevant as trying to find an adequate test of mental competence for patient-subjects.

In his review of the literature on this subject, Beecher concluded by suggesting that consent might not be the only or the most important issue (53, pp. 289–290):

> An even greater safeguard for the patient than consent is the presence of an informed, able, conscientious, compassionate, responsible investigator, for it is recognized that patients can, when imperfectly informed, be induced to agree, unwisely, to many things. . . .
>
> A considerable safeguard is to be found in the practice of having at least two physicians involved in experimental situations, first there is the physician concerned with the care of the patient, his first interest is the patient's welfare; and second, the physician-scientist whose interest is the sound conduct of the investigation. Perhaps too often a single individual attempts to encompass both roles.

Beecher is not clear as to which of these two physicians he would have negotiate informed consent. The Declaration of Helsinki requires the following (Principle I.10):

> When obtaining informed consent for the research project the doctor should be particularly cautious if the subject is in a dependent relationship to him or her or may consent under duress. In that case the informed consent should be obtained by a

doctor who is not engaged in the investigation and who is completely independent of the official relationship.

In its report on IRBs, the Commission suggests in its commentary under Recommendation 3D that the IRB should be aware of the advantages and disadvantages for patient-subjects of having one individual perform the dual role of physician-investigator. At its discretion, the IRB may require that a ''neutral person'' not otherwise associated with the research or the investigation be present when consent is sought or to observe the conduct of the research. This neutral person may be assigned to play a role in informing subjects of their rights and of the details of protocols, assuring that there is continuing willingness to participate, determining the advisability of continued participation, receiving complaints from subjects, and bringing grievances to the attention of the IRB. In its report on those institutionalized as mentally infirm, the Commission suggests the need for ''independent clinical judgment'' when persons propose to play the dual role of physician-investigator (cf. Chapter 11). Federal regulations do not reflect these considerations.

IRBs commonly approve research activities in which the investigators have ongoing physician-patient relationships with the subjects. In some situations in which prospective subjects are extremely vulnerable, IRBs have imposed requirements for the involvement of third parties having the same or similar functions as those prescribed by the Commission for the neutral person.

More extensive discussion may be found in Chapters 4, 11, and 12 of factors that tend to diminish autonomy and procedures that may be used to protect those who are vulnerable by virtue of having seriously impaired autonomy. In this chapter, the issues discussed in Elements 1 to 3 and particularly 8, 11, and 12 are designed to enhance the subject's free power of choice.

Consent

Ordinarily, after the processes of informing, assurance of comprehension, and assurance of autonomy have been completed to a mutually satisfactory extent, the prospective subject signifies his or her willingness to become a subject by consenting. Although this process commonly involves the signing of a consent form, in some cases written documentation may be unnecessary or undesirable (*infra*). Thus, it may be said that at this point there is established a contract—not in the legal sense of the term but more in the senses of agreement or commitment. This agreement differs in several important respects from the common commercial contract; it is a commitment to form a relationship resembling a fiduciary relationship. The investigator is obligated, ethically and legally, to renegotiate this agreement from time to time and in certain specified ways during the conduct of the research.

When protocols require a major commitment of the investigator's time or resources to a particular subject, as the research progresses, the investigator's motivation increases to persuade the subject to continue. Investigators must be aware of their motivations and take care not to subjugate the subject's will to their own.

Barriers to Informed Consent

Throughout this chapter and indeed throughout most of this book I have called attention to a large number of barriers to achieving the goals of informed consent. Now I wish to mention some that, although very important, do not seem to fit conveniently in any other section. These obstacles are rooted in the ethos of the medical profession as it exits in the United States. It is this ethos that determines who doctors are and who patients are and what they expect of each other. Although I focus on the medical profession, to some extent this discussion is relevant to other professional-client relationships.

In this section I shall draw most explicitly on Jay Katz's excellent book, *The Silent World of Doctor and Patient* (343). This is largely a theoretical analysis grounded in the experiential world of a practicing psychoanalyst who is also a distinguished legal scholar. I shall also draw, sometimes implicitly, on the empirical and analytical work of Renée Fox; her book *Experiment Perilous*, although written in 1959, provides penetrating and still valid insights into what it is like to be a part of a metabolic research ward as a research physician and as a patient-subject (238). I shall also draw on the fine book she wrote with Judith Swazey, *The Courage to Fail*, which examines human interaction in the development of organ transplants and dialysis (240).

Katz provides"contemporary and historical evidence that patients' participation in decisionmaking is an idea alien to the ethos of medicine. The humane care that physicians have extended to patients throughout the ages rarely has been based on the humaneness of consensual understanding; rather, it has been based on the humaneness of services silently rendered. It is the time-honored professional belief in the virtue of silence, based on ancient notions of a need for faith, reassurance, and hope, that the idea of informed consent seeks to question" (343, pp. xvi–xvii).

Katz argues that the time has come to dispel this silence. He calls for conversations,—searching conversations, between doctors and patients in which they reveal their knowledge, their values and, to the extent they can, their motivations; to do this they must learn to trust each other. In his view, respect for autonomy creates an obligation for conversation, the goal of which is to share the burdens of decision-making. These conversations should no longer presume an identity of interests or "consensus on goals, let alone on which paths to follow. . . . Two distinct and separate parties interact with one another—not one mind (the physician's), not one body (the patient's), but two minds and two bodies. Moreover, both parties bring conflicting motivations and interests to their encounters. . . . Conflicts within and between the parties are inevitable. Silent altruism alone cannot resolve these conflicting tensions" (343, p. xviii).

He acknowledges that conversations of this sort rarely take place. He identifies three major obstacles to achieving his goal of shared decision-making: authority, autonomy, and uncertainty. In my view, these obstacles must be considered not merely as an exercise in thinking about how the doctor-patient relationship ought to be in the future; they also stand as important barriers to informed consent as it is and ought to be conducted today. Katz and I may disagree on how we should deal with these barriers; we do not, however, dispute their importance.

Authority   Sociologists of medicine have identified two criteria as central to the definition of a profession. The first is the possession of esoteric and abstract knowledge over which they have a monopoly. The second is self-regulation of the profession by the profession with freedom from lay control. Patients are expected to trust physicians to make the correct decisions for them. Moreover, they are expected to be compliant with "doctors' orders." Physicians claim that this trust is justified by the altruistic traditions of the profession, which acts only for the benefit of the patient.

Physicians, by contrast, are not necessarily expected to trust patients. Physicians are taught carefully in medical school how to detect malingering and patient noncompliance, resorting to deceptive strategies if necessary for the "good of the patient" (cf. Chapter 9). More importantly, they do not trust patients to grasp the knowledge necessary to participate in decision-making. Katz argues that doctors and patients must learn to trust each other and to share authority.

Autonomy   In order to achieve the goal of shared decision-making in medical practice, Katz argues, we must revise our understanding of the concept of autonomy. He begins with a radical rejection of the concept of autonomy as envisioned by Kant and his followers. Kant is selected for attack because of his dominant influence on the literature of bioethics. Here is a very small sample of Katz's rejection (343, p. 108):

> Immanual Kant, in restricting his conception of autonomy to capacities to reason, without reference to human beings' emotional life and their dependence on the external world, projected a vision of human nature that estranged his principle from human

beings and the world in which they must live. That Kant did so deliberately and with full awareness, because he wished to isolate and abstract a single aspect of human psychology, i.e., rationality, into its pure form, is a separate matter. What matters here is the Kantian principle of free will, since it is based on this single aspect of human psychology, makes, if applied to actual situations, demands for human conduct that human beings cannot fulfill.

Thus, it is Katz's view that the prevailing concept of autonomy held by American philosophers and lawyers is destructive to shared decision-making because it does not take into account the psychological realities of persons, both patients and physicians. It presumes that decision-making is, or should be, both conscious and rational. The psychological reality, according to Katz, is that "the conscious and unconscious, rational and irrational forces . . . shape all thoughts and actions. . . ." (343, p. 129). Thus, a comprehensive definition of individual autonomy must take this into account.

Uncertainty "Medical knowledge is engulfed and infiltrated by uncertainty" (343, p. 166). Of the three barriers to communication between doctors and patients, it seems to me that this is the most intractable. This may come as a surprise to the lay reader. Has not the introduction of the scientific method into the development of medical knowledge done much to dispel uncertainty? Well, yes and no.

Although the scientific method yields facts about the nature of some things, it is a commonplace that the more we learn, the more we become aware of how much there is to learn. Science thus reveals to us our ignorance. Life must have been easier for the followers of Benjamin Rush who *knew* that the *only* way to treat a large number of diseases was blood-letting (352, p. 233). But of course it was not; they had other uncertainties to deal with.

According to Katz, it is only within the last 150 years that physicians have

> begun to acquire the intellectual sophistication and experimental tools to distinguish . . . systematically between knowledge and ignorance, between what they know, do not know, and what remains conjectural. These so recently acquired capacities have permitted physicians to consider for the first time whether to entrust their certainties and uncertainties to their patients. . . . Without the emergence of medical science, the legal doctrine of informed consent probably could not have been promulgated (343, xvi).

Following Fox, Katz identifies

> three basic types of uncertainty. . . . The first results from incomplete or imperfect mastery of available knowledge. No one can have at his command all skills and all knowledge of the lore of medicine. The second depends upon limitations in current medical knowledge. There are innumerable questions to which no physician, however well trained, can as yet provide answers. A third source of uncertainty derives from

the first two. This consists of difficulty in distinguishing between personal ignorance or ineptitude and the limitations of present medical knowledge (343, p. 171).

Katz's proposal that we must share these uncertainties with patients must strike terror into the heart of the average physician. Many doctors remain unconvinced that they should inform patient-subjects in randomized clinical trials that their therapy will be decided by chance. They believe that the uncertainty revealed by such a disclosure would cause most sensible patients to leave to find a doctor who knows what he or she is doing (cf. Chapter 8).

Physicians fear acknowledging uncertainty to themselves, to their colleagues, and to their patients. In Katz's view the problem is not uncertainty itself. Rather,

> the problem posed by uncertainty of knowledge for mutual decision-making is how to keep the existence of uncertainty clearly in mind and not replace it with certainty whenever one moves from theoretical to practical considerations. Put another way, the problem is not uncertainty of medical knowledge but the capacity to remain aware of, and the willingness to acknowledge, uncertainty (343, p. 171).

Lidz and his colleagues have observed decision-making in several hospital settings (446). Here they observed several significant barriers to the participation by patients in decision-making.

> These barriers include the fact that treatment decisions take place over a long period; there are often too many decision to be made; although patients want information about treatment, they typically believe that decision-making is the physician's task; physicians do not understand the rationale for the patient's role in decisions; the medical decision-making process often involves so many people that the patient does not know who is responsible.

Jonsen call attention to a "profound moral paradox" which pervades medicine (335). "That paradox arises from the incessant conflict of the two most basic principles of morality: self-interest and altruism." A careful reading of this important paper will facilitate further reflection on barriers to informed consent.

Standards for Informed Consent to Research and Practice

Informed consent to research is generally viewed as something very different from informed consent to medical practice. At an operational level, this perception is correct. There are major differences that reflect important distinctions between the nature and purpose of research and practice (cf. Chapter 1). However, at a more basic level, there is much more similarity than difference, because the fundamental purpose of informed consent is the same in research, practice, or any other context—to show respect for persons. In this discussion of the similarity and differences between research and practice, I shall speak in general terms about the nature of these activities, or what they are commonly thought to be, without addressing the many exceptions to these generalizations.

Informed consent to research is a very formal process. Investigators must develop a comprehensive account of the elements of informed consent for presentation to prospective subjects. Before contacting any prospective subject, the investigator must submit these plans to an IRB for its review and approval. All of the consent information must be incorporated into a consent form. The prevailing view is that the entire transaction is accomplished at a single meeting. For some complicated protocols, the subject may be advised to take a copy of the consent form; peruse it at his or her leisure; perhaps discuss it with family, friends, or personal physician; and then return to complete the process of consent.

By contrast, in the practice of medicine, informed consent is generally considered very informal. There are no regulations stipulating the elements of information and no review of the plans by any committee. In most medical practice, there is little documentation of informed consent except for the rather short and uninformative permission forms used for surgery, autopsy, and some other invasive procedures. Discussions leading to agreements about the overall plans for interaction between physician and patient are not generally thought to occur at a single encounter. Rather, at each meeting there is a little informing and a little consenting, although it is usually not called that.

The research program is viewed as one designed to serve the interests of society. Investigators invite subjects into the research program to serve those interests that are commonly portrayed as being in conflict with those of the subject. Subjects are asked if they are willing to be used as means to the ends of others. As Hans Jonas puts it, individuals are conscripted to sacrifice themselves in the service of the collective (cf. Chapter 4). Thus, if one accepts this view of the investigator-subject relationship, it is necessary to keep in mind that respect for persons requires that we leave them alone—not touch them or encroach upon their privacy unless they agree to such touching or encroachment.

By contrast, patients generally take the initiative in seeking out physicians to serve their own interests. No other interests are seen as inherently or inevitably standing in opposition to those of the patient. Through their seeking out of the physician, patients convey the clear message that they do not wish to be left alone; they expect the physician to touch them and to probe into their private lives. These understandings are commonly implicit, and we accept the fact that they are usually not negotiated clearly. This is not a relationship conducted at arm's length.

The relationship between the investigator and the subject is generally conceived of as a brief one, measured in hours or weeks, and ending when the investigator has accomplished his or her purpose. There is no time and, perhaps, little motivation for the investigator to get to know the subject to the extent that he or she can evaluate and be responsive to the subject's personal style of expressing choices. Thus, investigators are required to plan carefully for informed consent, to include all information that might be considered material by the reasonable person, and to use language that can be understood by the average layperson.

In the practice of medicine, however, we presume a prolonged relationship measured in terms of years. The physician does not speak in a language appropriate for the average layperson. Rather, he or she speaks to the unique individual who is the patient. More importantly, the physician listens and responds to that unique person. The physician evaluates the patient's remarks, not necessarily in the paternalistic sense of deciding whether these remarks are rational or appropriate, but more in the sense of whether it is authentic for that person (cf. Chapter 11). For example, "I have heard Susan's choice, but that is not like the Susan I know; perhaps we should discuss this matter further."

Investigators generally avoid recruiting as research subjects persons who have limited autonomy. Because limited autonomy makes one vulnerable and because research is generally regarded as imposing a burden on subjects, we attempt to protect vulnerable persons from bearing the burdens entailed in service as a research subject. Through our ethical codes and regulations we define ideal research subjects as highly autonomous and then proceed to develop rules to assure that we treat them as such.

By contrast, in medical practice we recognize that many patients have diminished autonomy, particularly at times when they are most in need of medical care. Indeed, Eric Cassell has argued persuasively that the primary function of medicine is to preserve or restore autonomy (119). When the patient is relatively incapacitated by illness—incapacitated in the sense of having reduced autonomy—it seems fitting to show respect in the second of the senses identified by the Commission, by providing protection against harm. This is not an all-or-none phenomenon in the sense that patients are either autonomous or they are not. To the extent that they are autonomous, physicians may respond by offering the patients information on which they may base their choices. To the extent that patients are not autonomous, however, physicians will commonly accept responsibility for making certain therapeutic decisions for them. The range of decisions that physicians will make for patients is, or should be, limited, and the delegation of decisionmaking authority should usually be explicit (430).

The goal of research—the development of generalizable knowledge—is advanced by working according to a detailed protocol. One can and should specify the characteristics of the subjects and precisely what will be done with each of them, as well as the amount of variation that will be tolerated. Thus, in the interests of good science, one develops a relatively inflexible protocol; this, in turn, enables the formulation of plans for informed consent and its documentation that would be reasonably suitable for all subjects within a protocol.

The practice of medicine is entirely different. One cannot specify who the patients will be. Persons with attributes that would provide grounds for their exclusion from a research protocol must be included. Similarly, one cannot specify the content of the physician-patient relationship. Even if we narrow our focus to the use of various therapeutic maneuvers and modalities, these must be selected to be responsive to the needs of individual patients. In research, an appropriate question might be, "What is the antihypertensive effect of administration of this beta-receptor blocker in a specified dose range for 8 weeks to patients with moderately severe hypertension?" In medical practice, this is never a suitable question. Instead one asks, "What is the best way to control the blood pressure of this patient, who not only has moderately severe hypertension but also has asthma, congestive heart failure, and recently lost his job?"

Because the fundamental values we wish to uphold through the devices of informed consent are the same in both research and practice, other things being equal, one might reasonably anticipate parallel developments of regulations for informed consent in the two enterprises. However, because research is conducted according to protocol, the development of detailed regulations for informed consent not only can be done, it has been done. Because medical practice cannot be conducted according to a protocol, it is impossible to develop detailed regulations for informed consent to medical practice; I believe this is the main reason that such regulations have not been developed.

It is commonly observed that, although there are many cases in which physicians have been found negligent for having failed to provide full disclosure in the context of medical practice, there is but one case in which an investigator was found negligent on the same grounds; this is *Halushka v. University of Saskatchewan*. Yet, examination of several medical malpractice cases found against physicians for failure to provide full disclosure reveals that what was not fully disclosed is that the procedure used was experimental, novel, or innovative, e.g., *Slater v. Baker and Stapleton, Natanson v. Kline*, and *Fiorentino v. Wenger*; these cases are each excerpted by Katz (341).

Supervision of the Negotiations

The Commission recommended that in certain situations there should be a third party in addition to the investigator and subject involved in the consent negotiations; in some circumstances, they should also be involved in the continuing negotiations during the course of the research to see whether the subject wishes to withdraw from the protocol, among other reasons. These third parties are variously called "consent auditors" (Chapters 10 and 11), "advocates" (Chapter 11), "neutral persons" (this chapter), and so on. Except for some types of research on those institutionalized as mentally infirm, the Commission recommends that that need for such third parties be a discretionary judgment of the IRB (Chapter 11). The policy of the National Institutes of Health (NIH) Clinical Center for "research with subjects who are impaired by severe physical or mental illness," also discussed in Chapter 11, reflects the Commission's recommendations on this point.

In this discussion of third parties, I shall use the term "trusted advisor" as a generic for those who act in an advisory capacity and who are consulted or not according to the wishes of the prospective subjects or persons authorized to speak for them. "Overseer" is the term I shall use for agents whose employment is required by the IRB and who are empowered to prohibit the initial or continuing involvement of any particular subject.

The circumstances in which it might be helpful to involve a trusted advisor and the types of persons who might be consulted are discussed under Element of information 12. Suggesting consultation with a trusted advisor is quite a different matter from commanding the presence of an overseer. The requirement for an overseer should never be imposed frivolously. It is an invasion of privacy. The magnitude of the invasion can be reduced by allowing the prospective subject to select the overseer. Moreover, the imposition of such a requirement is tantamount to a declaration to the prospective subject that his or her judgment, ability to comprehend, ability or freedom to make choices, and so on is to be questioned. However, in some cases this will be necessary.

I shall now present an example of a case in which the IRB at Yale Medical School imposed an overseer requirement. In this case, two types of overseer were each assigned specific functions rather than global authority to monitor the consent negotiations. Moreover, they were selected so as to minimize invasions of privacy.

The research maneuver was catheterization of the coronary sinus to sample blood for assay of a chemical made in the heart. Subjects were to be men who had been admitted to a coronary care unit very recently with known or suspected myocardial infarctions, including only those who required catheterization of the right side of the heart for medical indications. To further minimize the risk of traumatic injury to the heart, if the catheter did not pass easily into the sinus on the first attempt, there would be no second attempt. No catheter was to remain in the coronary sinus more than 10 minutes. It was agreed that the most serious remaining risk was a small probability of arrhythmia. The facilities available for monitoring for the occurrence of this harm were optimal; interventions available to counteract arrhythmias had a very high likelihood of success.

No direct benefits to the subjects were intended or expected. There was a moderately large probability of a very large benefit to the population of indi-

viduals having myocardial infarctions. The investigators were attempting to identify a chemical which might contribute to the morbidity and mortality of this
disease. If they were able to establish the relationship, it would be possible to
intervene directly; means for blocking the formation as well as the effects of
this chemical already existed. Prior animal studies indicated a reasonably high
probability of associating the chemical with the disease.

Physicians generally assume that any stress may be harmful to individuals
with myocardial infarctions. Thus, great care is taken to avoid physical and
psychological stress, particularly during the early stages of treatment. Parenthetically, it is not proved that this has any beneficial effect. Nonetheless, patients
with known or suspected myocardial infarctions commonly are treated with
narcotics to diminish pain and anxiety. Thus, most individuals whom the investigators would approach to negotiate informed consent for this research would
have received recent treatment with a narcotic.

In the judgment of the IRB, the negotiations for informed consent would create
anxiety. This, in turn, might jeopardize the physical and psychological well-
being of the prospective subject. (As noted earlier, this could not be proved.)
Further, it was judged that the proposed research population was especially
vulnerable for either of two reasons. 1) Some of these individuals might perceive
themselves as in the process of dying; in some cases this perception might be
in accord with facts. 2) Some of these people might have had their abilities to
make rational judgments impaired by virtue of having received a narcotic; i.e.,
a state of inebriation would have been induced.

Thus, the IRB imposed a requirement for involvement in the consent process
by two different types of overseers:

1. Before any approach to the prospective subject was made, the investigator
would review the proposal (all elements of information) with the next of kin.
The purpose of this discussion was to determine if, in the view of the next of
kin, this was the sort of thing to which the prospective subject might be expected
to consent. If in the judgment of the next of kin, the answer was no, no invitation
would be offered to the prospective subject.

2. A physician not connected with the research and who, by virtue of his or
her relationship to the prospective subject, had the best interests of the prospective
subject in mind would be called upon to determine that the patient's physical
and psychological condition were such that he was not likely to be unduly
threatened or harmed by the consent negotiations and the patient's cognitive
function had not been impaired to the extent that he could not understand the
information.

The physician-overseer was to be the one who had most recently examined
the patient sufficiently thoroughly to make these determinations. When possible,

this would be the patient's personal physician, who had established a physician-patient relationship prior to the onset of the current illness. When a physician meeting this description was unavailable, the resident physician in the coronary care unit would be called upon. Personal physicians who were also members of the research team were disqualified as overseers.

In this case, the investigators were cardiologists who were highly skilled and experienced not only in catheterizations of the heart but also in the management of patients with myocardial infarction in the coronary care unit. Thus, it was judged unnecessary to have any other physicians in attendance during the process of negotiating for informed consent or during the brief period during which the research maneuver was performed. In the event the subject wanted to see his own physician during this period, his access to that physician would be no more limited than it would ordinarily be in the usual conduct of activities on a coronary care unit. The circumstances of the research were such that it would not be in the scientific interests of the investigators to continue the research if something went wrong.

Frank and Agich have reported a somewhat similar case that differed sufficiently in two particulars to justify a different resolution of the problems (243). In this case investigators wished to draw 10 ml of blood "as soon as possible after cardiac arrest and again approximately 10 minutes later." Because many of these patients would already have intravenous access routes available either before the cardiac arrest or shortly thereafter, additional vein punctures would seldom be necessary. The IRB decided against having consent negotiated in advance for individuals who were likely to become eligible to be subjects, even though it knew that at the time the blood was to be drawn the patient would be unable to consent and there would almost never be time to consult with the next of kin. "First, the potential risk of anxiety associated with informing a large number of patients regarding the possibility of their sustaining a cardiac arrest, even though most would not, was thought to be significant." In their judgment "the overall risk to patients would be minimized by enrolling a small number of patients who did experience a cardiac arrest if the informed consent requirement could be waived." Most importantly, in their view, "it was doubtful that truly informed consent could be obtained" from the sorts of patients who would have to be recruited prospectively for this study, who were "all seriously ill or at high risk."

Instead, they required that "the chief resident or head nurse be informed of the study in advance and asked to serve as an ombudsman for his or her patient." Further, "the individual withdrawing the blood would not be involved in the resuscitation effort and would in no way interfere with it." The decision to permit the research was to be made by the physician in charge of the resuscitation team and not by the investigator. The ombudsman was charged "to assure that the patient's welfare was not jeopardized and that even the vaguest expression of reluctance by the patient . . . would be respected."

Earlier in this chapter I suggested that there is a tendency to consider the investigator-subject relationship a fiduciary relationship. I commented on some regulatory requirements that were incompatible with this view. Authors who recommend separation of the roles of health professional and investigator and those who claim there is a need for overseers are denying implicitly that this is a fiduciary relationship. Dyer is explicit on this point (200). He recommends that in some situations there is a need for a "patient's advocate" to whom he also refers as "trustee." He describes the role of this agent as creating a fiduciary relationship with the subject. The investigator, by contrast, is portrayed as being in an "adversary relationship" with the advocate or trustee.

Documentation

According to the President's Commission (571), "Ethically valid consent is a process of shared decision-making based upon mutual respect and participation, not a ritual to be equated with reciting the contents of a form that details the risks of particular treatments." Thus far we have been considering informed consent, a process designed to show respect for subjects, fostering their interests by empowering them to pursue and protect their own interests. Now we turn to the consent form, an instrument designed to protect the interests of investigators and their institutions, and to defend them against civil or criminal liability. I believe that one of the reasons that there has been so little successful litigation against investigators, as compared with practicing physicians, is the very formal and thorough documentation of informed consent on consent forms. Consent forms may be detrimental to the subject's interests not only in adversary proceedings; signed consent forms in institutional records may lead to violations of privacy and confidentiality (594, pp. 74–86).

DHHS requires in Section 46.117 the use of one of two types of consent forms for all types of research not explicitly excluded from the documentation requirement (*infra*).

1. A written consent document that embodies the elements of informed consent required by Section 46.116. This form may be read to the subject or the subject's legally authorized representative, but, in any event, the investigator shall give either the subject or the representative adequate opportunity to read it before it is signed.

2. A "short form" written consent document stating that the elements of informed consent required by Section 46.116 have been presented orally to the subject or the subject's legally authorized representative. When this method is used, there shall be

a witness to the oral presentation. Also, the IRB shall approve a written summary of what is to be said to the subject or the representative. Only the short form itself is to be signed by the subject or the representative. However, the witness shall sign both the short form and a copy of the summary and the person actually obtaining consent shall sign a copy of the summary. A copy of the summary shall be given to the subject or the representative, in addition to a copy of the "short form."

As far as I can see, there is no advantage to any party to the negotiations to the short form. Considering the purpose of the form, what might one omit from the summary? Why have a witness? I have already discussed the consequences of unwarranted intrusions of third parties into the consent negotiations. I shall not discuss the short form further.

In general, it should be kept in mind by all concerned, but particularly the investigator and members of the IRB, that consent forms can never be constructed so as to anticipate all of any particular prospective subject's wishes to be informed. The consent form is most effective when it is viewed by the investigator as an instrument designed to guide the negotiations with the prospective subject. The consent form should contain at least the minimum amount of information and advice that should be presented during the negotiations. If any substantive new understandings are developed in the process of negotiations that have any bearing on the prospective subject's willingness to participate, these should be added to the consent form signed by that particular individual.

The consent form should present an adequate coverage of each of the elements of information that are germane to the proposed research. The fact that the consent form is considered a guide to the negotiations is reflected as indicated in Element 11, which is the offer to answer any inquiries concerning the procedures. It should be made clear that this is, in fact, an offer to elaborate on any of the elements of information to the extent desired by the prospective subject.

The use of general consent forms designed to document consent for research generally or even for several categorically related research protocols should be avoided. At the very least, a consent form should be designed to meet the specifications of a particular protocol. A fully satisfactory document designed to meet the needs of a protocol may, as discussed above, undergo further modification to reflect alterations negotiated with particular subjects. At times it is necessary to design more than one consent form for use within a single research protocol. For example, when a protocol is designed to conduct the same maneuvers on two distinctly different populations of subjects (e.g., diabetics and healthy volunteers), it is ordinarily appropriate to have separate consent forms for each class of subject. Many of the elements of information presented to the diabetics will differ form those presented to the healthy volunteers. For example, Element 3 would contain very different information as to why the prospective subject has been selected. It might also contain very different sorts of information regarding pretests to determine eligibility, the consequences of failure to pass the examinations, and the nature of the hoped-for benefits.

Many institutional guidelines prescribe the style of language for the consent form. Most commonly, they require that the form be worded in language that the average lay person could be expected to understand. This suggestion has provoked considerable controversy. For example, what is an average lay person? And what can he or she be expected to understand? Many protocols are designed to involve subjects who differ in some substantial way from average. For example, some studies are designed to involve subjects who have serious chronic diseases in whom standard and accepted therapeutic measures have failed. Commonly, such individuals have sophisticated understandings of the technical language used to describe their disease, various means of diagnosis and therapy, and the harms that may occur as a consequence of therapy or of not being treated. On the other hand, some protocols are designed to involve naive subjects who might have little schooling, who might have primary languages other than English, and so on. Thus, it seems more appropriate to suggest that the consent form be presented in language that the prospective subject population could be expected to understand. Protocols involving diverse populations might require more than one consent form.

Some institutions provide further recommendations on the language of the consent form, e.g., that the entire consent form should be worded in the first person. Thus, the various elements of information begin, "I understand that the purpose of the study is to . . ." or "I hereby agree to have Dr. Jones draw 10 cc (2 teaspoons) of blood. . . ." Other institutions have recommended that consent forms be worded in the second person. For example, "You are invited to participate in a research project designed to . . . (accomplish some purpose)." "If you agree to participate there is a small possibility that you might develop a rash." "The purpose of this research is not to bring direct benefit to you, but rather to develop information which might help us design better diagnostic methods for persons like you in the future."

I prefer consent forms that present the elements of information in the second person. After the elements of information have been presented there may be a statement worded in the first person, as follows:

> *Authorization*: I have read the above and decided that . . . (name of subject) will participate in the project as described above. Its general purposes, the particulars of involvement and possible hazards and inconveniences have been explained to my satisfaction. My signature also indicates that I have received a copy of this consent form.

Presentation of the information in the second person followed by an authorization written in the first person best conveys the sense of negotiation. I am aware of no empirical evidence that one style of language is to be preferred to another.

At the conclusion of the consent form, there should be a line provided for the signature of the consentor. There should further be some means of specifying how the consentor is related to the subject (self, parent, or guardian). It should also specify the date on which the form was signed. Some individuals may wish also to record the date on which the form was first presented to the prospective subject if it differs from that on which it was signed. When the consentor is the legally authorized representative, another space may be provided when appropriate for the signature of the subject. This signature may indicate, depending upon the situation, the actual subject's consent (which may or may not be legally valid), assent, or, perhaps, merely awareness that somebody is consenting to something on his or her behalf.

There should also be a space for the signature of the person who negotiated consent with the subject, as well as any overseers who were present. Ordinarily, a witness is not needed except, of course, when one uses the short form or when a witness is required by state or local law. Unless otherwise stipulated by local law or institutional policy, the only role of a witness is to certify that the person's signature is not a forgery (299). Holder points out that there are no reported cases in which either a patient or a subject claimed forgery of a signature on a consent form in an action against a physician or an investigator (299).

Finally, there should be clear instructions (including telephone numbers) about who to contact in case of injuries not anticipated during the initial consent discussions or for answers to questions.

Prohibition of Exculpatory Clauses

"No informed consent, whether oral or written, may include any exculpatory language through which the subject or the representative is made to waive or appear to waive any of the subject's legal rights, or releases or appears to release the investigator, the sponsor, the institution or its agents from liability for negligence" (DHHS, Section 46.116). Although it seems clear that this is an exhortation to remain silent on this matter, Bowker reports having seen such statements as (73), "by signing this consent form, I have not waived any of my legal rights or released this institution from liability for negligences (*sic*)." In another form he found, "This consent form includes no exculpatory language through which the subject is made to waive any of his legal rights or release the institution from liability for negligence." One wonders what, if anything, the authors of these consent forms were thinking of.

In the context of medical practice, Sprague has found a most remarkable consent form (666). To fully appreciate this consent form one should read all of it. I offer one sentence as an appetizer, "I assume any and all risk in connection with said blood transfusion and covenant with the parties designated above that I will never sue for or on account of my contraction of hepatitis or on account of any untoward reaction resulting from said blood transfusion and that this instrument may be pleaded as a defense to any action or other proceedings that may be brought, instituted or taken by me against the above designated parties.'' *Res ipsa loquitur*!

Readability of Consent Forms

Several authors have expressed great concern over the readability of consent forms. Using the Flesch Readability Yardstick, Gray et al. assessed consent forms designed for 1526 protocols in 61 institutions (273). Readability scores were very difficult (similar to scientific or professional literature) for 21%; difficult (similar to scholarly or academic publications) for 56%; fairly difficult (similar to *Atlantic Monthly*) for 17%; standard (similar to *Time*) for 5%; and fairly easy or easy for less than 2%. They further found that IRB review had no effect on the readability of consent forms. Grundner reported substantially similar results from his assessment of surgical consent forms using the Fry Readability Scale as well as the Flesch Yardstick (277). Grundner proposes that IRBs routinely should use the Flesch and Fry tests—and in some cases, the SMOG Grading System—to rewrite consent forms at the desired level (usually tenth grade or lower) (278). My inclination is to strive for a reasonably readable consent form. In some cases the following statement in the consent form is called for. "In the preparation of this consent form it was necessary to use several technical words; please ask for an explanation of any you do not understand." Parenthetically, the sentence I just recommended is very high in the college reading level according to the Fry Readability Scale and, according to the Flesch Readability Yardstick, it is in the upper range of difficulty for academic or scholarly prose. The statement may be reworded: "Some arcane words are on this page. I'll construe them as you wish." According to the Fry test, this is suitable for a first grader; Flesch rates it as easier than "pulp fiction."

Baker and Taub analyzed the readability of consent forms at a Veterans Administration (VA) Medical Center over a 5-year period (1975–1979) in which there were changes in DHHS and VA policy; although these policy changes had no effect on readability scores, there was a significant increase in the length of the forms (34). Thus, in their judgment the overall effect of policy changes was to increase the difficulty of reading consent forms. Because they agree with me that the essence of informed consent is in the negotiations rather than in the documentation, they were not alarmed by their findings. Meisel believes, as I do, that all too often, consent forms are used in place of informed consent (487); thus, he argues that, "Instead of improving the readability of consent forms, we should abolish consent forms—prohibit them by legal fiat as necessary." Roth et al. offer this reason for improving the readability of consent forms (616):

"Paradoxically, the existence of signed forms . . . that are difficult to understand by many patients and which may also be difficult to understand by jurors should the occasion arise may someday serve to incriminate rather than protect the clinicians or investigators."

Who Keeps Consent Forms?

DHHS requires the IRB (Section 46.115) to maintain records on each protocol including "copies of . . . approved sample consent forms . . . for at least 3 years after completion of the research." Various commentators have proposed that the IRB should also keep copies of the consent forms signed by each subject. Such a practice would be detrimental to the interests of many subjects.

Ordinarily, the investigator should assume the responsibility to retain the signed consent forms and to safeguard the confidentiality of these forms when appropriate and to the extent necessary. To some degree the mere fact that an individual has agreed to serve as a subject of a particular research activity is private information. For example, the consent form might specify, "You are invited to participate in this research project because you have cirrhosis of the liver."

While the regulations require the retention for 3 years of sample consent forms, it is not clear how long the signed forms should be retained. Considering that the primary purpose of the form is to defend the investigator and the institution from civil or criminal liability, the form should be held long enough for that purpose. Ordinarily, signed forms should be retained for a sufficient period longer than the statute of limitations as usually advised for physicians who are concerned with potential malpractice litigation; this is usually about 6 months, but in states having discovery statues, it may be forever (297, p. 321).

Many institutions require that signed consent forms be made part of the permanent record of the institutions. This is based upon three assumptions. 1) If there is any litigation against the investigator, the institution is likely to be named as a codefendant. 2) Institutional administrators commonly believe that the record keeping systems of the institution are superior in various ways to those of investigators. 3) Investigators may leave the institution. Each of these assumptions is at least partly correct. The only one that might be seriously challenged by some investigators is the second. Some investigators are more adept at keeping orderly records than almost any institution. The smaller number of records they keep as well as the smaller number of individuals who have access to these records each contribute to their lesser likelihood of losing a record.

On the other hand, individual investigators or small groups of investigators are much more likely than large institutions to be able to assure confidentiality.

In general, when confidentiality is a significant issue, the investigator should have the responsibility for keeping signed consent forms.

DHHS requires that "A copy shall be given to the person signing the form" (Section 46.117a). The primary purpose of the form notwithstanding, it can and should be designed to be helpful to the subjects. Having a copy of the form will afford them an opportunity to continue to get more information as additional questions occur to them. It will also be available as a constant reminder of their freedom to ask questions, to withdraw without prejudice, and so on. Moreover, the form should provide the name(s) and telephone number(s) of individuals they might wish to contact as necessary.

Who Should Negotiate with the Subject?

"The duty and responsibility for ascertaining the quality of the consent rests upon each individual who initiates, directs, or engages in the experiment. It is a personal duty and responsibility which may not be delegated to another with impunity" (Nuremberg 1). Some institutions have policies prescribing more or less specifically who may negotiate with the prospective subject for informed consent. For example, some hospitals require that, if the principal investigator is a physician, the responsibility for negotiating for informed consent cannot be delegated to a nonphysician. Some commentators on informed consent have suggested that its quality might somehow be reduced by virtue of the fact that a physician-investigator delegated responsibility for informed consent negotiations to a research nurse or, perhaps, even a nonprofessional (271, pp. 212 et seq.).

In general, as one examines the various component processes of the negotiations for informed consent, it might be assumed that some types of professionals might be better equipped than others to accomplish each of these purposes, either by virtue of their training or expertise or by virtue of their motivations. The principal investigator ordinarily will be better informed than other members of the research team about the technical aspects of the research. Thus, if he or she is willing, the principal investigator will be more capable of responding to detailed questions about risks, benefits, and alternatives than most others. Equally certified professionals might be expected to have approximately the same amount of detailed information at their disposal.

On the other hand, in the context of medical practice, particularly in large institutional settings, patients tend to be better informed when a professional who is not a physician is assigned responsibility for their education. It seems that such professionals as clinical pharmacists, physician's assistants, and nurse practitioners are generally more oriented toward patient education than most physicians. There are, of course, clear individual exceptions to this generalization. However, there is no *a priori* reason to assume that subjects will be generally better informed merely because it is the principal investigator who undertakes this responsibility.

Much has been written about barriers to comprehension that are created when the individual negotiating with the prospective subject is of a very much different social class or has a very much higher degree of education than the prospective subject. It seems reasonable to suppose that the more equal the negotiators are in these two categories, the more likely there is to be comprehension. In addition, relative equals would probably be more capable in perceiving nonverbal manifestations of noncomprehension. Persons who have assumed the sick role, as have many prospective subjects, may find it difficult to refuse to cooperate with physicians (cf. Chapter 4).

State medical practice statutes authorize physicians to delegate responsibilities to persons who are, in the physician's judgment, sufficiently trained and competent to discharge these responsibilities; the physician is legally accountable for the actions of these "physician's trained assistants" (624). I suggest that investigators, including nonphysicians, should have similar authority to delegate consent negotiating responsibility.

The individual(s) assigned this responsibility within a research team should be the one(s) who seems most capable of and interested in performing this role. In the event this is a person having professional qualifications or certification lower than that of the principal investigator, this should be made clear during the consent negotiations. In such cases it is advisable to name the principal investigator or one of the equally certified co-investigators on the consent form and to provide instructions as to how he or she might be made available to respond to questions that in the view of the prospective consentor cannot be answered adequately by the delegate. The principal investigator should be held accountable for the actions of all individuals to whom he or she delegates responsibility. The selection of the individual(s) who will function as negotiator(s) for informed consent in any particular research project should be determined by the principal investigator. The IRB should review this designation and make recommendations for modification as it sees fit.

In their survey of informed consent to biomedical research, the IRB at the University of Pittsburgh found that information about the research and the request to sign the consent form were usually handled by either one individual or small groups (444). In 78% of the protocols this was the principal investigator or a coinvestigator, fellow, resident, or attending physician. In 22% of cases, these actions were performed by specially trained nurses, social workers, or laboratory technicians.

Timing of the Negotiations for Informed Consent

DHHS requires: "An investigator shall seek such consent only under circumstances that provide the prospective subject or the representative sufficient opportunity to consider whether or not to participate and that minimize the possibility of coercion or undue influence" (Section 46.116). Subjects should be allowed sufficient time to weigh the risks and benefits, consider alternatives, and ask questions or consult with others. Whenever possible, one should avoid seeking consent in physical settings in which subjects may feel coerced or unduly influenced.

Often the prospective subjects will necessarily be relatively incapacitated, distracted, or preoccupied at the time the research will be done. For example, when one plans to evaluate an innovative approach to delivering a baby, it is possible to identify a suitable population of prospective subjects weeks or months before the procedure will be performed on any particular subject. Informed consent should be negotiated during or after a routine prenatal visit with the obstetrician rather than when the woman is experiencing the discomfort and anxiety of labor and perhaps, already has been treated with drugs that might influence her cognitive function. In general, the obstetrician's office is an environment more conducive to rational decision-making than is the labor or delivery room (cf. Gray, 271). Also, prospective subjects contacted during a routine prenatal visit will have the time to accomplish the various functions specified in the preceding paragraph.

Similar considerations are often in order when prospective subjects are to be invited to participate in research to be done during elective surgery, predictable admissions to intensive care units, terminal phases of lethal diseases, or periods of intermittent inability to comprehend (cf. Chapters 4 and 12).

In situations in which informed consent has been negotiated well in advance of the proposed research, the prospective subject should be reminded of the right to withdraw shortly before the research maneuver is initiated.

In some other circumstances, it will be impossible to conduct the consent negotiations at a time and in an environment conducive to high quality decision-making; see for example, the illustrative example provided in the section on supervision of the negotiations. Chapter 8 contains a discussion of situations in which deferred consent may be contemplated.

Robertson has proposed that there should be a mandatory waiting period between the investigator's invitation to a subject to participate and the signing of a consent form or entry into the study (608). "Separating the explanation from the decision by a period adequate for the person to reflect and consider what exactly is being proposed is likely to enhance the quality of the subject's choice.'' In the artificial heart program at Utah, the IRB imposed a mandatory waiting period of 24 hours (739).

In biomedical research at the University of Pittsburgh, in 57% of cases the giving of information to the subjects and reading and requesting a signature on the consent form were accomplished at the same session (444). ''In another 16% about 24 hours elapsed between telling the subject about the research and requesting the signature. In the remaining (26%) there were more than 24 hours and often as long as 2 weeks between informing the subject and requesting the signature.''

Conditions Under Which Consent Negotiations May Be Less Elaborate

Thus far, this chapter may have created the impression that the average negotiation for and documentation of informed consent must be an extremely elaborate process. This false impression derives from the fact that I have attempted to present a comprehensive account of the various factors that must be considered by and the various procedures available to individuals who are planning negotiations for informed consent. In fact, in most cases, most of these factors and procedures will be inappropriate or unnecessary. Each negotiation for informed consent must be adjusted to meet the requirements of the specific proposed activity and, more particularly, be sufficiently flexible to meet the needs of the individual prospective subject. By way of illustration, most consent forms approved by the IRB at Yale University School of Medicine contain less than two pages of single-spaced text.

In an important article entitled *The Unethical in Medical Ethics*, Ingelfinger expressed alarm over the increasing apparent need for review of research by various types of committees, excessive formality and documentation of informed consent, and so on (325). He expresses his alarm with the:

> . . . dilution and deprecation of the important by a proliferation of the trivial. The patient, asked to sign countless releases or consents, may respond with a blanket refusal or with a pro forma signature. The physician, immersed in a profusion of unimportant detail will lose sight of, and respect for the important issues. Perhaps he will feel compelled to practice defensive ethics—no more honorable than defensive medicine. For medical ethics, in short, trivialization is self-defeating.

Thus, he draws on the experience of observing the behavior of physicians in reaction to their awareness of ever-increasing possibilities for malpractice litigation. Physicians now obtain many more diagnostic tests than are necessary.

This custom, commonly referred to as the practice of defensive medicine, is enormously expensive for all concerned. The analogy to which he calls attention, defensive ethics, includes highly formal negotiations for informed consent to research procedures involving minimal involvement, risk or discomfort to the subject.

Ingelfinger earlier expressed his view that, particularly in biomedical research, consent is generally informed only technically but virtually never educated (323). He argues powerfully that the individual who has assumed the sick role is virtually never capable of either thorough understanding or total freedom of choice. He sees the relationship that might be created by educated consent as described by, for instance, Ramsey as a "... convenantal bond between consenting man and consenting man (that) makes them ... joint adventurers in medical care and progress ..." as essentially unattainable and utopian ideals. He suggests that it is worth striving toward these ideals, but that we should understand that they never will be reached. He acknowledges that:

> The procedure currently approved in the United States for enlisting human experimental subjects has one great virtue: patient-subjects are put on notice that their management is in part at least an experiment. The deceptions of the past are no longer tolerated. Beyond this accomplishment, however, the process of obtaining "informed consent," with all its regulations and conditions, is no more than an elaborate ritual, a device that, when the subject is uneducated and uncomprehending, confers no more than the semblance of propriety on human experimentation. The subject's only real protection, the public as well as the medical profession must recognize, depends on the conscience and compassion of the investigator and his peers.

Benjamin Freedman agrees, but without dismay, that fully informed consent is unattainable and suggests that striving for it is, in most cases, undesirable (246). He proposes that it might better serve the purposes of all concerned to negotiate for *valid consent* rather than for fully informed consent. He concludes:

> . . . that valid consent entails only the imparting of that information which the patient/subject requires in order to make a responsible decision. This entails, I think, the possibility of a valid yet ignorant consent . . . the informing of the patient/subject is not a fundamental requirement of valid consent. It is, rather, derivative from the requirement that the consent be the expression of a responsible choice. The two requirements which I do see as fundamental in this doctrine are that the choice be responsible and that it be voluntary.

He sees responsibility as a dispositional characteristic which can be defined only relatively and conditionally in the context of the totality of one's knowledge of an individual. Similarly, he claims that "voluntarism" and reward can be evaluated only in the context of the prospective subject's total environment.

Perhaps germane to the argument presented earlier in this chapter, that it is a serious step for an IRB to impose a requirement for supervision of the negotiations, are some of the points Freedman makes in his discussion of the right to consent.

> From whence derives this right? It arises from the right which each of us possesses to be treated as a person, and in the duty which all of us have, to have respect for persons, to treat a person as such, and not as an object. For this entails that our capacities for personhood ought to be recognized by all—these capacities including the capacity for rational decision, and for action consequent upon rational decision. Perhaps the worst which we may do to a man is to deny him his humanity, for example, by classifying him as mentally incompetent when he is, in fact, sane. It is a terrible thing to be hated or persecuted; it is far worse to be ignored, to be notified that you "don't count."

With these arguments in mind, the opposing arguments having been sampled earlier in this chapter, let us now consider some circumstances under which the negotiations for and documentation of consent should or may be much less elaborate than those presented in the comprehensive model.

Program Modification and Evaluation

> An IRB may approve a consent procedure which does not include, or which alters, some or all of the elements of informed consent . . . or waive the requirement to obtain informed consent provided the IRB finds and documents that:
>
> (1) The research or demonstration project is to be conducted by or subject to the approval of state or local government officials and is designed to study, evaluate, or otherwise examine: (i) programs under the Social Security Act, or other public benefit or service program; (ii) procedures for obtaining benefits or services under these programs; (iii) possible changes in or alternatives to these programs or procedures; or (iv) possible changes in methods or levels of payment for benefits or services under these programs; and (2) The research could not practicably be carried out without the waiver or alteration (DHHS, Section 46.116c).

Section 46.101(b)(6) exempts from coverage by DHHS regulations a class of activities that is nearly identical to those described in Section 46.116c. Thus, unless an institution has promised in its multiple project assurance to apply 45 CFR 46 to all research regardless of funding source, it seems that these requirements need never be followed; that is, not unless the Secretary exercises his or her authority to determine whether any particular activity is covered by the regulations (Section 46.101c). For a look at the fascinating history of the changes in this regulation, see Appendix D to the President's Commission's *Implementing Human Research Regulations* (574).

There seems to be no problem in waiving the requirements for some or all aspects of informed consent in most studies designed to evaluate ongoing programs if certain limits are observed. The procedures performed in the interests of evaluation should not present any burden greater than mere inconvenience and secure safeguards of confidentiality should be developed as necessary. When the IRB is in doubt about the acceptability of proceeding with evaluation of ongoing programs without consent, community consultation may help resolve the doubt.

Proposals to introduce substantive modifications in the system will present greater problems to the IRB. Innovations in the system should be considered just like any other nonvalidated practices (392). Examples of manipulations of the health care delivery system include the introduction of health maintenance organizations (HMOs), innovative social service programs, multiphasic screening, computerized medical record systems, computerized approaches to diagnostic and therapeutic decision-making, coronary care ambulances, innovative communications networks, and the introduction of new types of health professionals such as physician's associates. When such manipulations are introduced, they ought to be considered as nonvalidated practices; research should be done to learn of their safety and efficacy. The IRB should consider, in each case, whether it is appropriate to waive or alter the requirements for informed consent.

Proposals to modify the system are particularly problematic when a subject population is designated arbitrarily and offered no options. This is commonly a problem in social policy research, an activity that I shall not discuss further because it is outside the scope of this book. For an extensive discussion of the problems in this field, see Rivlin and Timpane (601). The relevance of these considerations to manipulations of the health care delivery system is exhibited with particular thoroughness in Robert Veatch's chapter in that book (694).

Waivers and Alterations

An IRB may approve a consent procedure which does not include, or which alters, some or all of the elements of informed consent . . . or waive the requirements to obtain informed consent provided the IRB finds and documents that:

1) The research involves no more than minimal risk to the subjects; 2) The waiver or alteration will not adversely affect the rights and welfare of the subjects; 3) The research could not practicably be carried out without the waiver or alteration; and 4) Whenever appropriate, the subjects will be provided with additional pertinent information after participation (DHHS, Section 46.116d).

These provisions are derived from the Commission's report on IRBs, specifically Recommendation 4H and the commentary under Recommendation 4F. The relevant passage from 4F is reproduced in Chapter 9; Recommendation 4H will be discussed in Chapter 7. DHHS departed substantially from the Commission's intent.

First, let us examine some of the problems with the conditions specified by DHHS. Note that all four provisions are linked by the conjunction "and." Requiring both Criteria 2 and 3 means that if one could "practicably" carry out the research without the waiver, one must do so even though the waiver does not affect adversely the rights and welfare of the subjects. Thus, all elements of informed consent must be disclosed whether or not they are material to the prospective subject's capacity to make a decision. In my view it is never appropriate to require the disclosure of information that has no bearing on the subject's capacity to protect his or her rights or welfare, even in the absence of Criteria 1, 3, and 4.

In Criterion 1, the word "procedure" should be inserted after "research." Many research activities involve multiple components of which one or more presents more than minimal risk. If withholding or altering information about those components that present no more than minimal risk can be justified, it should be authorized even in the context of a protocol that also includes more risky procedures. For an example, see the discussion of compliance monitoring in Chapter 9.

In Criterion 3, the word "practicably" will require interpretation. As I discuss in Chapter 7, most research on medical records could be carried out with full informed consent from each and every subject. However, this would create great expense and inefficiency without materially furthering the goal of showing respect for the patients whose records we examine.

Criterion 4 is concerned with "debriefing;" references to "alterations" in the information are directed at research based upon deceptive strategies. These matters both are discussed in detail in Chapter 9.

The IRB at Yale was presented with a protocol that required it to analyze and interpret Section 46.116d in relation to a proposal to do research involving deception (428). The IRB required the investigator to minimize deception and replace it with a negotiation for consent to incomplete disclosure (Element of information 14) to the extent she could without compromising the validity of the research design. It was essential that subjects be unaware of the purpose of the research until they had completed their participation. Subjects were to be asked to complete questionnaires concerned primarily with various details of their sexual experience. Their responses would enable the investigator to discern whether various antihypertensive medications caused sexual dysfunction in women; such linkages already had been established for men.

1. Does this research involve no more than minimal risk? The IRB reasoned that the DHHS definition of minimal risk depends on a determination that the risks "are not greater . . . than those encountered . . . during the performance of routine physical . . . examinations." In the course of routine medical examinations, it is customary to collect information of a highly personal nature; breaches of confidentiality can result in serious social injury. For this reason one might consider research of this type as presenting minimal risk, given adequate assurances of confidentiality.

2. Does this waiver or alteration adversely affect the rights and welfare of the subject? Given the assurances of confidentiality provided by the investigator, it seemed extremely unlikely that any subject could be injured socially, less likely than in routine medical practice (cf. Chapter 7). Moreover, the subjects would be fully aware of the nature of the information they were providing and its social implications. Failure to disclose the purpose of collecting this information, thus, would not in and of itself be detrimental to the rights or welfare of the subjects. However, if one considers that subjects have a "right" to be fully informed—to receive the information prescribed by each of the elements of informed consent—then any waiver or alteration must be construed as depriving them of their rights. This, of course, would make it impossible to justify the waiver or alteration. In the judgment of the IRB, this could not be the intent of the regulations.

In order to determine whether plans to withhold disclosure of the purpose of research can be justified ethically, it is necessary to decide whether such disclosure is material in that the reasonable prospective subject might choose to consent or refuse to consent based upon knowledge of the purpose (cf. Chapter 9). In the judgment of the IRB, there was no reason to believe that reasonable prospective subjects might object to the development of information on the adverse effects of antihypertensive drugs.

3. Could this research have been practicably carried out without the alteration? Although the IRB accepted the investigator's judgment that it could not, it decided that the proposed letter of invitation was entirely too vague and that such vagueness could be detrimental to achieving the goals of the research. As one IRB member put it, "The letter . . . must appear to the prospective subject to be a somewhat aimless probe into female sexual behavior. Under such circumstances, one might anticipate frivolous answers or a high rate of rejection. A frank statement of aims might, on the other hand, engender more numerous and more serious responses. . . ."

The final letter read, in part: "I . . . invite you to participate in a study of women being treated for high blood pressure. . . . The effects of some of the common drugs used for treating high blood pressure on women's sexual function are uncertain. Whether these drugs increase, decrease, or have no effect on sexual function needs further evaluation so that the medical community can better counsel patients receiving these medications."

This letter could be interpreted as somewhat deceptive in that the hypothesis was that the drugs would either decrease sexual function or have no effect. However, it was possible that successful treatment of a chronic disease could be associated with enhanced sexual function.

4. Were there adequate plans to provide additional pertinent information after participation? The investigator persuaded the IRB that no debriefing was nec-

essary. Responses from the control group were to be anonymous. In order to debrief these women it would be necessary to keep identifiers, which would create a possibility of a breach of confidentiality. The investigator further argued successfully that debriefing the women with hypertension also might create more problems than it was worth. The research was designed to see whether there was a general tendency toward sexual dysfunction in such women. If such a general trend were to be identified, this would form the basis of doing more careful studies of the effects of particular drugs in order to determine specific cause-effect relationships. It seemed likely that most women with hypertension would, during a debriefing session, be most interested in learning whether the specific drugs they were taking had any effects on their own sexual function. Although in the course of this research some temporal correlations might be identified in individual women, it would not be possible to advise them of cause-effect relationships until follow-up studies had been done.

Further discussion of debriefing and alterations in the information may be found in Chapter 9. Further discussion of circumstances in which informed consent may be waived or replaced by a right of notice is found in Chapter 7.

Emergency Exception and Therapeutic Privilege

"Nothing in these regulations is intended to limit the authority of a physician to provide emergency medical care, to the extent the physician is permitted to do so under applicable federal, state, or local law" (DHHS Section 46.116f). Implicit in this rule is a recognition of two exceptions to the legal requirement for informed consent—the emergency exception and therapeutic privilege. For a recent authoritative commentary on these two exceptions, the reader is referred to Appendix L of the President's Commission's report, *Making Health Care Decisions* (571, Appendix I, pp. 199–201):

> In an emergency, when the need for immediate rendition of medical care exists, a doctor is excused from having to make disclosure and obtain consent. As simple as the rule is to state, it is equally difficult to apply. Medical care does not fall neatly into two precise categories of 'emergency' and 'nonemergency.' The degree of urgency with which a procedure needs to be performed can vary substantially. As a result, in some instances a physician may have sufficient time to seek a patient's (or a surrogate's) permission to perform a procedure, but inadequate time to make the ordinarily required disclosure without seriously jeopardizing the patient's life or health. . . . Although many courts have recognized the existence of the emergency exception, few if any have had occasion to undertake a thorough discussion of how an emergency ought to affect physicians' dual obligations. . . .
>
> Therapeutic privilege is undoubtedly the most discussed and debated of the . . . exceptions . . . [to the disclosure rule.] The exception is formulated by courts and legislatures in a variety of ways, so that it is difficult, if not impossible, to capture

the essence of the privilege succinctly. The formulations range from the fairly vague to the specific. In Rhode Island, "The physician is required to disclose . . . unless the doctor makes an affirmative showing that non-disclosure was in the best interests of the patient." More specific is the formulation employed in Kansas: "If a complete disclosure of all facts, diagnoses, and possible alternatives would endanger the recovery of the patient because of his existing physical or mental condition, the physician may withhold such information." These formulations, and others, focus on the anticipated effect that ordinarily required disclosure would have on the patient, and reflect the importance ascribed to the physician's "primary duty to do what is best for the patient," as the Supreme Court of North Carolina put it. . . .

Other courts

have focused instead, or in addition, on the effect that the disclosure of frightening information might have on the patient's decision-making capabilities. Thus, in addition to other reasons, the Minnesota Supreme Court concluded that the therapeutic privilege is applicable "if disclosure of the information would . . . cause such emotional distress as to preclude a rational decision" by the patient. This formulation is most in keeping with the dual goals of informed consent of maximizing the patient's well-being and encouraging autonomous decisionmaking . . . (571, Appendix L, p. 201).

Finally, other courts, contrary to these goals of informed consent, have held that a physician has a privilege to withhold information that "risks frightening the patient out of a course of treatment which sound medical judgment dictates the patient should undertake. . . ." This and similar formulations of the privilege ignore the caution issued by the Court of Appeals for the District of Columbia in the leading informed consent case, *Canterbury v. Spence*: "the physician's privilege to withhold information for therapeutic reasons must be carefully circumscribed, however, for otherwise it might devour the disclosure rule itself" (571, Appendix L, pp. 201–202).

Thus, the law on therapeutic privilege varies substantially from state to state. For a table showing the highlights of various state laws on these matters, the reader is referred to Appendix L provided by the President's Commission (pp. 375–414).

Virtually all commentators agree that emergency exceptions and therapeutic privilege have no relevance to protocols that have no therapeutic components and that its scope should be generally more limited with nonvalidated that it is with validated therapies. How much more limited? I am aware of no court case on the point.

According to FDA regulations (Section 50.23):

A. The obtaining of informed consent shall be deemed feasible unless, before use of the test article (except as provided in paragraph (b) of this section), both the investigator and a physician who is not otherwise participating in the clinical investigation certify in writing all of the following:

1. The human subject is confronted by a life-threatening situation necessitating the use of the test article.

2. Informed consent cannot be obtained from the subject because of an inability to communicate with, or obtain legally effective consent from, the subject.

3. Time is not sufficient to obtain consent from the subject's legal representative.

4. There is available no alternative method of approved or generally recognized therapy that provides an equal or greater likelihood of saving the life of the subject.

B. If immediate use of the test article is, in the investigator's opinion, required to preserve the life of the subject, and time is not sufficient to obtain the independent determination required in paragraph (a) of this section in advance of using the test article, the determinations of the clinical investigator shall be made and, within 5 working days after the use of the article, be reviewed and evaluated in writing by a physician who is not participating in the clinical investigation.

C. The documentation required in paragraph (a) or (b) of this section shall be submitted to the IRB within 5 working days after the use of the test article.

In my view, Section 50.23 is defective in several important respects (418). Firstly, these rules seem to reflect a lack of confidence in the validity of the federal drug approval process. If the physician knows that there is not available any alternative method of approved therapy that provides an equal or greater likelihood of saving the life of the subject, then, most commonly, this test article should already have been approved for marketing in the United States. In the view of many physicians, the drugs of choice for many disorders are not approved by the FDA for use in those disorders. This is exemplified by the therapeutic orphan problem discussed in Chapter 10. This also accounts for the fact that so many test articles are used according to compassionate INDs, as discussed in Chapter 14.

A careful reading of the phrase "because of an inability to communicate with, or obtain legally effective consent from, the subject" could be construed as permitting, to some extent, application of the therapeutic privilege exception. However, it could equally be read to recognize only the emergency exception. In Chapter 8 there are some examples of how some IRBs have used combinations of both exceptions to justify waiver or postponement of part or all of the disclosure requirements.

Limitation of the application of this rule to life-threatening situations is much too narrow. It should apply also in situations in which the subject is confronted by a situation presenting the threat of irreversible harm. If taken literally, this regulation would preclude timely use of an investigational therapy which might prevent loss of a limb or a kidney.

Finally, in my view the requirement for disinterested confirmation of the investigator's judgment as well as the requirements for excessive documentation are unfortunate. They appear to reflect a presumption that in these situations investigators may have important conflicts of interest. This may be correct in a small number of cases. However, in most situations covered by Section 50.23, the investigator's only motivation is to serve the health interests of the subject.

Such subjects, who are really patients, usually do not meet the requirements for inclusion in protocols designed to meet FDA requirements to gain approval for commercial distribution of test articles.

For further discussion of this rule and other alarming trends in proposed federal legislation, see Levine (418).

Conditions in Which Consent Need Not Be Documented

In its report on IRBs, the Commission recommends (4G):

> Informed consent will be appropriately documented unless the (IRB) determines that written consent is not necessary or appropriate because (I) the existence of signed consent forms would place subjects at risk, or (II) the research presents no more than minimal risk and involves no procedures for which written consent is normally required.

DHHS regulations approximate the Commission's intent except in one important respect (Section 46.117c).

> An IRB may waive the requirement for the investigator to obtain a signed consent form for some or all subjects if it finds either:
> 1. That the only record linking the subject and the research would be the consent document and the principal risk would be potential harm resulting from a breach of confidentiality. Each subject will be asked whether the subject wants documentation linking the subject with the research, and the subject's wishes will govern; or
> 2. That the research presents no more than minimal risk of harm to subjects and involves no procedures for which written consent is normally required outside of the research context. In cases where the documentation requirement is waived, the IRB may require the investigator to provide subjects with a written statement regarding the research.

Subparagraph 1 presents a problem by stipulating that the signed form must be the *only* record linking the subject and the research. In my opinion, the existence of records that constitute a threat to a person's privacy does not justify the imposition of a requirement to create yet another.

It is my wish to abstain from sarcasm. But how else does one discuss a requirement to ask a subject, "Do you wish to create a threat to your privacy? Do you wish to provide me with defense against any future allegation on your part that I wronged you or harmed you?"

In very careful studies, Singer has demonstrated that the fact that many persons simply do not like to sign forms may cause them to decline participation in research for illegitimate reasons (653). Three groups of subjects were each presented with the same information. Of those not asked to sign forms, 71% consented. The refusal rate was 24% higher among those asked to sign consent forms before participating in the study (an interview) and essentially the same in the group that was told they would be asked to sign forms after the interview.

The IRB may properly require providing subjects with documents that truly serve their interests if and when they seem appropriate (cf. the section on who keeps consent forms).

The FDA's documentation requirements (Section 56.109c) are essentially the same, except that they provide for no waiver in the interests of confidentiality. This seems reasonable, because the FDA does not regulate research of the sort for which this exemption is intended.

Epilogue

In the course of his extensive survey and analysis of informed consent, a book that provides a particularly comprehensive coverage of empirical studies of informed consent, Barber observes (36, p. 76):

> Satisfactory informed consent obviously requires effective communication among the many different participants in . . . the complex subsystems of medical relationships, relationships often thought to include only doctor and patient or researcher and subject but actually including families, friends, medical colleagues, paramedicals, administrators, and other professionals such as social workers and genetics counselors.

In 1974, Congress charged the Commission to consider "The nature and definition of informed consent in various research settings." The Commission, in turn, asked me to write a paper on this topic (404). I conclude now as I did then. The nature and definition of informed consent cannot be described definitively in the abstract. Functionally relevant definitions can be developed only in relation to specific proposals. In each case, the investigator should draft the proposal based on his or her knowledge of all aspects of the research and of the prospective subjects. The IRB should review the plans and negotiate with the investigator the minimum standards for that particular project. The investigator should then proceed to negotiate with prospective subjects with an awareness that the plan he or she has agreed to with the IRB often will have to be supplemented or modified to meet the needs of particular subjects. Thus it is that the nature and definition of informed consent are and ought to be continually negotiated and renegotiated.

Chapter 6

Compensation for Research-induced Injury

In 1982, the President's Commission concluded its deliberations on *Compensating for Research Injuries* with these statements (570):

> A social policy experiment is needed to see whether compensation programs might provide a feasible means further to reduce the risk of unremedied injury to subjects and to avoid the occurrence of events that might needlessly tarnish the reputation of research.

> Accordingly, the Commission recommends that the Secretary of Health and Human Services conduct a small-scale experiment in which several institutions would receive Federal support over 3 to 5 years for the administrative and insurance costs of providing compensation on a nonfault basis to injured research subjects.

As far as I know, this is the only "action" taken in the federal policy arena since 1981. In the first edition of this book, I stated that it appears likely that a norm calling for compensation for research-induced injury will be added to federal regulations. Because it seems much less likely at the time of this writing, this chapter is presented with only minor revisions. For further information on this topic, the reader is referred to the report of the President's Commission and its Appendix.

During the 1970s, commentators on the ethics of research reached a consensus that subjects who are injured as a consequence of their participation in research are entitled to compensation. The ethical arguments to support this entitlement are grounded in considerations of compensatory justice (132, 205). Compensatory justice consists in giving injured persons their due by taking account of their previous conditions and attempting to restore them. Sometimes it is possible to literally restore injured persons to their previous conditions, e.g., through medical therapy for the research-induced injury or illness. On other occasions, when literal restoration is not feasible, a monetary substitute is about the best

155

we can do. Most discussions of compensation for research-induced injury focus on the provision of monetary substitutes in cases in which there is temporary or permanent disability or death.

Childress analyzes the issue by examining two other situations in which society encourages its members to take risks to serve society's interests; these are the provision of veteran's benefits for those who are disabled in military service and compensation for "good Samaritans" who are injured while aiding others, trying to prevent a crime, or assisting the police in apprehending suspected criminals (132). Childress identifies at least three attributes that seem to give rise to a societal obligation to compensate for injuries. 1) The injured party either accepts or is compelled to accept a position of risk ("positional risk"). By accepting the position, the injured party is exposed to objective risks that he or she would not have encountered otherwise. 2) The activity is for the benefit of society. Parenthetically, any particular individual's motives may not be to benefit society. Thus, we distinguish the general aim of the activity from the aims of any particular participant, whether researcher or subject; for these purposes it is important to stress the objective end or function of the activity. 3) The activity is either conducted, sponsored, or mandated by society through the government or one of its agencies. The third criterion provides some more or less official verdict that the social practice is important for society apart from any particular individual's judgment of its value. Thus, a societal obligation to compensate for injuries exists in activities that involve positional risk for participants who objectively (if not subjectively) act on behalf and at the behest of society.

The Functions of Compensation

The development of a system of compensation for research-induced injury, in addition to fulfilling an ethical obligation of those who conduct or support research, may be expected to have several salutary consequences. Most important, such a development would change the nature of the cost-benefit calculations conducted by investigators and sponsors of research (8, 100, 285). If the investigators and sponsors of research were obliged to contribute to the costs of compensation, this would add to the cost side of the cost-benefit calculations in direct proportion to the probability and magnitude of the injuries to be expected. This would tend to discourage the conduct of excessively dangerous research, to enhance the motivation of investigators to be attentive to the subjects' safety in both the design and conduct of research activities, and to reward the timely termination of research activities when actual risks are found to exceed expectations. Robertson further predicts that the development of compensation systems would encourage persons to volunteer to become research subjects by limiting the potential costs to them of their participation (603).

Common Law Remedies for Research-Induced Injury

Adams and Shea-Stonum have surveyed the remedies for research-induced injury that are now available or that could be developed under common law principles (8). They conclude that none of them are adequate to meet the requirements of compensatory justice. For example, malpractice actions based upon claims of negligence provide no damages unless the physician is found at fault. By contrast, compensation would be required for all research-induced injuries whether or not the investigator is at fault.

At common law, an individual's consent is ordinarily construed as an agreement to assume the burdens of those risks that are disclosed during the negotiations for informed consent (8). This means that the individual agrees to assume the financial burden of paying for medical therapy for any such injuries; additionally, if these injuries result in loss of time from work, the consenting individual has also assumed responsibility for this financial loss. Most commentators agree that this arrangement is unsuitable for research-induced injury. One of the purposes of establishing a compensation system is to encourage individuals to volunteer to take certain sorts of risks of injury to serve the interests of society. Thus, the notion of assumption of risk through consenting makes no sense. Childress analogizes this to a public policy that would provide veterans benefits for those who were drafted while denying the same benefits to those who volunteered for military service (132). Moreover, participation in research designed to assess the safety or efficacy of a nonvalidated practice presents the possibility of injury of an unknown nature. One of the purposes of such testing is to identify the nature of the injuries that might result. Thus, it is impossible to disclose the probability and magnitude of an injury the nature of which is unknown. Finally, Robertson argues that the lack of availability of compensation for research-induced injury of itself presents a material risk: "... one can easily envisage a successful suit against the investigator . . . and the institution for medical expenses, lost earnings, and pain and suffering resulting from a non-negligently caused research injury, in which the subject claims that if he had been informed that injuries are not compensated, he would have not participated" (603). As we shall see, in the absence of a system of no-fault compensation, the Department of Health and Human Services (DHHS) decided that the unavailability of compensation must be disclosed.

Public Policy

As far as I can determine, the first public policy document that presented the concept of no-fault compensation for injury as an ethical requirement for research or practice was the report on the International Conference on the Role of the Individual and the Community in the Research, Development, and Use of Biologicals (155). This report resulted from a conference convened by several organizations, including the World Health Organization, the World Medical Association and the Centers for Disease Control of the United States Public Health Service.

> The basis of plans currently operating in six countries for compensating victims of injuries from immunization which is obligatory or recommended by health authorities is redress for having rendered benefit to the community by participating in a vaccination programme. . . . The view was strongly expressed that compensation of persons injured as a result of participation in field trials (of vaccines) was as important as compensation for injury from licensed products and such systems should be expanded to include research subjects.

This report also presented a set of "Criteria for Guidelines," including the following:

> National and international bodies in recognizing the special characteristics of biologicals research, development, and use should take into account. . . . Social and legislative action . . . to provide for the needs of subjects in biologicals research and recipients of biologicals in general use who suffer from disabling adverse effects.

Thus, with Childress, this international conference recommended compensation for injuries sustained by individuals through their participation in activities on behalf and at the behest of the community.

At about the same time (1975–1976), the Health, Education, and Welfare (HEW) Secretary's Task Force on Compensation of Injured Research Subjects conducted an extensive study of this issue. In its report, the Task Force recommended (183):

> Human subjects who suffer physical, psychological, or social injury in the course of research conducted or supported by the U.S. Public Health Service should be compensated if 1) the injury is proximately caused by such research, and 2) the injury on balance exceeds that reasonably associated with such illness from which the subject may be suffering, as well as with treatment usually associated with such illness at the time the subject began participation in the research.

In a letter to the HEW Secretary (June 29, 1977), the Commission supported the Task Force's recommendation very strongly and urged that "The goal of a fully operational compensation mechanism should be attained as soon as possible." However, in only one of its reports—*Research Involving Prisoners*—did the Commission recommend that the availability of compensation should be a necessary condition for the justification of research. Recommendation 4B assigns to the Institutional Review Board (IRB) the duty to ". . . consider . . . provisions for providing compensation for research-induced injury." The commentary under this recommendation states:

> Compensation and treatment for research-related injury should be provided, and the procedures for requesting such compensation and treatment should be described fully on consent forms retained by the subjects.

HEW decided not to implement this recommendation as a regulation ''. . . until the Department's Task Force Report on the subject has been thoroughly evaluated'' (184, p. 1052).

In Recommendation 4F in the IRB Report, the Commission states that during the negotiations for informed consent subjects should be informed ''. . . whether treatment or compensation is available if harm occurs . . . and who should be contacted if harm occurs. . . .'' (527). This recommendation was promptly implemented by HEW as an ''interim final regulation'' (186). The revised DHHS regulations specify as an element of informed consent (Section 46.116a):

> For research involving more than minimal risk, an explanation as to whether any compensation and an explanation as to whether any medical treatments are available if injury occurs and, if so, what they consist of, or where further information may be obtained. . . .''

Although there is broad consensus that no-fault compensation systems should be established for research-induced injury, very few institutions have developed such systems. The HEW Secretary's Task Force was able to identify only one, at the University of Washington (183). Their system, which is patterned after Workmen's Compensation, is described in Appendix B to the Task Force's Report. There seem to be formidable obstacles to the development of compensation systems. For various practical reasons, insurance companies seem unwilling to underwrite such insurance policies (285). Further, there is no general agreement on which injuries sustained in the course of research designed to test the safety and efficacy of nonvalidated practices should be compensable. Thus, most institutions are confronted with the problem of informing prospective research subjects that in the event of physical injury no compensation is available (411).

Policy at Yale Medical School

Yale University School of Medicine has established a policy on the provision of medical therapy for research-induced physical injury (410). This policy was designed to be in compliance with the 1978 ''interim final regulation'' which limited to *physical* injuries the requirement for disclosure of the availability of therapy and compensation. The final regulation does not specify the nature of the injury. Yale has not decided whether to modify its policy. Although it seems imprudent to publish a policy promising free medical therapy for psychological injury, Yale's IRB decides on a protocol-by-protocol basis which psychological injuries, if any, should be treated as if they were physical injuries in that the investigator should explicate for such injuries the offer of free medical therapy. Moreover, it is Yale's custom, although not a published policy, to provide care for minor anxieties and stress reactions experienced by some research subjects.

This policy is also responsive to the intent of the report of the Task Force providing medical therapy for injury that

> on balance exceeds that reasonably associated with such illness from which the subject may be suffering, as well as with treatment usually associated with such illness. . . .

> Yale University School of Medicine will assure that, for physical injuries sustained by human subjects as a consequence of their participation in research, medical therapy will be provided at no cost to the subjects. As elaborated subsequently, this policy is designed to cover only those costs of medical therapy made necessary by virtue of participation in research—not those ordinarily expected in the course of customary medical practice in the absence of research.

> At present, the School of Medicine is unable to provide compensation for disability occurring as a consequence of participation in research. If and when it becomes possible to provide such compensation, this policy will be changed accordingly.

> Guidelines for Preparation of Protocols for Review by the Human Investigation Committee are *amended* as follows:

> (In the appropriate section of the protocol) specify the risks, if any, of *physical* injury. If none, state: There are no risks of physical injury. If there are risks of physical injury there should be for each injury a careful estimate of its probability and severity as well as of its potential duration and the likelihood of its reversibility.

> There should be a statement as to whether these risks are presented by: 1) a procedure or modality performed or administered solely for purposes of developing or contributing to generalizeable knowledge (research), or 2) a procedure or modality performed or administered with the intent and reasonable prospect of yielding a direct health-related benefit to the subject (investigational practice).

> Ordinarily, the patient (or third-party payer) is expected to assume the costs of medical therapy for physical injuries sustained as adverse effects of noninvestigational diagnostic, prophylactic or therapeutic maneuvers. The intent of this Medical School policy is to provide medical therapy at no cost to subjects for physical injuries to which the subjects would not be exposed in the course of customary medical practice. Therefore, investigators should specify which of the risks of research or investigational practice are incremental in that they are either: 1) qualitatively different from those presented by noninvestigational alternatives; or 2) substantially more likely to occur or likely to be substantially more severe with the investigational practice than with noninvestigational alternatives. In addition, in the course of studying an investigational practice, new types of physical injuries may be discovered; medical therapy for these will be provided at no cost to subjects.

The source of funds, if any, to cover the costs of medical therapy for injuries should be specified; if none are available to the investigator, this should be stated. When the protocol is designed to test an investigational practice and there is an industrial sponsor (e.g., a drug company), the investigator is required to secure from the sponsor a written commitment to pay for the costs of medical therapy for physical injuries; a copy of this commitment must be attached to the protocol.

Consent forms: The following amendments to Guidelines for preparation of consent forms apply only to those protocols in which there is a risk of *physical injury*.

A. For protocols in which the risks of physical injury are presented by procedures or modalities performed or administered solely for research purposes (not investigational practice), state in language appropriate for the proposed subject population:

Medical therapy will be offered at no cost to you for any physical injury sustained as a consequence of your participation in this research. Federal regulations require that you be informed that—in the event of injuries—no additional financial compensation is available.

B. For protocols in which the risks of physical injury are presented by investigational practices, state in appropriate lay language:

If you participate in this study, you will be exposed to certain risks of physical injury in addition to those connected with standard forms of therapy; these include: (provide a complete description). In addition, it is possible that in the course of these studies, new adverse effects of (name of drug, device, procedure, etc.) that result in physical injury may be discovered. Medical therapy will be offered at no cost to you for any of the aforementioned physical injuries. You or your insurance carrier will be expected to pay the costs of medical care for physical injuries and other complications not mentioned in this paragraph since these are either associated with your disease or commensurate with the complications or risks expected of the usual therapies for your disease. Federal regulations require that you be informed that—except as specified above—no financial compensation for injury is available.

Chapter 7

Privacy and Confidentiality

Privacy is "the freedom of the individual to pick and choose for himself the time and circumstances under which, and most importantly, the extent to which, his attitudes, beliefs, behavior and opinions are to be shared with or withheld from others" (349). Because this is the definition used in this chapter, some matters considered under the rubric of privacy by the law are excluded, e.g., the right to abortion and contraception. In general, in clinical research we do not condone intrusions into individuals' privacy without their informed consent (Chapter 5). When an informed person allows an investigator into his or her private space, there is no invasion.

"Confidentiality" is a term that is all too often used interchangeably with "privacy." "Confidentiality" refers to a mode of management of private information; if a subject shares private information with (confides in) an investigator, the investigator is expected to refrain from sharing this information with others without the subject's authorization or some other justification.

The ethical grounding for the requirement to respect the privacy of persons may be found in the principle of *respect for persons*. The ethical justification for confidentiality, according to Sissela Bok (68), is grounded in four premises, three of which support confidentiality in general; the fourth supports professional confidentiality in particular.

First and foremost, we must respect the individual's autonomy regarding personal information. To the extent they wish, and to the extent they are capable

163

of doing so, they are entitled to have secrets. This facilitates their ability to live according to their own life plans.

Closely related is the second premise, which recognizes not only the legitimacy of having personal secrets but also of sharing them with whom one chooses. This premise embodies an obligation to show respect for relationships among human beings and respect for intimacy. Bok illustrates this premise by pointing to the marital privilege upheld in American law, according to which one spouse cannot be forced to testify against the other.

The third premise draws on the general requirement to keep promises. Thus, a pledge of confidentiality creates an obligation beyond the respect due to persons and existing relationships. Once we are bound by a promise, we may no longer be fully impartial in our dealings with the promisee.

According to Bok, these three premises, taken together, provide strong *prima facie* reasons to support confidentiality. Thus, these are binding upon those who have accepted information in confidence unless there are sufficiently powerful reasons to do otherwise—e.g., when maintaining confidentiality would cause serious harm to innocent third parties.

Bok's fourth premise adds strength to the pledge of silence given by professionals. The professional's duty to maintain confidentiality goes beyond ordinary loyalty "because of its utility to persons and to society. . . . Individuals benefit from such confidentiality because it allows them to seek help they might otherwise fear to ask for; those most vulnerable or at risk might otherwise not go for help to doctors or lawyers or others trained to provide it."

Investigators, of course, are not necessarily professionals to whom individuals (research subjects) turn for professional help. Thus, only part of Bok's fourth premise applies to investigators, that part that grounds the justification and requirement for confidentiality in its social utility. If investigators violated the confidences of their subjects, subjects would refuse to cooperate with them. This, in turn, would make it difficult, if not impossible, for many investigators to contribute to the development of generalizable knowledge.

In her book, *Secrets*, Bok presents an excellent and thorough examination of the topic of confidentiality and its limits (67). Parenthetically, much of what professionals are required to keep confidential is not, strictly speaking, secret.

Over the millenia, the professions have viewed the obligation to maintain confidentiality as very important. The Hippocratic Oath requires, "What I may see or hear in the course of the treatment or even outside the treatment in regard to the life of men, which on no account one must spread abroad, I will keep to myself holding such things shameful to be spoken about."

Thomas Percival's *Code of Medical Ethics*, upon which was based the first code of ethics of the American Medical Association, incorporated the following requirement (564): "Patients should be interrogated concerning their complaints in a tone of voice which cannot be overheard."

In all states, the law not only recognizes the obligations of physicians and many other professionals to maintain confidentiality, it requires it. In many states, there are statutes granting testimonial privilege to information secured by physicians from patients in the course of medical practice (85). Testimonial privilege means that physicians cannot be compelled to disclose such information even under subpoena. Although the United States Supreme Court refused to extend constitutional protections of privacy to physician-patient communications (*Whalen v. Roe*), the Federal Rules of Evidence, promulgated by the Judicial Conference, deferred to state laws on physician-patient privilege (85).

Most state laws on physician-patient privilege contain various exceptions, including mandatory reporting of information regarding battered children, various communicable diseases, and gun-shot wounds and certain proceedings concerned with health issues including workers' compensation and insurance claims (85).

In recent years, there has been a general movement to place further limits on the scope of testimonial privilege accorded to the physician-patient relationship. Consider, for example, some recent developments in the field of psychiatry, the medical specialty which in most states has been most successful in securing testimonial privilege. In 1974 in the case *Tarasoff v. Board of Regents*, the Supreme Court of California ruled that a psychiatrist has a duty to warn of the threat of violence if it is likely that such a threat would be carried out (313). A more recent ruling by the Massachusetts Supreme Judicial Court is likely to be even more problematic to psychiatrists (167). This case (*Commonwealth v. Kobrin*) involved a Medicaid fraud investigation. The state's attorney general subpoenaed the complete medical records for all of the defendant's Medicaid patients. The defendant refused, pointing to the state statute that protects psychiatrist-patient communications. According to Culliton, the court ruled (167): "Those portions of the records . . . which reflect the patients' thoughts, feelings, and impressions, or contain the substance of psychotherapeutic dialogue are protected and need not be produced." Apparently, all other components of the records must be revealed including ". . . notations that the patient has suffered any of the following . . . disturbance of sleep or appetite; . . . impaired concentration or memory; hopelessness; anxiety or panic; dissociative states; hallucinations. . . ."

Medicine is not the only profession that regards confidentiality as among its most important duties to clients. As Bok discusses, many professions are engaged in a constant struggle to enhance the legal protection of their authority to maintain clients' confidences; in general, lawyers seem most successful in this regard. Lebacqz has written a very fine book on professional ethics; although addressed to the clergy, it is highly relevant to other professions (376). The centrality of confidentiality to professional ethics is highlighted by Lebacqz's focusing the entire book on an analysis of whether a minister ought to keep confidential the secret plans of a teenager to seek an abortion. For an excellent discussion of threats to privacy and the major consequences of invasion of privacy, see Kelman, who concludes (349):

> Social and psychological research generates three kinds of concerns about invasions of participants' privacy: that public exposure of their views and actions may have damaging consequences for them; that the procedures used to elicit information may deprive them of control over their self-presentation; and that research may probe into areas that constitute their private space, overstepping the customary boundary between self and environment.

Most direct social injuries to research subjects result from breaches of confidentiality. An investigator may identify a subject as a drug or alcohol abuser; as a participant in various deviant sexual practices, as having any of a variety of diseases that may be deemed unacceptable by his or her family or social or political group, as having a higher or lower income than acquaintances might have predicted, as a felon, and so on. If such information becomes known to certain individuals, it might cost the subject his or her reputation, job, social standing, credit, or citizenship.

Threats to Confidentiality

In modern society it is becoming increasingly difficult to maintain confidentiality. Almost everywhere we look we find evidence of nearly universal surveillance. Cameras seem to be focused on us everywhere as we cash our checks, enter public buildings, and board airplanes. It seems that almost every day we read about surveillance activities that we did not even know about by public health officials, potential creditors, and agents of the Federal Bureau of Investigation and other members of the intelligence community. Increasingly, much personal and private information and misinformation is being entered into computers, the security of which is limited, even in medical practice (503). It is debatable as to whether we are approaching or surpassing the universal surveillance envisioned by George Orwell in *1984*. Siegler's reflections on confidentiality in medicine caused him to conclude that it is a "decrepit concept," particularly in large hospitals where so many people have legitimate access to confidential medical records (650). It is against this backdrop that we turn to a consideration of threats to confidentiality that may arise in the course of research involving human subjects.

Research records may be subject to subpoena. Although communications between physicians and patients are accorded limited testimonial privilege, it is not at all clear that such communications will be regarded as immune to subpoena when the physician is working in the role of investigator or physician-investigator. Holder reviewed the cases in which the records of epidemiologists and other investigators have been subpoenaed (311). She notes that such subpoenas are becoming increasingly frequent. "In most . . . cases, the ultimate disposition of the issue has been that all identifiers of research subjects have been protected and not released to the opposing party, but the data . . . have been held to be a proper subject for discovery." Brennan reviewed the special problems associated with doing research on toxic substances (85). He concluded that it is necessary to inform potential subjects/workers that research records could possibly be used as evidence in a compensation hearing or trial.

Plans to publish case studies may present formidable problems if the subject of the report either has a very rare disease or is very well known (236). At times it may be difficult or impossible to disguise the subject without distorting scientific content. In some cases, it may be appropriate to offer the subject the right to review and approve the manuscript before it is submitted for publication.

Plans to publish photographs present even more vexatious problems. Murray and Pagon review the precautions that may be considered in preserving patients' privacy and in assuring that one gets truly valid authorization from patients for publication of pictures and other data that might lead to their identification (517). They cite several cases in which physicians were found liable for publication without adequate consent of recognizable data not only in the lay press but also in medical books and journals.

In the past, it was a routine procedure at many colleges to take posture pictures of all entering freshmen. The students posed nude for lateral and posterior photographs that were developed as negatives. The purpose of these pictures was to identify those with poor posture; they were then required to enter a basic motor skills class. Dieck described the ethical and bureaucratic problems she encountered in securing such pictures in order to do research on the pathogenesis and epidemiology of disorders of the lower back (194).

In some social groups it is very difficult to maintain privacy or confidentiality. For an interesting analysis of the challenges presented by attempts to do research on one such group, see Nelson and Mills (535). They present a sophisticated resolution of some difficult problems in maintaining the privacy of subjects, cloistered religious celibates, in a study designed to evaluate a screening test for carriers of gonorrhea.

The name Barney Clark is a household word in the United States. This calls attention to an occasional but important problem. The Working Group on Mechanical Circulatory Support of the National Heart, Lung, and Blood Institute observed (740, p. 29):

> Dramatic therapeutic innovations such as the artificial heart have a tendency to be turned into media events. Efforts must be made to balance the public's interest and

patient-subject's right to choose his or her self-presentation. To the extent that damage to the latter can be anticipated or cannot be avoided, the prospective subject must be made aware of what to expect so that this consideration may enter into his or her "risk-benefit calculation" in deciding whether to consent. No patient should be excluded from selection on the basis of willingness to make public personal or private information, and the patient should be informed that this choice is available.

The Working Group recommends (740, p. 35):

In spite of public interest, the privacy of the patient and family must be respected. Institutions involved have a responsibility to help patients and families in this respect. Participation in media relations should not become a precondition for participation in clinical investigative use of mechanical circulatory support systems.

At times, the media event cannot be anticipated. This was illustrated dramatically in 1983 when a patient with leukemia petitioned a medical center to learn the name of a woman who could serve as a bone marrow donor for a transplant that he needed desperately (110, 176). The Institutional Review Board (IRB) chairperson contacted the prospective donor, Mrs. X, who said she was not willing to serve as a donor unless it was for a relative. The patient believed that Mrs. X might be more willing to donate bone marrow if she knew there was a specific patient who might be helped by her donation. When the institution and the IRB refused his demand, he retained a lawyer and requested a court order commanding the IRB to try once again to recruit Mrs. X. Such an order was granted by the Iowa District Court; on appeal, however, the Iowa Supreme Court reversed that decision. The Supreme Court's decision relied on a technical interpretation of the state statute defining "hospital record."

Experienced IRB members learn from time to time of a variety of other unanticipated threats to confidentiality. For example, investigators carrying private information have been robbed. A temporary secretary learned of her daughter's abortion when she was assigned to type the results of research. Investigators have inadvertently enclosed copies of raw data from which identifiers had not yet been stripped with their progress reports to funding agencies. Such surprises suggest that we should promise to use due care in maintaining confidentiality rather than making exorbitant promises of absolute confidentiality.

Sometimes a breach of confidentiality can occur as a consequence of elaborate plans to protect the subjects' privacy. Angoff describes one such case in which survey research was designed to develop information on various attributes of narcotic addicts and their families (19). Questionnaires were sent to individuals who were being treated at a local substance abuse treatment center; these questionnaires solicited information that most persons would find highly sensitive and personal. In order to protect the subjects' privacy, any connection between participation in the project and affiliation with the substance abuse treatment center had to be disguised. Among the elaborate precautions taken were these: the investigators used stationery that did not identify their affiliation with the substance abuse treatment center. Subjects were asked to mail completed questionnaires to a neutral-sounding title at an address of a box in the university mailroom. The subjects' names were affixed to completed questionnaires with a peel-off label. The plan was to remove this label immediately upon receipt and replace it with a code number.

Unfortunately, when the first questionnaire was returned, the mailroom staff did not make the connection between the neutral-sounding study title and the assigned mailbox. Because the address on the envelope identified no person and no department, the mailroom personnel could not identify the intended recipient. Consequently, the envelope was first delivered to the wrong department where it was passed from one person to another until "three investigators not associated with this study had opened the envelope, read portions of the questionnaire, and concluded that it did not belong to them or their studies." Finally, the IRB was consulted; because it was aware of all plans for this study, it was able to return the envelope to the investigator and make plans to prevent recurrence of this particular problem.

At times, individuals who are not engaged in research decide to publish what they learned while serving in other roles. Pattullo describes the experience of a college undergraduate who worked as a volunteer in a prison social work program (557). He decided to organize his experiences in order to develop an honors thesis examining the prison's power structure. By the time he decided to do this, many prisoners had revealed to him personal and private information that if it became generally known within the prison could put their lives in jeopardy. The IRB approved the student's plans to write the honors thesis, noting that there was little they could do to get valid informed consent after the "research" was done. Pattullo subsequently debated with Rolleston (611) the propriety of the IRB's disposition of this matter.

Investigators must give careful consideration to the circumstances under which they may either choose or be required to breach confidentiality. Generally, they can accomplish this by considering the sorts of information they are pursuing. Are they likely to learn about child abuse, intended violence to third parties, facts material to workers' compensation or litigation, or communicable diseases? If so, they may be required by law to share such information with others.

Professionals are often presented with the necessity to make difficult judgments about whether it would be in the best interests of the client to "betray" his or her confidences. Both Bok (68) and Lebacqz (376) illustrate this problem by presenting cases of young pregnant teenagers who disclose to professionals their intentions to have abortions without informing anyone—not their parents, not their boyfriends, not anyone. What is the professional to do?

Professionals who serve in the dual roles of physician and investigator may be presented with very difficult conflicting obligations. Investigators are expected to pursue the development of generalizable knowledge. However, when they are also physicians, they are expected to intervene or at least to offer to intervene in ways that would benefit the patient-subject. The most satisfactory way to deal with such conflicting obligations is to anticipate them, to decide in advance how one is likely to act in the event such a conflict should arise, and to discuss such plans forthrightly with prospective subjects (cf. Chapters 3 and 5).

Department of Health and Human Services (DHHS) and Food and Drug Administration (FDA) regulations require that research records be made available for inspection by agents of the government (*infra*). There are also proposed regulations requiring that such records be made available to the sponsors of FDA-regulated research.

Serious problems may be presented by investigators who seek to enter the private spaces of individuals without their informed consent or by use of deception (Chapter 9).

Protection of Confidentiality

". . . (T)he IRB shall determine that . . . Where appropriate, there are adequate provisions to protect the privacy of subjects and to maintain the confidentiality of data" (DHHS, Section 46.111a). This Regulation is in accord with Recommendation 4I of the Commission's Report on IRBs. In the commentary under this recommendation, the Commission provides suggestions for safeguards of confidentiality. Depending upon the degree of sensitivity of the data, appropriate methods may include the coding or removal of individual identifiers as soon as possible, limitations of access to data or the use of locked file cabinets. Researchers occasionally collect data that if disclosed, would put subjects in legal jeopardy. Because research records are subject to subpoena, the Commission suggests that when the identity of subjects who have committed crimes is to be recorded, the study should be conducted under assurances of confidentiality that are available from DHHS as well as from the Department of Justice.

Reatig has reviewed federal regulations on the privacy of research subjects and the confidentiality of data gathered from or about them (589). DHHS Regulations (42 CFR Part 2) require that the records of patients and research subjects who are involved in programs concerned with the study or management of problems related to alcohol and drug abuse be kept confidential and disclosed only 1) with the written consent of the patient (subject) or 2) pursuant to an authorizing court order based upon a finding of good cause. The regulations further provide that in certain limited circumstances (e.g., medical emergency or the conduct of research, audits, or evaluations), these records may be used without either consent or a court order. Although these regulations provide for penalties for those who violate confidentiality requirements, they do not grant immunity from subpoena. Special regulations that authorize confidentiality for patient records maintained by methadone treatment programs have been promulgated by the FDA (21 CFR 310.505g).

Three Federal agencies have the authority to grant confidentiality certificates which provide immunity from the requirement to reveal names or identifying information in legal proceedings. For a detailed survey of the provisions of these certificates along with instructions on how and to whom to apply for them, see Reatig (589).

DHHS Confidentiality Certificate

DHHS may authorize persons engaged in research on mental health, including research on the use and effect of alcohol and other psychoactive drugs, to protect the privacy of individuals who are involved as subjects in such research (42 CFR Part 2a). These certificates, which provide immunity from subpoenas, are available on application to the Director of each of the three institutes in ADAMHA, which are Mental Health, Alcoholism and Alcohol Abuse, and Drug Abuse. Applicants need not be recipients of DHHS funds.

Grants of Confidentiality

The Department of Justice may authorize investigators to withhold the names and other identifying characteristics of subjects of research projects involving drugs or other substances subject to control under the provisions of the Controlled Substances Act (21 CFR 1316.21). These confidentiality certificates, which also grant immunity from subpoena, may be obtained through applications submitted to the Administrator, Drug Enforcement Administration (DEA), Department of Justice. In order to be eligible, research projects need not be funded by the Department of Justice; however, they must be of a nature that the Department is authorized to conduct.

Privacy Certification

This is unlike the two other provisions for confidentiality in that it imposes positive duties and obligations on recipients of funds from the Law Enforcement Assistance Administration (LEAA) in the Department of Justice. According to the provisions of the Omnibus Crime Control and Safe Streets Act, no agent of the federal government or any recipient of assistance under the provisions of this statute may use or reveal any research or statistical information furnished by any person or identifiable to any specific private person for any purpose not specified in the statute. Privacy certification also provides immunity against subpoena. LEAA regulations (28 CFR Part 22) require that each applicant for support must submit a Privacy Certificate as a condition of approval of the grant or contract whenever there is a research or statistical project component under which information identifiable to a private person is to be collected. Penalties for violation of provisions of the Privacy Certificate include fines not to exceed $10,000.

The field of protection of privacy and confidentiality has become very complicated in recent years. For additional information see Privacy Protection Study Commission (578) and Reiss (594). Boruch and Cecil provide a general overview of the field with particular concentration on the development of ingenious statistical methods for making the responses of any particular subject uninterpretable even if the research records are subpoenaed (70).

Confidentiality Assurances

DHHS requires: "A statement describing the extent, if any, to which confidentiality of records identifying the subject will be maintained" (Section 46.116a). Statements about confidentiality of research records should not promise more than the investigators can guarantee. For most protocols in which the private information to be collected is not especially sensitive, it suffices to state that the investigators intend to maintain confidentiality, that they will take precautions against violations and that all reports of the research will be in the form of aggregated data and devoid of identifiers. When dealing with more sensitive information, it may be useful to specify some of the precautions; e.g., videotapes will be destroyed within 60 days or, if the subject so requests, earlier; data will be kept in locked files; individuals will be identified by a code and only a small number of researchers will have access to the key that links code numbers to identifiers.

Plans to incorporate data in the subject's medical record should be made explicit. In general, when these data are relevant to patient management they should be incorporated unless the subject objects. Incorporation of the results of nonvalidated diagnostic tests may lead to false diagnostic inferences with adverse consequences to the patient's medical care or insurability. Most people do not understand the full implications of signing forms that release their medical records to insurance companies (650). In some studies of the nature of the doctor-patient relationship, it is essential to keep the patient-subject's responses from the personal physician (Chapter 3); in fact, in some studies, to minimize potentials for intimidation, it may be necessary to prevent the physician from learning whether the subject even consented to participate.

It is essential to disclose all serious threats of confidentiality breaches which can be anticipated. Following are some examples of disclosures that have been approved by the IRB at Yale University School of Medicine.

From a study on the toxic effects of perchlorethylene involving as subjects individuals who were exposed to this substance in their jobs (85):

> Because the results of this study are directly relevant to your medical problem, they, in summary form, will be entered in your medical chart. You have the option of requesting that the results not be entered in your chart. In this case, they will be kept in a separate, confidential research file. At any rate, the results will be discussed with you and transmitted to anyone you designate. All possible steps will be taken to assure confidentiality of the results of this study. . . . However, you should know that a court of law or the Workers' Compensation Board could require access to any of your medical records if they chose. This may either help or hinder your chances for compensation.

From a study of devices designed to encourage young children who are thought to have been abused sexually to speak freely of their experiences, the following was incorporated in the permission form for the normal control subjects:

> Because this study is not intended to be related to either diagnosis or therapy for you or your child, you are entitled to decide whether the information obtained during this study shall be entered into the medical record. If, however, during your child's interview, his or her behavior raises a concern of sexual abuse, a clinical evaluation of your child will be performed. . . . If after that evaluation the suspicion of sexual abuse persists, the case will be reported as mandated by law.

In many studies it seems pointless to discuss confidentiality with prospective subjects. Commonly, the data developed by investigators are of no interest to anyone but them and their colleagues, and then only in aggregated form. I wonder how many prospective subjects have pondered, "Why is this investigator promising not to tell anybody my serum sodium level? Is there something more I should know about this?"

Until recently, I considered it unnecessary to discuss confidentiality with prospective subjects of most research designed to evaluate nonvalidated therapies when the results were to be entered in their medical records. Physicians are required by state laws to maintain confidentiality of these records and subjects presume that they will. Unfortunately, such discussions are now required in protocols subject to FDA regulations.

> FDA requires: A statement describing the extent, if any, to which confidentiality of records identifying the subject will be maintained and notes the possibility that the Food and Drug Administration may inspect the records (Section 50.25a).

This regulation is related to other provisions in FDA regulations. Section 312.1, describing form FD-1573 for INDs, requires the clinical investigator to make available to FDA employees all records pertaining to a clinical investigation; this includes case histories. Under certain specified circumstances, patients' names must be divulged. Similar requirements applying to investigators, sponsors, and IRBs are in the medical device regulations (Section 812.145). The FDA has published a proposal to revise this requirement and make it applicable to all research covered by FDA regulations; proposed Section 56.15a authorizes FDA agents (225):

> To copy such records which do not identify the names of human subjects or from which the identifying information has been deleted; and, (to) copy such records that identify the human subjects, without deletion, but only upon notice that (FDA) has reason to believe that the consent of human subjects was not obtained, that the reports submitted by the investigator to the sponsor (or to the IRB) do not represent actual cases or actual results, or that such reports or other required records are otherwise false or misleading.

This set of requirements has been criticized as unwarranted and unworkable (e.g., imagine stripping medical records of identifying information) (301, 415).

Similar regulations have been proposed requiring monitoring by industrial sponsors (21 CFR 52). Although these regulations are only in proposed form, sponsors have already begun to comply with them. Another agency that frequently requires access to patient records in the course of their monitoring and auditing activities is the National Cancer Institute. Thus, with time, increasing numbers of agencies and agents are required to look at research records. In much research involving the study of investigational therapies or what FDA calls "test articles," the use of the investigational therapy is an integral part of the patient-subject's medical care. Accordingly, the progress of such activities and the results customarily are entered into the medical record. The medical record, of course, also contains much information—often of a highly personal and sensitive nature—unrelated to the clinical investigation. For this reason, some institutions have begun to require that separate research records be kept; these records satisfy the record-keeping requirements of FDA regulations without containing other private information unrelated to FDA interests. Some institutions (e.g., Yale) recognize that in some cases a requirement for the keeping of separate medical and research records may impose an unwarranted burden on physician-investigators. Such a requirement could also be hazardous to patient-subjects; information that may be essential to patient care could be entered into the research record and omitted from the medical record. For these reasons, our policy does not require the maintenance of separate research records; the development of separate records is instead "strongly encouraged."

Many industrial sponsors print on the forms they provide for reports by investigators to sponsors spaces for subjects' names (or initials) and hospital unit numbers. Many IRBs require that this information not be sent to the sponsors. If necessary, a research identification number may be substituted.

Access to Prospective Subjects

In some protocols, the prospective subject population will be one to which the investigator does not have access in the course of his or her customary professional activities. The investigator may wish to secure the names of prospective subjects through contact with practicing professionals (e.g., physicians or social workers) or through the records of physicians, hospitals, schools, or welfare agencies. When this mode of access is to be used, it is ordinarily necessary to involve the practicing professional or the institutional record system in the consent process in a way that minimizes the potential for violations of confidentiality or coercion.

In some situations, the prospective subject may be the client of a professional whose overall care of the client depends upon his or her awareness of all activities that might have any influence on the health of the client. In such circumstances, no matter how access to the prospective subject is established, negotiations for informed consent should not proceed until the investigator secures the approval of the practicing professional. In such cases, the prospective subject should be informed that the research is proceeding with the awareness of the practicing professional who will be kept informed of any consequential findings, adverse reactions, and so on. In most hospitals, there are policies requiring that the personal physician approve the involvement of each patient in research.

Let us now consider research activities in which the practicing professional need not necessarily be aware of the activities of the researcher because the researcher has no plans to do anything likely to have an effect on the client's health. Most research activities in the field of epidemiology can be so characterized. Kelsey describes four methods that can be used by investigators to establish access to research subjects (350). A summary of these four methods in descending order of their protectiveness of the prospective subjects' privacy and confidentiality interests follows. Method 1, although most protective of these interests, is as we shall see least efficient from the perspective of the investigator.

In Method 1, the investigator provides the practicing professional with a form letter addressed to suitable prospective subjects. This letter describes the proposed research in general terms and suggests to the recipients that those who are interested should contact the investigator. At that point, negotiations for informed consent begin as in any other research context. When appropriate, the prospective subject should be advised in the initial communication that not only is there no need to inform the practicing professional whether the client consents to become a subject, but also (when appropriate) the investigator will not reveal to the practicing professional whether an investigator-subject relationship has been established. Thus, to the extent necessary, it is possible to minimize potential for violations of confidentiality, coercion, and duress. The latter potential presents itself most significantly in research designed to develop new knowledge on some characteristic of the professional-client interaction.

Following these procedures affords a high level of protection of the interests of prospective subjects. However, it does introduce into the system a good deal of inefficiency. Insistence upon all of these procedures can impede seriously the conduct of some types of research in which it is necessary to recruit large numbers of subjects, particularly when a high percentage of consentors is essential to the validity of the results, e.g., studies in epidemiology (350).

Many personal physicians who are asked to cooperate in providing access to prospective subjects refuse. Their reasons are often irrelevant to the purposes of involving the physicians in the first place, e.g., they simply do not wish to invest a lot of time and energy in activities in which they have no interest. Often it is possible to overcome this obstacle through providing investigators direct access to medical or other records. (Research use of medical records is discussed subsequently in this chapter.) In this way the investigators can themselves assume the tedious burden of identifying prospective subjects.

Once prospective subjects are identified, any of these four methods can be followed to establish contact with them. In choosing a method, its costs and benefits should be kept clearly in mind. Consider, for example, if the nature of the study (or the population of prospective subjects) is such that it merits *that* much protection at *that* cost (e.g., dollars, professional time, scientific validity)?

In Method 2, the investigator drafts a letter, as described in Method 1. The letter may be signed either by an employee of the hospital or by the investigator. The letter asks the prospective subject to return an enclosed self-addressed postcard indicating whether he or she would or would not like to participate. Alternatively, the prospective subject may be invited to telephone the signer.

This method assures that subjects will not receive unwelcome intrusions by investigators. However, many persons who might be quite willing to participate in the research once they learned more about it apparently do not respond to these invitations. Thus, the method is inefficient owing to the frequency of uninformed refusal.

Requiring first contact with a hospital employee rather than the investigator allows the prospective subject to respond without revealing his or her identity to a person who might be an outsider to the hospital system. This could have the effect of encouraging contact. In some cases, hospital employees are actually used to identify prospective subjects through examining, for instance, discharge diagnoses; ordinarily, they are much less skillful than investigators in identifying suitable prospective subjects. Apparently, in some other cases, investigators actually identify prospective subjects from medical records and the letter is signed by the hospital employee in order to create the semblance of a higher degree of confidentiality than actually exists. The propriety of such behavior is dubious. In any event, if the prospective subject is instructed to contact a hospital employee rather than the investigator, the former will generally be much less capable of responding to questions about the nature of the study.

In Method 3, participation is invited by letter as in Method 1; however, the recipient is asked to return a postcard or make telephone contact only if he or she does not wish to participate. The letter goes on to say that if there is no communication within, for example, 3 weeks, the person will be contacted by an investigator for purposes of negotiating informed consent.

In this method there is much less uninformed refusal than in Method 2. In Kelsey's experience, however, about 15% of people return postcards indicating that they do not wish to participate; she presents evidence that this often reflects an uninformed refusal on the part of individuals who, upon receiving more complete information, seem quite pleased to participate. As in Method 2, the contact person may be either the investigator or a hospital employee.

In Method 4, the letter sent to prospective subjects is similar in most respects to the other letters. However, there is no invitation to refuse either by telephone or by mail. Rather, addressees are notified that within 2 or 3 weeks they will hear from an investigator who will explain the study in more detail and then ask whether or not they wish to get involved. In Kelsey's view, which I share, this is in all respects the most efficient method and, for many studies, it is adequately respectful of the prerogatives of prospective subjects.

Which of these methods one is likely to select depends on several factors, some of which I have already identified. The method chosen also depends upon one's view of how much weight should be assigned to the right to privacy if it is balanced against institutional needs to accomplish goals efficiently. Pattullo (559) and Veatch (700), in their debate on this point illustrate vividly the gulf that exists between those holding differing perspectives. Cann and Rothman review the ways in which excessive protection of privacy interests can be highly destructive of research in epidemiology (103). They argue that most research involving interviews, because it is exempt from coverage by DHHS regulations, should not be reviewed by IRBs at all. Hershey rebuts their arguments on several grounds; one of the most important is that institutions have a legal responsibility to safeguard the confidentiality of records (293).

In survey research, it is quite common that researchers and subjects never meet face to face. All communications may be conducted by mail or by telephone.

Singer and Frankel have published an excellent discussion of informed consent in telephone interviews (654). Their empirical studies yield the conclusion that varying the amounts of information provided about the purpose or the content of the interview had very little effect on either the rate of response or refusal. They identify several factors that modestly influenced either the rate of agreement to be interviewed or the quality of responses provided by interviewees.

Right of Notice

. . . (I)nformed consent is unnecessary, where the subject's interests are determined (by the IRB) to be adequately protected in studies of documents, records or pathological specimens and the importance of the research justifies such invasion of the subject's privacy (527, Recommendation 4H).

In its commentary under this recommendation, the Commission indicates that such proceedings without consent should be conducted according to the conclusions of the Privacy Protection Study Commission (586). In particular, it calls attention to the Privacy Commission's Recommendation (10c) under "Record Keeping in the Medical-Care Relationship;"

. . . (N)o medical care provider should disclose—in individually identifiable form, any information about any . . . individual without the individual's explicit authorization, unless the disclosures would be: . . . (c) for use in conducting a biomedical or epidemiological research project, provided that the medical-care provider maintaining the medical record:

(i) determines that the use or disclosure does not violate any limitations under which the record or information was collected; (ii) ascertains that use or disclosure in individually identifiable form is necessary to accomplish the research . . . purpose . . . (iii) determines that the importance of the research . . . is such as to warrant the risk to the individual from the additional exposure of the record or information . . . (iv) requires that the adequate safeguards to protect the record or information from unauthorized disclosure be established . . . and (v) consents in writing before any further use or redisclosure of the record or information in individually identifiable form is permitted. . . .

The Commission further elaborates (527):

When the conduct of research using documents, records or pathology specimens without explicit consent is anticipated, incoming patients or other potential subjects should be informed of the potential use of such materials upon admission into the institution or program in which the materials will be developed, and given an opportunity to provide a general consent or object to such research.

Informing potential subjects of the possible use of documents, records, or pathology specimens and affording them an opportunity to provide a general consent or objection is, by no stretch of the imagination, informed consent. Yet, in the Commission's judgment, it seems to be an adequate expression of *respect for persons* in these conditions. Holder has proposed that such expressions be construed as responsive to a right of notice, which differs from the right of consent held by prospective subjects of most research (303).

The following notice is presented to all patients at Georgetown University Medical Center at the time of admission or initial treatment (707):

> As you may already know, the Georgetown University Medical Center is a teaching institution. Thus, the care you receive here is part of an educational process. As part of this process, the medical record which is kept concerning your treatment may be used for research purposes intended to benefit society through the advancement of medical knowledge. If information contained in your medical record is used for medical research, your anonymity will be carefully protected in any publication which is based on information contained in the record.

Appelbaum et al. analyze various approaches to affording patients adequate opportunities to protect their own interests in regard to use of their medical records for research purposes (24). They conclude that the right of notice is inadequate. The procedure they recommend is one they call "access with patients' general consent obtained at the time of admission." Upon admission to the hospital, each of their patients is presented with a written statement calling attention to the fact that their records might be "reviewed by approved researchers for purposes of screening you or your child as a potential subject in a . . . research program. Whether or not you or your child decide(s) to participate in a research project is, of course, up to you." This statement is followed by a signature block where patients indicate they have read the form and boxes to check indicating whether or not they are willing to have their records screened.

Although there is no requirement in federal regulations for patients' consent to use of their medical records for research purposes, Hershey cautions that there are regulations in some states requiring explicit consent (291, 292). As he points out, state statutes and regulations may be vaguely worded, and their interpretation may be difficult. He cautions investigators and IRBs to be sure that they are operating in accord with authoritative interpretations of such laws.

As we consider permitting investigators to examine medical records for research purposes, it is important to recognize that the hospital or the practicing physician is responsible for safeguarding their confidentiality. Many institutions have developed policies that are relatively generous with regard to permitting access to such records when the investigators are subject to the disciplinary procedures of the institution. Institutional guidelines at Yale University School of Medicine, for example, identify as "authorized individuals" the hospital medical staff, members of the faculty of the School of Medicine and the School of Nursing, and students in those two schools working with faculty sponsors. All other persons are identified as "nonauthorized." Nonauthorized investigators are not permitted access to medical records unless some member of the faculty of the School of Medicine or School of Nursing assumes responsibility for maintaining confidentiality of the records. Unlike some other institutions, Yale does not require that a member of its faculty be a coinvestigator in the project. In this regard, Yale is concerned with confidentiality, not credit for research productivity.

For further discussion of privacy and confidentiality issues in research involving medical records, see Cowan and Adams (164) and Gordis and Gold (268). Cowan also describes tumor registries, mandated by state laws, and their use for research purposes (163).

Holder and Levine recommend the following additions to hospital forms used to authorize surgery or autopsy (314):

> . . . the customary practices of the hospital (should be) made explicit on the consent forms. Thus, the standard forms for consent to autopsy might indicate—when appropriate—that it is customary practice . . . to remove and retain some organs, tissues, and other parts as may be deemed proper for diagnostic, research, or teaching purposes. Similarly, the . . .form for . . . surgery might indicate that it is customary to deliver the removed part to the surgical pathologist for diagnostic testing and that in some cases it might be retained for research or teaching purposes before it is destroyed.

Holder and Levine also recommend criteria for determining when notices are inadequate; when both attributes are present, informed consent is necessary (314):

> 1) The information to be obtained by the research procedure either will or might be linked to the name of the individual from whom the specimen was removed. If there is no way to link the information to the name of the individual there seems to be no possibility of putting that individual or the next of kin at risk.

> If, and only if, the first attribute is present the proposed research should be reviewed for the possible presence of the second.

> 2) The proposed research may yield information having diagnostic significance. In this regard, one should be particularly concerned if the presence of the diagnostic information might expose the individual to liability for criminal action (e.g., detection of alcohol or abuse drugs), or might jeopardize his (or her) insurance or Workmen's Compensation status or create the potential for civil litigation. Additionally, one should be concerned about the possibility that the information might significantly change the way in which the individual is perceived by self, family, social group, employers, and so on (e.g., detection of venereal disease).

There are several other commonly performed types of research activities (e.g., study of leftover blood, urine, or spinal fluid collected for diagnostic purposes) that may be assessed similarly to those performed on specimens secured at surgery or autopsy. In general, the same two attributes will suffice to identify those activities for which informed consent is necessary.

As discussed in Chapter 5, there is a good deal of controversy over whether subjects should be entitled to a share of the profits from marketable products developed in the course of research using their cells or tissues. In some cases, it may be prudent to add statements regarding such entitlements or lack thereof to standard forms used to authorize surgery or autopsy. In the event such statements are necessary, it seems essential to make these on forms to be signed by persons authorized to do so rather than to rely on notices published in patient information brochures.

Acquired Immune Deficiency Syndrome

Acquired immune deficiency syndrome (AIDS) is the special case of the 1980s, the one that focuses most sharply and urgently the need to balance the privacy and confidentiality interests of prospective subjects against the need to develop sound information promptly. The stakes on both sides are very high indeed.

Although AIDS was first identified in 1981, the earliest cases have been traced back to 1979. Since then its spread may be characterized as a geometric progression; for the first 5 years, the number of afflicted people doubled about every 6 months. By August 1983, over 2,000 cases of AIDS as defined by the Centers for Disease Control (CDC) had been reported; of these patients, 80% died within 2 years of diagnosis (171). By 1985, the doubling time had increased to approximately 11 months, and by the end of the year, the total number of cases in the United States had reached 16,000. Landesman et al. estimate that in 1985 the direct costs to society of the AIDS epidemic will exceed $500,000,000 (365). Others have estimated total direct costs to society for the first 10,000 cases will be $6,300,000,000. Suffice it to say that in 1983 Edward Brandt, Assistant Secretary of Health, identified AIDS as the "number 1 health priority" of the U.S. Public Health Service (82).

Research on AIDS has resulted in the identification of a virus as its probable cause; it is called human T-cell lymphotropic virus (HTLV-III) or lymphadenopathy-associated virus (LAV or ARV). Tests have been developed to identify antibodies to HTLV-III in human blood; these tests are being used to identify persons who have been exposed to the virus. The diagnosis of AIDS cannot be established based upon a finding of antibodies to HTLV-III in a person's blood; a positive test establishes only the fact that a person has been exposed to the virus. Upon hearing this statement, many persons who do not know what a biologist means by "exposed" develop incorrect beliefs. "Exposed" in this sense means that the virus has been in the person's blood; in the case of HTLV-III, it almost certainly means that the virus is in the person's body and probably that it is alive and capable of causing infection.

Because the virus has been identified, optimism regarding the eventual development of a vaccine seems warranted. We cannot be certain about that, however. Because this virus, like the influenza virus, has frequent genetic mutations, development of a vaccine may be rather difficult.

We can also share Brandt's hope that a cure might be developed even though most viral diseases have proven refractory to efforts to cure them with chemotherapeutic agents (82).

While awaiting the development of vaccines and cures, we need careful epidemiologic studies to determine the natural history of AIDS. We need refined information on how AIDS is spread. This will enable us to advise people on behavioral changes that might reduce their chances of becoming infected. It will also facilitate reaching sound judgments on such controversial matters as whether children with AIDS should be excluded from schools.

We need information on what the odds are that a person with HTLV-III antibodies or viruses in his or her blood will go on to develop AIDS, AIDS-related complex (ARC), or, for that matter, any other disease. This will permit more adequate counseling regarding prognosis than physicians are now able to give.

In passing, it is worth noting that many different conditions are being referred to by physicians and others as AIDS. The fully developed syndrome that has such a high fatality rate has been defined carefully by CDC (549). Some individuals who do not have the fully developed syndrome are referred to as having ARC. There are other individuals who have no abnormality other than a positive antibody test, some who have only generalized lymphadenopathy (enlarged lymph nodes), and some who have only the abnormalities in ratios of T-lymphocytes associated with fully developed AIDS. We have little reliable information at this point to provide to individuals with any of these conditions, other than fully developed AIDS, as to the prospects for their future health. Given this situation, individuals who learn that they have positive antibody tests or any of the other conditions I mentioned may be terrified—probably unnecessarily—about the prospects for the future, particularly if all of their information on prognosis is based on newspaper accounts, which, in turn, are generally based upon reports of patients with AIDS as defined by CDC.

Much of the necessary research can only be done by epidemiologists gathering data of a most sensitive nature from persons in at-risk groups. Among the groups identified by the U.S. Public Health Service as at-risk are homosexual men, intravenous drug (e.g., heroin) abusers, individuals with hemophilia, sex partners of AIDS patients, and some others (82, 384, 543).

Among the at-risk groups identified by the U.S. Public Health Service, two are conspicuously vulnerable to adverse consequences of breaches of confidentiality; these are male homosexuals and users of illegal intravenous drugs. Any breach of confidentiality regarding their personal data could result in grievous losses, such as imprisonment, loss of jobs, and loss of insurance. Even becoming identified as a subject in certain sorts of epidemiologic research could place these people in grave jeopardy. For this reason, many representatives of the gay community expressed great anxiety about participating in epidemiologic research. Although they recognized that the efficient and successful conduct of such research was greatly needed and that they would be among the greatest beneficiaries of the research, they expressed concern about the possibility that research conducted or supported by the federal government could result in great social harm to them (547). Some others were skeptical about the likelihood that agents of the federal government who had access to their private data would show more loyalty to them than they would to their colleagues in other branches of the federal government (e.g., the Department of Justice) or to the local police.

Boruch provided evidence suggesting that such concerns should be taken seriously; as he observed, the CDC has potentially conflicting missions: the conduct of epidemiologic research and law enforcement (69). Thus, he argues that we should consider carefully the proposition that in research on AIDS, even that conducted or supported by the federal government, private agencies and not the federal government should be the custodians of personal identifiers.

Currently, millions of persons are having blood tests for HTLV-III antibodies. All persons who offer to donate blood, for example, are told that such a test will be done and that they will be informed of positive results. Those who have false positive results will be frightened unnecessarily. The way these tests are being done now, false positives should be rare. False negatives are more of a problem owing to the lag time between infection and the development of antibodies. Those who have false negative results may experience inappropriate relief of anxiety unless they receive very careful explanations of the meaning of a negative test. Although in most states those who are tested can now be offered the option of not having the results entered in their medical records, some states require mandatory reporting of positive tests to health departments for surveillance purposes (384). Individuals now may choose not to be tested; they can refuse to donate blood or to participate in research. However, some states are considering requiring HTLV-III testing before issuing marriage licenses (365). These are among the continuing conflicts between the interests of the state and those of the individual as played out in society's struggle to contain the AIDS epidemic.

Guidelines for the conduct of research on AIDS have been published recently by several agencies, including the U.S. Public Health Service (549) and a working group convened by the Hastings Center (46). In short, they urge that all of the threats to privacy and confidentiality and precautions used to defend against these threats that I have discussed in this chapter be taken especially seriously in the conduct of research on AIDS. For further reading on AIDS, I recommend a collection of scholarly articles, public documents, and newspaper stories assembled by Watkins (716) and two recent symposia (384, 543).

Chapter 8

Randomized Clinical Trials

The randomized clinical trial (RCT) is a device used to compare the efficacy and safety of two or more interventions or regimens. The RCT was designed primarily to test new drugs (294); however, over the last 20 years, as its popularity has increased, it has been applied to the study of old drugs, vaccinations, surgical interventions, and even social innovations such as multiphasic screening. The RCT has four main elements. 1) It is "controlled," i.e., one part of the subject population receives a therapy that is being tested while another part, as similar as possible in all important respects, receives either another therapy or no therapy. The purpose of simultaneous controls is to avoid the fallacy of *post hoc ergo propter hoc* reasoning. 2) The significance of its results is established through statistical analysis. Commonly, in order efficiently to generate sufficiently large numbers for statistical analysis, it is necessary to involve several hospitals in the study; consequently, the design of the protocol and the evaluation of the results are often conducted at a central office remote from that in which any particular patient-subject is being studied. 3) When it is feasible, a double-blind technique is employed. That is, neither the investigator nor the subject knows until the conclusion of the study who is in the treatment or control group. The purpose of double-blinding is to overcome biases on the part of both subjects and investigators; it also tends to mitigate the Hawthorne effect (cf. Chapter 9). 4) It is randomized, i.e., the therapies being compared are allocated among the subjects by chance. This maneuver also is designed to minimize bias, as when investigators assign those patients whom they think have the best prognosis to what they believe to be the superior therapy.

In this discussion, the term "bias" is used in two different senses. In the first, "bias" means a prepossession with some point of view that which precludes impartial judgments on matters relating to that point of view; double-blinding is designed to correct for bias in this sense. "Bias" is also used in the sense in which statisticians use the word, meaning any tendency of an estimate to deviate in one direction from a true value. Randomization is among the strategies employed to prevent bias in this sense. The sense intended with each use in this chapter will be apparent from the context in which bias is used.

Let us consider some data that suggest the dimensions of the RCT issue (245). In fiscal year 1975, the National Institutes of Health (NIH) spent 114 million dollars on RCTs. The total cost of these trials from their beginnings through the anticipated dates of their completion was estimated at 650 million dollars. In 1975 there were over 750 separate protocols involving over 600,000 patient-subjects. These numbers are for NIH-sponsored trials only; many additional RCTs are conducted or sponsored by drug companies or with funding from other sources.

In the years 1975–1979, funding for RCTs remained nearly constant when expressed as a percentage of the total NIH research budget; by contrast, in the Veterans Administration (VA), the cost of RCTs has increased progressively from 3.1% of the total research and development budget in 1970 to 7.1% in 1981 (485). In 1979, the median cost of RCTs expressed in terms of dollars per patient per year was $1657 for the National Heart, Lung and Blood Institute and $603 for the National Cancer Institute.

In the course of planning and conducting RCTs, one encounters a vast range of ethical problems encompassing nearly all of those presented by all research involving human subjects. In the negotiations for informed consent for participation in a RCT, nearly all the traditions and motivations of the physician must be suspended. Similarly, the effects on the expectations and wishes of many of the patients may be devastating. There are those who argue that the entire process is so technological and dehumanizing that its use should be curtailed sharply (253). On the other hand, there are scientists (294) and physicians (128) who contend that the power of the RCT to develop sound information on the relative merits of therapies is so great that it would be unethical to introduce a new therapy without this sort of validation.

In this chapter I shall examine some ethical problems that are more or less peculiar to RCTs. Many other ethical problems presented by RCTs, as well as other research designs, are discussed in other chapters. More extensive discussions of problems presented by RCTs have been published by Fried (253), in the Proceedings of the National Conference on Clinical Trials Methodology (615), and by others to whom I shall refer in passing. I wish to emphasize that the focus of this chapter is on ethical problems presented by RCTs rather than on their obvious virtues when used intelligently and in appropriate circumstances.

Justification to Begin

It is now generally accepted that the ethical justification for beginning a RCT requires, at a minimum, that the investigators be able to state an honest null hypothesis. It is necessary, then, to state that there is no scientifically validated reason to predict that therapy A will be superior to Therapy B. Further, there must be no Therapy C known to be superior to either A or B unless there is good cause to reject Therapy C, e.g., the population of research subjects will consist of either those in whom Therapy C has been tried and failed or individuals who are aware of Therapy C and, for various reasons, have refused it (cf. Chapter 3).

Shaw and Chalmers recognize the possibility that reasonable physicians may differ on whether an honest statement of a null hypothesis may be made (648). Consequently, they focus on whether any particular clinician can justify his or her collaboration in a RCT.

> If the clinician knows, or has good reason to believe, that a new therapy (A) is better than another therapy (B), he cannot participate in a comparative trial of therapy A versus therapy B. Ethically, the clinician is obligated to give therapy A to each new patient with a need for one of these therapies.

> If the physician (or his peers) has genuine doubt as to which therapy is better, he should give each patient an equal chance to receive one or the other therapy. The physician must fully recognize that the new therapy might be worse than the old. Each new patient must have a fair chance of receiving either the new and, hopefully, better therapy or the limited benefits of the old therapy.

Shaw and Chalmers note that early usage of a new therapy commonly is performed in an uncontrolled manner. This often leads to false impressions about its safety and efficacy. Poorly controlled pilot studies generate inadequate information that tends to inhibit the subsequent conduct of adequate RCTs. For example, the results of an early pilot study may show that the new therapy (A) seems far superior to the standard (B); thus, the statement of the null hypothesis to justify a comparison between A and B will be challenged inappropriately by IRB members and others. To avoid such problems, Shaw and Chalmers encourage the use of RCTs as early as possible in the course of testing a new therapy. In fact, in a subsequent publication, Chalmers et al. urge randomization with the very first use of the new therapy; i.e., the first patient to receive a new drug should have a 50–50 chance of receiving either it or the standard alternative drug (128).

It is not customary in the United States to introduce new therapies by random allocation with the very first patient. Most clinical investigators demand some preliminary evidence that the new therapy will have at least some beneficial effect (436). For an example of an argument over whether this should be the case, see the exchange between Hollenberg et al. (315) and Sacks et al. (623). In my view, early demonstration of some beneficial effect is essential in most cases. However, I concede the point that the availability of information derived from preliminary and uncontrolled studies complicates subsequent defense of null hypotheses. For an example of how the conduct of a scientifically inadequate early trial seems to have inhibited the conduct of subsequent adequate RCTs, see the discussion of an RCT proposed to evaluate the effects of folic acid in preventing serious neural tube defects (Chapter 13). Preliminary data suggesting the efficacy of folic acid caused three IRBs to disapprove a RCT designed to evaluate its effect thoroughly (202).

All too often, the null hypothesis used by investigators to justify a RCT consists of a statement that certain gross measures of outcome will be the same, e.g., that 5 years after the initiation of either course of treatment, the probability that the patient will still be alive is equivalent (441). Such statements of medical equivalence based upon gross measures of outcome may be inadequate in at least two respects.

First, the apparent medical equivalence may not always obtain in consideration of any particular patient (253, pp. 52–53):

> Consider, for instance, the choice between medical and surgical intervention for acute unstable angina pectoris. I would suppose that a group of patients could be so defined that the risks and benefits of the two available courses of action were quite easily balanced. But, when a particular patient is involved, with a particular set of symptoms, a particular diagnostic picture, and a particular set of values and preferences, then one may doubt how often a physician carefully going into all of these particularities would conclude that the risks and benefits are truly equal.

Physicians often state that treatment decisions must be based at least in part on their intuitive grasp of what is needed based upon their experience in responding to minor variations in the clinical picture; consequently, the physician may have grounds for judging that even though two therapies are medically equivalent in terms of the long range outcome for a *group* of patients, one may be superior for any *particular* patient. In that case, the two treatments cannot be judged equivalent for that particular patient.

Fried identified a second reason for challenging claims of equivalency between treatments; patients differ in their values and preferences and may find the quality of life offered by one or the other treatment to be as important to them as the eventual outcome. Thus, although two treatments may be medically equivalent in terms of life expectancy, they may be vastly different in terms of the value that the prospective subject may place upon the quality of life that he or she may enjoy while awaiting the outcome (cf. Chapter 5).

A dramatic illustration is provided by consideration of a RCT performed to compare the results of radical mastectomy with wide excision in the treatment of early breast cancer (29). The two treatments were judged to be potentially medically equivalent in terms of such gross outcome measures as life expectancy and probability of recurrence of cancer. However, one of the alternative treatments was, to quote the investigators, of a "mutilating character." Consequently, from the perspective of the patient-subjects, the two approaches to therapy can not be considered equivalent. For other examples of this problem, see Fried (253, pp. 141–148).

In some circumstances, the conduct of a RCT may be justified even when there is strong reason to predict that the new therapy will prove superior to standard approaches to therapy (621). When a new therapy first becomes available, there is often not enough to go around. One of the fairest ways of distributing a scarce good or benefit is through a lottery (cf. Chapter 4). Thus, in many cases it may be appropriate to recruit a pool of prospective recipients of the new therapy that is twice as large as the supply can accommodate. Those who will receive the new therapy are chosen by chance and the others may serve as controls receiving standard therapy. One of the RCTs justified and conducted in this fashion was the trial of streptomycin in the treatment of pulmonary tuberculosis (294). Parenthetically, this RCT is of historical significance in that it is generally recognized as the first RCT conducted according to modern standards.

Scientists usually are tentative in their acceptance of the results of research until they have been confirmed in a subsequent investigation conducted by a new group of investigators. Insistence upon such independent confirmation of the results of an RCT is especially problematic. When the null hypothesis has been disproved in an adequate RCT, defense of the same null hypothesis in order to justify a second RCT is usually a travesty. Thus, when such independent confirmation is necessary, one should conduct the two (or, if necessary, more) RCTs concurrently.

Occasionally, we find that two apparently reproachless RCTs, conducted concurrently and even published simultaneously, yield opposite conclusions. Bailar has written an excellent and concise editorial on how we might react to such distressing happenings (32). He suggests that we should, in general, examine the RCTs carefully to see if we can discern reasons for the discrepancies. If this fails, he has no concrete advice on what should be done next. Neither do I.

At times strong pressures are brought to bear on medical researchers to conduct RCTs to evaluate therapies that, in their view, are worthless, such as laetrile. This problem is assessed by Cowan, who concludes (162): "No physician should sanction such an activity or be part of it. . . . A physician may feel that he cannot prevent a patient from resorting to quackery. He should not, however, legitimize quackery by incorporating it into a clinical trial regardless of the procedures used to obtain informed consent. There can be no informed consent to fraudulent treatment."

Cowan concedes that a RCT designed to evaluate laetrile might be permissible, if at all, "in patients no longer responsive to conventional therapy." In my view, this presents two problems. First, proponents of laetrile therapy would rightly claim unfair bias; the cards would be stacked against laetrile owing to selection of patients with poor prognoses. Secondly, if physicians are not permitted to participate in such RCTs, who would take care of the patient-subjects?

Cowan notes without comment the fact that the National Cancer Institute, in response to heavy political pressure, conducted clinical trials of laetrile. In my view, it is permissible for physicians to participate in such RCTs if, and only if, they believe that such RCTs can be begun with a valid null hypothesis. A second necessary condition is this: the patient-subjects should be only those who are believers in laetrile to the extent that they would accept no other form of cancer therapy currently "known" to be effective for their diseases. The second condition probably would result in a requirement that the RCT involve either placebo or no-treatment controls. It is possible, of course, that some believers in laetrile might accept assignment to either laetrile or some standard cancer treatment as a condition of enrollment in a RCT; for the true believer, this must be considered an act of extraordinary altruism.

Riskin contends that "therapist-patient sexual relations should be regarded by professional organizations as ethically justifiable . . . but only if performed pursuant to a research protocol reviewed and approved . . ." by an IRB (600). In his view, such sexual encounters for the "benefit of the patient" must be regarded as nonvalidated practices. In response, Culver argues that there is already sufficient evidence to reject the null hypothesis (168). In his view, such encounters are much more likely to be harmful than beneficial. Riskin remains unpersuaded.

Subject Recruitment

In the conduct of a RCT is it essential to recruit large numbers of subjects in order to generate sufficient data for statistical analysis of the results (204; 652, pp. 116–122). It is necessary to recruit a sufficient number of subjects to avoid Type I errors, which may be defined as incorrect rejection of the null hypothesis. Such errors result in invalid declarations that (e.g.) Therapy A is superior to Therapy B. To say that the results of a RCT are significant at a level of $P = 0.05$ means that the probability of having committed a Type I error is 5%. It is also necessary to avoid Type II errors, which are failures to reject incorrect null hypotheses. Such errors result in incorrect declarations that Therapy A is no better or worse than Therapy B. Commonly, acceptable levels of risk for Type II errors negotiated by those who plan RCTs inflate about four-fold the size of the population of subjects required to avoid only Type I errors.

Those who conduct RCTs commonly refer to subject recruitment as patient accrual. According to Ellenberg (204),

> Patient accrual is nearly always a problem in RCTs. The number of patients estimated to be available for a particular trial is often far greater than the number that appears to be available once the trial is underway. Low rates of accrual result in abandoned studies, studies with power so low as to limit the inferences that can be made, and studies that continue accrual over such a long period that the treatments being compared may no longer be of interest by the time the study results are available.

The need for large numbers of subjects has implications for both subject recruitment and informed consent (*infra*). In the interests of efficiency, there is, perhaps, too great a tendency to capitalize on the administrative availability of various institutionalized populations. (The concept of administrative availability is elaborated in Chapter 4). Various critics have strongly condemned such practices as unfair.

> The main participants . . . are now primarily recruited from the temporary or permanent inmates of public institutions, our Veterans Administration Hospitals, our mental institutions, and, often, our prisons. This means that it is mainly the underprivileged of our society who perform these services . . . (749).

> There are certain injustices that are so gross that few would try to defend them. That RCTs should be performed on, say, the poor . . . or those who happen to come for treatment to a Veterans Administration Hospital, while others received the benefit of a fully individualized treatment, would be grossly unfair (253, p. 61).

Levine and Lebacqz have examined the charge that the use of patients in VA hospitals as subjects in RCTs violates the principle of justice, particularly the requirement that we reduce the level of burdens imposed on persons who are less advantaged in relevant ways (441). The results of this analysis bear on considerations of the fairness of conducting RCTs in various dependent populations and in devising means to enhance the fairness of such use of dependent populations.

There is at least one condition of being a patient in a VA hospital that tends to render one more vulnerable and less advantaged than other patients, i.e., the condition of having limited options. Private patients may choose another physician or, in most cases, another hospital setting if they prefer not to participate in a RCT. Veterans wishing to exercise such an option, however, often will be obligated to assume large financial burdens. In order to receive the benefits due to them as veterans, they must use certain specified facilities; their options for receiving medical care thus are usually more limited. To this extent, they may be considered more vulnerable and less advantaged than other patients with similar conditions receiving private care. Thus, one could argue that in general, they should receive more benefits and bear fewer burdens of research.

The extent to which participation in a RCT represents a burden or a benefit is often debatable. However, Fried has argued that there is one type of burden that is imposed by participation in most, if not all, RCTs: the subject is deprived of the relationship with a physician that is characteristic of medical practice, a physician whose only professional obligation is to the well-being of the patient, not complicated by competing obligations to generate high quality data. In Fried's words, the subject is deprived of the "good of personal care" (253, p. 67).

Thus, it seems that VA hospital patients are less advantaged, and it further seems that participation in a RCT may be considered, at least in one respect, a burden. This seems to establish a *prima facie* case for claiming that involvement of VA hospital patients in RCTs must be considered unjust. However, there are ways to modify the design or implementation of the RCT that are directly responsive to the grounds for these charges. Levine and Lebacqz conclude that if these modifications are implemented, the use of VA hospital patients in RCTs need not be unjust.

First, one could take steps to minimize the burdens imposed by the RCT through the loss of the personal physician-patient relationship. For example, the subjects might be afforded the opportunity to maintain a physician-patient relationship with a physician not involved in the RCT but sufficiently familiar with it to facilitate the integration of its components and objectives with those of personal care. For an example of such involvement of personal physicians, see the discussion of the Coronary Primary Prevention Trial (*infra*).

Second, steps might be taken to minimize the disadvantaged position of the VA hospital patient. For example, they may be given the option of personal care or participation in the RCT, thus increasing their range of options to a level similar to that enjoyed by private patients. Alternatively, veterans refusing participation might be referred for personal care to a private facility at the expense of the VA.

Finally, a judgment that participation in a RCT is a burden (or, for that matter, a benefit) is a value judgment, and it is possible that patients' value judgments might differ from those of investigators or Institutional Review Board (IRB) members. When in doubt as to whether a prospective subject population considers itself disadvantaged or whether it considers participation in a RCT more of a benefit or a burden, community consultation may be helpful in resolving the uncertainties (cf. Chapter 4). Coulehan et al. have reported on the successful performance of a placebo-controlled RCT of the efficacy of vitamin C in the prevention of acute illnesses in Navajo school children performed after consultation with the community (158).

The results of survey research by Cassileth et al. provide interesting insights into whether prospective subjects consider participation in RCTs burdensome and, if so, who should shoulder such burdens (123). These studies were conducted in the aftermath of a series of newspaper articles that gave high visibility to reports of misconduct and abuses in clinical research. Respondents, who were assured anonymity, included 104 patients with cancer, 84 cardiology patients, and 107 members of the general public. Responses were similar regardless of subgroup or demographic considerations.

Of the respondents, 71% said patients should serve as subjects, giving as their primary reasons their belief that such service would result in potential benefit to others and an opportunity to increase scientific knowledge. When asked what their main reasons for participation in medical research would be, 52% responded that it would help them get the best medical care available. However, when asked which patients get the best medical therapy, 36% reported that it was those patients who were treated by private physicians who knew them, 13% said participants in medical research, 25% said there was no difference, and an additional 25% reported that they did not know. Cassileth et al. provide explanations for these apparent inconsistencies.

Mattson et al. surveyed the attitudes of subjects who were actually participants in two large scale RCTs (473). In these subjects' view, the advantages outweighed the disadvantages of participation in a RCT to the extent that a large majority said they would volunteer for similar studies in the future. Among the advantages identified by these subjects were additional medical monitoring and information, opportunities for second opinions, and reassurance and support received from RCT staff. Each of these advantages were rated as more important than actual feelings of physical improvement. Some subjects claimed altruistic motivations, and a few pointed to the availability of some free services. There was a low frequency of perceived disadvantages, consisting largely of transportation problems and clinic waiting times.

At first glance, the findings of Mattson et al. suggest a need for reconsideration of whether patient-subjects in RCTs are deprived of the "good of personal care." The RCTs they studied were not suitable for empirical testing of this question. More suitable RCTs to study for these purposes are those in which patient-subjects are faced with a high probability of death or disability in the easily foreseeable future, such as many protocols in the field of oncology.

In some RCTs, participating physicians are limited by the protocol design in the types of medical interventions and advice they can offer. Consider, for example, the Coronary Primary Prevention Trial (449).

> Intervention is restricted to prescription of the study medication and the LRC (Lipid Research Clinic) diet. The LRC staff neither encourages nor discourages changes in exercise, smoking or body weight. More restrictive diets and other cholesterol-lowering drugs are discouraged. Participants who develop morbid conditions during the trial are diagnosed and treated according to standard medical practice by their personal physicians, who are asked to refrain, if possible, from ascertaining blood lipid levels

and breaking the double-blind and to avoid prescribing other cholesterol-lowering medications or diets.

Yet the consent form for this protocol read, in part:

> Personal benefits obtained from the study are close medical surveillance of several aspects of health with physician contact and frequent blood tests. Although the study cannot take responsibility for general medical care, any markedly abnormal blood tests will be immediately reported to the participant's personal physician.

The importance of clarifying the role of the physician in the RCT cannot be overemphasized. Patients who have frequent contact with physicians who are apparently doing all of the things physicians commonly do (e.g., taking histories, performing physical examinations, drawing blood for tests, ordering electrocardiograms, writing prescriptions) are likely to presume that these physicians are performing as their personal doctors unless they are informed very clearly and reminded often that this is not the case. As discussed in Chapter 5, it may be most difficult for a physician to argue successfully (e.g., in court) that he or she does not have a fiduciary relationship with a patient-subject on grounds that he or she so informed the patient-subject.

Consent to Randomization

As a general rule, prospective subjects of RCTs should be informed that their therapy will be chosen by chance. The primary reason for this is that, although the therapies to be compared are considered medically equivalent by investigators, the composition of risks and benefits usually differs in ways such that reasonable prospective subjects might choose one or another based upon their personal values. Moreover, patients expect that physicians will provide advice based upon their personal knowledge of the patient, his or her ailment, and of therapeutics. If the state of relevant knowledge in therapeutics is such that choice of therapy by chance is justified, this fact, of itself, should be disclosed. For supporting arguments for these statements, as well as an assessment of the major counterarguments, see Levine and Lebacqz (441).

By the late 1970s, it seemed to me that consensus had been achieved on this matter, which in the 1960s and early 1970s had been the subject of sustained and heated controversy. In 1979, however, the arguments began anew. Reopening of the debate was occasioned by Zelen's proposal of prerandomization before negotiating informed consent (*infra*), a proposal grounded in the belief that a requirement for consent to randomization presented a major obstacle to subject recruitment.

Is consent to randomization such a great barrier to subject recruitment? Taylor et al. surveyed the surgeons participating in a national protocol designed to compare two forms of primary therapy for breast cancer (680). Specifically, they wanted to learn why there was a disappointingly low rate of patient accrual for this RCT. They found that only 27% of surgeons enrolled "all eligible patients." The others offered the following explanations: "1) concern that the doctor-patient relationship would be affected by a RCT (73%), 2) difficulty with informed consent (38%), 3) dislike of open discussions involving uncertainty (22%), 4) perceived conflict between the roles of scientists and clinician (18%), 5) practical difficulties in following procedures (9%), and 6) feelings of personal responsibility if the treatments were found to be unequal (8%)."

What they learned about was the surgeons' problems, not those of the prospective subjects. They did not learn whether there was any substance to the speculation that patients either cannot or will not accept disclosures that therapy will be decided by chance. They did not learn whether Chalmers, for example, was correct in his prediction in 1967 about patients who learned that their therapy was chosen by chance (441): "If they were in their right senses, they would find a doctor who thought he knew which was the best treatment, or they would conclude that if the differences were so slight . . ., they would prefer to take their chances on no operation." To be fair to Chalmers, he has long since changed his position (127); all prospective subjects of RCTs at his institution are informed that their therapy will be decided by chance, apparently without inhibiting patient accrual. What seems in order is not a new strategy to avoid disclosure of the fact of randomization, but rather a strategy to educate those physicians who doubt that this can be done without adverse consequences to any relevant interests.

Zelen's initial proposal of a prerandomization design was, in his view, "especially suited to a comparison of a best standard or controlled treatment with an experimental treatment" (751). This design entails randomizing patients first and then proceeding with negotiations for informed consent according to the standards of research only for those patients receiving experimental therapy. In his legal analysis of Zelen's proposal, Curran expresses several limitations, but concludes that it is worthy of serious consideration (174). Fost, on the other hand, rejects the proposal on ethical grounds (233); his specific reasons include most of those I have offered supporting the requirement to secure consent to randomization. Subsequently, nearly all commentators on the ethics of Zelen's initial proposal have rejected it on ethical grounds; for references to these articles, see Ellenberg (204). According to Ellenberg, "No major clinical trial has used or is using this randomization technique."

Zelen has responded by proposing a "double-consent randomized design;" this entails randomization before consent and then negotiating full informed consent by research standards with subjects assigned to each of the two therapies under comparison (752). This design has been employed in several major RCTs

(204). For a full analysis of this design and its variants and their implications for informed consent, see Kopelman (360). Ellenberg calls attention to some important ethical concerns (204). For example, although patients have the option of accepting or refusing the assigned treatment, they are not presented with the option of withdrawing entirely from the study. Moreover, the physician's knowledge of the assigned treatment "allows conscious or subconscious tailoring of the . . . presentation to predispose the patient to accept the assigned therapy." With Angell, "I find it difficult to justify strategies to increase the accrual of patients if the result is that people do not act as they would if full and necessarily neutral information had been provided before randomization" (17).

It is worth recalling that the purpose of introducing prerandomization strategies was to improve the efficiency of patient accrual. Ellenberg has surveyed the record in this regard (204). The most conspicuous success story is that of the National Surgical Adjuvant Breast and Bowel Project's Study of Total Mastectomy as Compared with Segmental Mastectomy With or Without Radiation Therapy; conversion to a prerandomization design resulted in a six-fold increase in accrual rate. This is the same study in which Taylor et al. surveyed surgeons' reasons for not enrolling eligible patients; it is difficult to imagine how or why conversion to a prerandomization design alleviated all of the important concerns expressed by the surgeons. Prerandomization has been employed in several other RCTs with more modest increases in accrual rates associated with substantial rates of refusal of the assigned treatment. In one RCT, the refusal rate was so high that the sponsoring group abandoned its use of the prerandomization design. As Ellenberg observes, in at least two studies, the increase in accrual has not compensated sufficiently for the inefficiency of the study design. Thus, even if one could justify on grounds of efficiency suspending the ethical requirement to inform patients that their therapy will be chosen by chance, it is not clear that increased efficiency can be expected to result.

Chalmers et al. have reviewed RCTs of the treatment of acute myocardial infarction in order to determine the importance of blinding the randomization process (129). "Blinded randomization" was defined as "assignment prearranged at random and communicated to the investigator only after the patient had been accepted for the study and informed consent had been obtained." "Unblinded randomization" was defined as any study in which the patients could "be selected or rejected after the physician knew the treatment assignment." When compared with blinded randomization, unblinded randomization was associated with nearly double the frequency of maldistribution of at least one prognostic variable and nearly triple the frequency of differences in case-fatality rates between treatment and control groups. "These data emphasize the importance of keeping those who recruit patients for clinical trials from suspecting which treatment will be assigned to the patient under consideration." This appears to be one further important reason to avoid prerandomization techniques.

Veatch has proposed a semirandomization design in which prospective subjects are offered the opportunity either to consent to randomization or to choose the therapy they will receive (702). I merely wish to call attention to this proposal and to the criticisms published simultaneously by Gordon and Fletcher (270) and by Lebacqz (375).

There are situations in which therapy must be administered promptly in order to be effective and in which neither the patient nor the next of kin is capable of expressing personal value choices. Fost and Robertson describe a protocol designed to test alternative therapies for patients with acute serious head trauma (237). There was no reason to suspect that either form of therapy would prove to be more advantageous to the patients, who were all expected to be incapable of consent. Moreover, it was necessary to begin therapy promptly in order for it to have any beneficial effect. Waiting for the next-of-kin might deprive patients of prompt therapy. In addition, in the judgment of their IRB, relatives who might be available in time would be experiencing such severe emotional distress that it would be inhumane to approach them soon after they had learned of the subject's critical illness. "It was considered unlikely that consent from such persons would be meaningful, given the difficulty of absorbing complex information in such circumstances." The IRB, therefore, authorized proceeding with the RCT using what they called "deferred consent":

> Families would be informed at the time of admission that their kin would be entered in an experimental project, but they would not be explicitly told that they could refuse, or be asked at that time to consent either orally or in writing. Within 48 hours, however, the investigators would have to obtain written consent from the family to continue. . . . It was assumed that by 48 hours after admission the family members would be under less stress and would be better able to assimilate the emotional and cognitive aspects of consenting to a research protocol. Initial notification gave an opportunity for withdrawal to those families who might feel especially strongly about the issue.

Fost and Robertson propose three conditions necessary for justification of deferred consent. 1) There must be a substantial reason for deferring consent, 2) consent should not be deferred any longer than is necessary to protect whatever interests deferral aims to protect, and 3) harm to the patient from deferring consent must be minimal.

Beauchamp agrees that the concept of deferred consent or various forms of nondisagreement are worthy of consideration by IRBs (48). However, he argues that the three conditions for justification as articulated by Fost and Robertson are too vague to provide clear guidance. In my view, in cases in which two approaches to therapy in emergency situations seem medically equivalent and no person is *reasonably* available to express values or choices on behalf of the patient, it seems reasonable to proceed with deferred consent. This seems to be consistent with Food and Drug Administration (FDA) regulations recognizing the place of emergency exceptions and therapeutic privilege in appropriate contexts (cf. Chapter 5). In the past, some investigators have abandoned attempts at RCTs in emergency situations owing to the impossibility of negotiating full informed consent (736).

Abramson et al. discuss RCTs on cardiopulmonary resuscitation (CPR) (3). In an emergency, the physician's duty is to do what is best for the patient. If there is a valid null hypothesis, however, the physician has no way to know which procedure is best for the patient. When CPR is required, there is no time to negotiate consent with the next of kin or legally authorized representative. Thus, they contend that it is permissible to include the patient in the RCT, performing the procedure selected by the randomization technique with no consent whatever. Their argument is grounded partially in the fact that this approach is consistent with the emergency exception to the informed consent standards in FDA regulations. In Chapter 5 there is a description of procedures designed to protect the rights and welfare of research subjects during cardiac arrest as reported by Frank and Agich; I prefer their approach to that proposed by Abramson et al.

IRBs are commonly presented with proposals to waive the requirement to inform prospective subjects that therapies which are not investigational will be distributed by chance. The argument may take this form: "Four different formulas are prescribed by pediatricians for newborns in this hospital. There are no rational grounds to select among these formulas. Therefore, infants are getting one or another according to a system of randomization depending upon which pediatrician is taking care of them. All we want to do is to systematize this randomization. Why bother the parents with making what appears to be a major decision when it is a trivial matter?" Similar requests also seem to come frequently from anesthesiologists and others. In general, at Yale, we require that such randomizations be disclosed in as innocuous terms as can be justified by the facts.

Anisfeld and Lipper studied the effects of early contact between mother and infant on maternal-infant bonding (21). Infants were assigned to early contact or delayed contact conditions by a process of randomization. Mothers were not informed of this until 1 day later. According to the authors,

> At the time we decided to study the effects of early contact, the delivery room procedures for postpartum contact . . . were in a state of transition. Some mothers got to hold their babies for varying lengths of time, and some not at all. Patients were not consulted about this. Hospital personnel were planning to introduce extended early contact as a routine procedure in the near future. We helped them standardize what they were already doing informally, so that at the conclusion of the study, they were able to implement the early contact procedures in an organized fashion.

There are other interesting features to this case which are not germane to the present discussion. Mahler has provided a detailed critique of this study (462).

Cello et al. report on their RCT designed to compare endoscopic sclerotherapy versus portacaval shunt in patients with severe cirrhosis and hemorrhaging esophageal varices (126). "If neither the patient nor the next of kin was able to give informed consent (15 patients), a patient advocate designated by the protocol (generally, the attending physician) was approached for informed consent. The protocol was approved by the Committee on Human Research, University of California." I have insufficient knowledge of the institutional context in which this procedure was approved to evaluate its ethical propriety. Given a similar protocol at Yale, I believe our IRB would have required use of some form of the research living will (cf. Chapter 11).

Preliminary Data and Emerging Trends

In the course of most RCTs, preliminary data are accumulated indicating that one of the therapies may be more effective or safe than the other. Another common occurrence is the discovery of a serious adverse effect which may or may not be caused by one of the therapies. Preliminary data present serious problems to the investigators. Under what circumstances must a RCT be terminated because continuation would be unethical? Must subjects be informed of emerging trends indicating superiority of one of the therapies although such superiority is not yet established at a statistically significant level? Must subjects be informed of the possibility of an injury when it has not been established that the therapy causes the injury?

Cowan has surveyed the literature on the statistical factors that must be considered in determining whether and when a RCT should be terminated prematurely (161). It is necessary to keep in mind that if a RCT is continued longer than necessary, it is not only the subjects enrolled in the group receiving the less advantageous therapy who will be harmed; other patients having the same disease who are not enrolled in the RCT will be deprived of the benefits of receiving the more advantageous therapy. Similarly, premature termination of a RCT may mean that data adequate to establish the safety and efficacy of the test therapies may never be developed. In general, it is wise to develop rules for stopping a RCT before the trial is initiated; in developing these rules, one should attempt to anticipate all conceivable contingencies.

Veatch has analyzed the issue of whether preliminary data reflecting trends in safety and efficacy that have not yet reached statistical significance ought to be disclosed to subjects in a RCT (697). For example, if Therapy A seems to be more advantageous than Therapy B and the probability that this could have occurred by chance is 0.07 (the level of 0.05 having been agreed upon for statistical significance), must this be disclosed to subjects? Must this be disclosed to new prospective subjects who are being recruited to enter the trial? Must this also be disclosed to those who are already enrolled in the RCT? Must the public as well as the medical profession be notified? Disclosure of the emerging trend would almost certainly disrupt the RCT; many subjects would decline participation in the RCT in order to gain access to what seemed to be the more advantageous therapy.

Veatch concludes that there are four possible solutions:

1. "Nondisclosure of preliminary data because the subject would not want to know." It seems to me that this solution would not be acceptable in most RCTs. There is no reason to presume that most reasonable subjects would not be curious about information that they might consider important to their well-being.

2. "Nondisclosure of preliminary data because benefits to society outweigh subjects' rights." This is a straightforward utilitarian approach. I am inclined to oppose such solutions except in extreme circumstances.

3. "Insistence on full disclosure." This solution would almost certainly render impossible any continuation of the RCT.

4. "Consent to incomplete disclosure." The fourth solution I find most acceptable for the following reasons. The generally accepted standard for determining whether any particular bit of information must be disclosed in the negotiations for informed consent is whether that particular bit of information is material to the prospective subject's decision (Chapter 5). I suggest that statements about the relative superiority of one form of therapy over another are not material before they are validated statistically.

Our culture seems to affirm statistics as one way of determining what is true, i.e., what is sufficiently true (or sufficiently proven) so that it may be revealed as an accepted fact, one that is worthy of disclosure. Editors of journals resist the efforts of scientists to publish data that have not been shown to be significant statistically. Without such demonstrations of significance, the FDA forbids changes in labeling of a drug and the Federal Trade Commission forbids advertising claims.

We have chosen arbitrarily to say that something is true when the probability is less than 0.05 that it could have occurred by chance. This choice, although arbitrary, reflects a consensus developed among scientists, statisticians, those responsible for developing federal regulations in relevant fields, and some others. There are those who argue, however, that some individuals would be willing to choose a therapy proved more advantageous at the 0.06 or 0.07 level; consequently, they contend that such information is material and ought to be disclosed.

There may even be others who argue that something ought to be disclosed at the 0.1 or 0.2 level. Each individual will use his or her intuition to select a cut-off number; it will always be arbitrary.

In many institutions, the informed consent discussions are used to educate the prospective subject on the initial null hypothesis that forms the ethical justification for the RCT. I suggest that we should go on to say that with time there will be increasing clarity that, e.g., A is superior to B. This superiority could be manifest as a decreased probability of serious harms or an increased probability of benefits. However, for good reasons, we have designed the RCT so that neither the investigator nor the subject will be aware of this disparity until such time as the arbitrarily preselected criteria of significance are achieved. Thus, what I would ask the subject to consent to is an acceptance of the standards of proof agreed upon within the community of professionals. This approaches the issue at the most fundamental level by calling upon the subject to accept or reject the values of the scientific community.

Although I find this resolution satisfactory for dealing with emerging trends relevant to the relative overall superiority of one therapy or another, I do not think it deals satisfactorily with the discovery of serious and unanticipated harms. Until it is established, preferably by prearranged criteria, that the unanticipated harm is caused by one of the therapies being tested, it may be appropriate to call upon subjects to consent to their nondisclosure. However, once it seems highly likely that the harm is caused by one of the therapies being tested, it seems to me that this must be disclosed not only to new prospective subjects, but also to those who are already enrolled in the RCT. This seems obligatory even though this might damage the validity of the RCT. I do not think that investigators or IRBs have the authority to withhold information about unanticipated serious harms because most subjects are likely to consider them material; it changes the composition of risks the subject must assume in order to pursue the hoped-for benefits. In cases in which there is serious doubt as to whether nondisclosure would be acceptable to subjects, such doubts may be resolved through consultation with the community (Chapter 4) or with a community of surrogates (Chapter 9).

Shaw and Chalmers describe vividly the threats to the integrity of the RCT if the investigators are allowed to be aware of emerging trends during the course of the RCT (639). They recommend keeping "the results confidential from the participating physician until a peer review group says that the study is over. Where appropriate, an understanding of the need for confidence until a conclusion is reached, fortified by an explanation of the built-in safety mechanisms, should be a part of the information imparted in obtaining informed consent." It is customary in large scale RCTs to have Data Monitoring Committees (254). These groups monitor data that are developed in the course of the RCT for purposes of determining whether the RCT should be discontinued or whether the RCT or the consent procedures should be modified; their deliberations are kept secret from investigators and subjects.

Confidentiality of preliminary findings is vital to the integrity of RCTs (269). In recognition of this, the Health, Education, and Welfare (HEW) Ethics Advisory Board (EAB) recommended a limited exemption from the Freedom of Information Act (208). The EAB concluded that confidentiality was justified if two criteria were satisfied. First, the protocol submitted for IRB review must state explicitly that interim results are to be held confidential. Secondly, prospective subjects must be informed that interim results are not to be revealed to them or to anybody else until the RCT had reached a conclusion as specified in the original protocol.

Schafer (626) and Marquis (468) call attention to another peril of allowing physicians in the RCT to learn of preliminary trends. Such physicians would be faced with the problem of permitting some of their patient-subjects to continue to receive what appears to be less than the best available therapy. Some physicians would feel obliged to withdraw from the RCT. If many of these physicians chose not to withdraw and chose to keep the patient-subject ignorant, this "would represent an important shift from a patient-centered to a social-welfare centered ethic" (626). For a dramatic example of the types of problems that could be presented to such physicians, see the case study, "The Last Patient in a Drug Trial" (663).

Placebo Controls

Until recently, most of the published commentary on the use of placebo controls in RCTs focused on the problem of deception. As discussed in Chapter 9, this is a misplaced concern; it is appropriate for medical practice, not for RCTs.

It is important to distinguish two classes of placebo-controlled RCTs. The distinction is based on the purpose of the active agent against which the placebo is to be compared. In the first class, the purpose of the active agent is to mitigate that component of a disease process that leads to lethal or disabling complications. In the second, the purpose of the active agent is to relieve symptoms without altering the rate of development of lethal or disabling complications.

The second class of RCTs is exemplified by comparisons between pain-relieving agents and placebo. This is the class of activities characterized by Beecher as presenting "no discernible risk" (53, p. 291). Although this class of placebo usage may present important ethical problems, I shall not discuss it further here.

In order to understand the problems presented by placebo usage in the first class of clinical trials, let us consider two diseases in which placebo-controlled

RCTs are done commonly: peptic ulcer and essential hypertension. For each of these diseases there are therapies known to be safe and effective; these therapies retard substantially the development of disabling or lethal complications. Complication rates for these diseases have been estimated as follows (436). Patients with untreated peptic ulcer will develop one of the three major complications, hemorrhage, perforation, or gastric outlet obstruction, at a rate of approximately 1.5 times per 100 patients per year. For essential hypertension, we have data available from a RCT. Men with initial diastolic blood pressures of 115 to 129 who are treated with placebo have an approximately 28% chance per year of having a severe complication such as stroke, malignant hypertension, heart failure, or death. Similar men who receive active treatment have an approximately 1.6% chance per year of a severe complication. For men having initial diastolic blood pressures of 105 to 114, the cumulative incidence of serious morbid events over a 5-year period was 55% in the placebo control group and 18% in those who received active therapy.

As noted earlier, a major ethical requirement for justification of a RCT is a bona fide null hypothesis. This requirement notwithstanding, placebo-controlled RCTs are generally not begun until there is some preliminary evidence that the agent to be compared with placebo is more effective than placebo in bringing about the desired consequence. There are important practical reasons for this. Not the least of these reasons is that RCTs are expensive and those who finance such trials do not, in general, wish to make major investments in studying agents that are not likely to be superior to a placebo. Moreover, in the period before beginning a RCT, studies are done that are designed not only to establish preliminary evidence of efficacy but also to determine the optimum dose to be studied during the RCT (124, p. 224).

Quite commonly, at the time an RCT is begun, the investigators have available quite a bit of evidence about the agents to be tested. In the United States, the results of extensive experience with new drugs obtained during clinical investigations or in medical practice in other countries are often available. Moreover, proposals to repeat RCTs in which the null hypothesis was rejected are not uncommon; usually in such cases the investigator or the industrial sponsor claims that either the design or the conduct of the first RCT was inadequate.

In a RCT in which a placebo is compared with an active agent, the purpose of which is to mitigate that component of a disease process that leads to lethal or disabling complications, the administration of placebo must be viewed as a nontherapeutic procedure. Attempts to justify placebo administration in this context in terms of the benefit it might yield for the patient-subject are generally without legitimate warrant. Such attempts usually cite experience with placebo usage derived either from clinical practice or from research conditions simulating clinical practice. I do not dispute the power of the placebo to produce salutary effects when it is administered as is customary in clinical practice. In RCTs, however, placebos are administered in ways that are designed to minimize their effects, especially their beneficial effects (Chapter 9).

Some commentators argue that placebo administration may benefit the subject indirectly. Lasagna, for example, observes: "Too often the placebo-treated patients turn out to be the lucky ones in such a trial, 'deprived' only of a toxic and ineffective chemical" (367). Such reasoning may be applicable at times but only in those unusual circumstances in which there is little or no preliminary evidence indicating the safety and efficacy of the active agent to be compared with placebo.

When we consider placebo-controlled RCTs in the treatment of patients with diseases such as peptic ulcer or essential hypertension, it is necessary to recognize that placebo usage is not only a nontherapeutic procedure but also one that presents the patient-subjects with more than minimal risk. In many cases, it is possible to develop reasonably accurate estimates of the risk involved. Thus, for example, if one wishes to justify starting a placebo-controlled RCT in patients with peptic ulcer, one must defend the claim that the knowledge to be gained is worth the risk of permitting one of the three major complications of peptic ulcer at a rate of 1.5 times per 100 patients per year. To put this in perspective, we commonly see heated controversies over the ethical justification of performing certain invasive diagnostic procedures for research purposes. This rate of serious complications with placebo usage in patients with peptic ulcer is substantially greater than the risk of similarly serious complications with, for example, liver biopsy (282, 283).

Justification of placebo-controlled RCTs of antihypertensive drugs should be even more difficult. The knowledge to be pursued must be worth the risk of permitting a severe complication at a rate of approximately 26 per 100 control subjects per year for patients having initial diastolic blood pressures of 115 to 129 or 37 per 100 control subjects per 5 years for patients having initial diastolic blood pressures of 105 to 114.

I do not mean to say that placebo-controlled trials of this sort can never be justified. I simply claim that their justification should be more difficult than it now appears to be. It is worth noting in this regard that the Chairman of the IRB at the University of California, San Francisco, has reported their disapproval of placebo-controlled RCTs designed to evaluate therapies for gastric ulcer in one case and, in another, moderately severe depression (717). For further discussion of the problem presented by this class of activities, see Levine (436) and Schafer (627).

Schafer claims correctly that a "substantial body of preliminary evidence" is available before beginning many RCTs, not just those that are placebo-controlled (627). In the name of consistency, therefore, he argues that placebo-controlled

RCTs should not be singled out for special attention. To that I respond there is an important and relevant difference. Placebo-controlled RCTs are designed to assess absolute efficacy, which may have been established or suggested strongly by preliminary studies. RCTs comparing two or more active therapies are different in that they are designed to assess comparative efficacy. The question is not whether the new drug is effective, but rather whether it is more or less effective than another drug, commonly one considered the best established therapy available for the patient-subjects' disease. With active controls, patient-subjects are not deprived of the benefits of any known effective therapy, as they so often are in placebo-controlled RCTs.

In the event any particular placebo-controlled RCT can be justified, patients should be informed forthrightly of the perils of withholding active therapy. To the extent that important preliminary evidence indicates efficacy of the active agent, there should be an increasing presumption of a requirement to inform prospective subjects of the preliminary evidence and its implications. In those cases in which there has already been one RCT which rejected the null hypothesis, this, too, should be disclosed. This assertion may seem inconsistent with the conclusion I reached earlier, that preliminary data developed during the conduct of a RCT may and should be withheld from subjects. There are, however, some relevant distinctions. Preliminary data of the sorts I am discussing now are generally known already to the physician-investigator and probably also to others. By contrast, preliminary data developed in the course of a RCT either is or should be kept confidential and not made available to anybody, including the physician-investigator.

Schafer argues that a policy of disclosing preliminary evidence would probably cause most subjects to decline participation in the RCT (627). In my view, if the preliminary evidence is so strong as to cause a high rate of refusal, one must question the ethical justification of having designed the RCT. Schafer counters, "Perhaps traditional physician ethics, with its highly individualistic commitment to patient welfare needs to be modified. A more socially oriented ethic might permit RCTs to proceed with a statistically adequate sample of patient-subjects. . . . If the traditional ethical rules governing physician-patient interaction are to be changed, then all parties should be made aware of this fact." I am not prepared to endorse the radical change in the doctor-patient relationship suggested by Schafer.

In recent years there has been mounting evidence that at least some patient-subjects and physician-investigators have been able to discern who was receiving placebo and who the active agent in some RCTs. In the Aspirin Myocardial Infarction Study (AMIS) comparing aspirin with placebo, 52% of the subjects correctly identified their study therapy and 28% mistakenly named the alternative treatment (319). According to the authors' formula for evaluating the patient blind, 24% of the subjects made informed guesses regarding their therapy. Some of the subjects figured out what treatment they were receiving by biting their capsules and tasting their contents. Others discerned their inclusion in the aspirin group through their experience of side effects. In the Beta-Blocker Heart Attack Trial (BHAT) comparing propranolol hydrochloride with placebo, before unblinding, subjects, physicians, and coordinators were asked to guess the treatment group assignment for each subject. Approximately 70% of all guesses were correct. Subjects assigned to propranolol treatment were most successful (79.9%) (99). Investigators for both AMIS and BHAT provide arguments to support their views that these correct guesses did not undermine the validity of the study.

By contrast, Brownell and Stunkard argue that correct guesses may undermine the validity of placebo-controlled RCTs (91). In their study comparing fenfluoramine with placebo, patients and physicians correctly identified medication assignments in 70% of the cases. They further found that correct identification of the active agent was correlated with therapeutic responses. In their view informed consent is the problem.

> The patients reported that the detailed consent forms alerted them not only to the possibility that they might receive a placebo but also to the specific side effects that they might encounter if they received the drug. With this information, it was relatively easy for them to detect their medication assignment, and a number of the subjects indicated that the procedures served as a challenge to do so.

According to FDA Guidelines (222):

> During all phases of clinical investigation the objective in using a placebo is to control the study adequately. It should be recognized that there are other methods of adequately controlling studies. In some studies, and in some diseases, the use of an active control drug rather than a placebo is desirable, primarily for ethical reasons. If a drug gives a positive dose response, this in itself may constitute adequate control in some studies. In some diseases or conditions where the natural course of the disease or the condition is predictable and in which objective measurements of therapeutic or prophylactic response can be made, carefully executed open studies may be compared to the historical data to provide acceptable evidence of efficacy. Some studies should be designed to ascertain the degree of safety and effectiveness of the investigational drug in comparison with one or more marketed drugs for the same indication.

> With the majority of investigational drugs, placebo and/or active drug controlled studies are necessary.

Dose-response designs may be used for drugs that have gradations of therapeutic effects depending upon the dose administered. In such a study, a high dose is compared against a modest dose of the drug. In my view, such studies present the same ethical problems as placebo-controlled studies. If a 2-mg dose is compared against a 4-mg dose, one is deliberately giving one group or the other either an excessive or inadequate amount of the drug.

Crossover placebo-controlled designs entail assignment of one group of patients to placebo and the other to active drug. Halfway through the study, the assignments are reversed; those who began taking placebo now receive active drug. At first glance, crossover designs seems responsive to the ethical problems presented by placebo controls. No patient-subject is deprived more than any other of the benefits of the active drug. However, in my view, the problem remains in that all patient-subjects are assigned to placebo for one-half of the study. For further discussion of crossover designs and the problems they present, see Louis et al. (455). See Chapter 3 for a discussion of add-on designs.

Excessive Reliance on RCTs

In recent years there have been several publications of harsh criticism of the American tendency to rely very heavily on the RCT, almost to the exclusion of other research methods, for validating new approaches to therapy. The most comprehensive and severe criticisms have been published by Fried (253) and by Freireich and Gehan (250); the latter publication is in the same book as one of the more comprehensive defenses of the RCT (62). Although a comprehensive survey and analysis of the arguments for and against use of the RCT are beyond the scope of this book, I shall identify some recent publications in which the arguments are presented in order to provide the reader a portal of entry into the literature on this debate.

Freireich and Gehan charge, that, although the RCT is a highly valuable instrument in appropriate circumstances, we tend to use it far more often than can be justified (250). They see the dominance of the RCT as a form of intellectual tyranny that stifles creative approaches to developing more efficient and less expensive techniques for validation of new therapies. They argue that this orthodoxy is supported by the FDA and by the National Cancer Institute, agencies that control the entry of new therapies into the practice of medicine, through their refusal to accept data developed with other methods. They then enumerate many ethical problems that arise in the course of conducting RCTs that could be avoided through the use of other methods.

Nelson has surveyed the various statistical problems that can occur in the course of planning, conducting and publishing the results of RCTs (536). He explains why many RCTs yield either false negative or false positive results and shows why, given the statistical rules for conducting RCTs, such artifactual results are inevitable. He then develops a statistically based argument that the entire literature based upon RCTs tends to indicate that drugs are in general more effective than they really are. Because editors tend to favor the publication of positive results, negative results tend not to get published.

Feinstein has written an excellent and thorough account of the limitations of RCTs (216); this article should be required reading for all persons who conduct or review plans to conduct RCTs. He identifies as a major problem the fact that there are conflicting schools of thought about how RCTs ought to be conducted; Feinstein refers to these as "pragmatic," which includes most clinical practitioners, and "fastidious," which includes most biostatisticians. The former "usually want the trial to answer pragmatic questions in clinical management. For this purpose, the plans would incorporate the heterogeneity, occasional or frequent ambiguity, and other 'messy' aspects of ordinary clinical practice. The advocates of the opposing viewpoint fear that this strategy will yield a 'messy' answer. They prefer a 'clean' arrangement, using homogeneous groups, reducing or eliminating ambiguity, and avoiding the spectre of biased results."

Currently, the fastidious viewpoint seems to dominate the conduct of RCTs. Consequently, according to Feinstein, RCTs are not good predictors of what will occur in practice of medicine. Among other reasons, the RCT, being a carefully contrived laboratory model in which the patient-subjects are carefully selected and the conditions of therapy rigidly controlled, does not simulate very well the practice of medicine. In addition, there is a tendency to overlook the acquisition and analysis of what Feinstein calls soft data. Soft data are concerned with symptoms caused and relieved by the new therapies; often these are most important in determining whether patients will accept a new therapy or consider it superior to an alternative (217).

Similarly, Murphy draws a contrast "between the literalism of most pragmatic scientists and the formalism of most statisticians. The terms and notions of the one may have imperfect correspondence with those of the other, or perhaps none at all" (515).

In a scathing editorial directed at the Beta-Blocker Heart Attack Trial, in which he takes exception to the authors' contention that the high rate of correct guesses about treatment assignment did not undermine the validity of the RCT, Morgan espouses the Feinsteinian perspective (513). In his view, the data suggest that the entire RCT may have had a placebo effect.

According to Feinstein, the therapy

> tested in a trial can have a remedial or prophylactic target. . . . Remedial targets have been the ones for which RCTs have been particularly successful. In a study of remedial therapy, the patients under investigation are relatively homogeneous because they all possess the same target to be altered; the target can be directly "measured" and observed for change; the change usually requires only a relatively short time to occur; and the compared differences in the measured changes can be statistically significant without requiring massive numbers of patients.

Prophylactic trials, designed to assess relative efficacy in preventing some future adverse occurrence, by contrast, "are the ones that have been notoriously expensive and that have often created, rather than clarified, controversy.'' This is because they lack the attributes that facilitate success in the remedial trials.

Sackett identifies three objectives of clinical trials: validity (its results are true), generalizability (its results are widely applicable), and efficiency (the trial is affordable and resources are left over for patient care and for other health research) (622). ''The first objective, validity, has become a nonnegotiable demand; hence the ascendency of the RCT.''

Sackett argues that many of the problems that have been identified by critics of RCTs reflect an inherent competition between these three objectives. For example, in the interests of efficiency, an RCT might enroll only subjects who are at high risk in order to show large risk reductions in short periods of time. Data developed from such a RCT are probably not generalizable to the usually much larger population of patients at average or low risk. For another example, some RCTs have analyzed data developed only in that subset of the subject group who were found to be highly compliant with the therapeutic regimen. This presents problems with validity in that highly compliant patients tend to have better prognoses than the average patient whether they receive active therapy or placebo.

Chalmers reviews the evidence that when RCTs develop compelling evidence to support the rejection of current medical practices, physicians often do not do so (127). In his view ''the reluctance of physicians to accept the results of clinical trials when the conclusions are contrary to conventional wisdom, no matter how well the trials have been designed, may gradually disappear as physicians become better educated in clinical trial methodology.''

Using Bayesian methods of statistical analysis, Diamond and Forrester examined the results of several major published RCTs and concluded that they were probably invalid (193). For example, they conclude that the University Group Diabetes Program probably did not establish that tolbutamide increases cardiac mortality, and that the Aspirin and Myocardial Infarction Study probably did not establish a protective effect of aspirin. For a balanced perspective on these findings, see Small and Schor (658).

Those who argue that RCTs are employed too frequently generally offer alternative research strategies that in their view are either more efficient or present fewer ethical problems. Freireich and Gehan have provided an account of such alternative strategies (250). The alternative to the RCT that is discussed most commonly in the literature on the ethics of RCTs is the historical control. In a clinical trial using historical controls, control data are derived from the experience of the institution with the treatment of the disease in question accumulated before the introduction of the new therapy. Variants on this approach involve the use of literature controls (control data derived from publications on the outcomes of treatment with the best available standard therapy) and the use of control data developed in the conduct of other clinical trials. The strongest defense of the use of historical controls can be made when the disease in question has a uniformly lethal outcome when untreated and for which there is no effective therapy. Thus, it is commonly argued that it would be unethical to establish a RCT designed to evaluate a new treatment for rabies if there were any cause to predict that it might be effective in even a small percentage of cases (621). When the disease is not uniformly lethal, there may be heated arguments over whether it is more appropriate to use a RCT or historical controls. For example, McCartney has argued that the placebo-controlled RCT designed to test the efficacy of adenine arabinoside in the treatment of herpes simplex encephalitis was unethical in that the disease was known to be fatal in 70% of cases and there was no known effective therapy (478). His argument was rebutted by the investigators (727); in the same issue of the same journal, McCartney rebuts the rebuttal.

Another strategy that merits consideration in selected cases is the randomized play-the-winner technique first proposed by Zelen in 1969 (718, 750). This method is particularly suited for studies in which the outcome of the intervention is known shortly after randomization. The first subject has an equal chance of random assignment to either of the two therapies being compared. If Therapy A is selected and is successful, the next subject is more likely to be assigned randomly to that therapy. If, however, it fails, the next subject is more likely to be assigned randomly to Therapy B. This selection process continues for each successive subject; the chance of each subject being assigned randomly to Therapy A or B is influenced by the outcomes of all of his or her predecessors. For an example of application of this technique to a study of extracorporeal circulation in the treatment of respiratory failure in newborns, see Bartlett et al. (41). In this study, one patient was assigned randomly to conventional treatment and died. The other 11 were assigned randomly to extracorporeal membrane oxygenation and all survived. Consider what the death rate might have been in a conventional RCT.

Observational case control studies seem particularly suited to the evaluation of nonvalidated practices that have become accepted in the community as the treatment of choice for very serious conditions (317). This design entails systematic observation of patients as they are treated by practicing physicians. Inclusion and exclusion criteria exactly like those used for RCTs are used to determine who will be observed. After the outcome has been established, patients are sorted according to outcome, e.g., fatalities and survivors. Causes of the outcome are evaluated as in case-control studies. For an example of application of this technique to studying therapeutic efficacy of lidocaine in preventing death from arrhythmia in patients with acute myocardial infarction, see Horwitz and Feinstein (317).

Epilogue

As I mentioned at the outset, the purpose of this chapter is to identify some of the major problems associated with the design and conduct of RCTs. I hope the reader now stands prepared to identify these problems when they are present and to recommend strategies for their elimination or mitigation. On this point, a word of caution is necessary.

IRBs are presented frequently with protocols describing RCTs designed by NIH or by one of the major cooperative groups, such as the Eastern Cooperative Oncology Group (ECOG). In the design of these RCTs, the sponsoring groups engage large numbers of experts including representatives of most, if not all, competing schools of thought on the subject. The final protocol reflects a compromise negotiated among these experts.

The IRB may have available to it one or two expert consultants in the field of study. These consultants may suggest changes in the design of the RCT. In considering these suggestions, the IRB must keep two points in mind. First, the positions recommended by their consultants probably have already been accommodated in the compromise negotiated by the panel of experts assembled by the sponsoring group. What the IRB is not hearing are the dozen or so other perspectives represented in the compromise. Secondly, these protocols are designed to be executed in precisely the same fashion in each of the cooperating institutions. Any deviations from the protocol render useless to the cooperative study the involvement of the institution in which such deviations occur. One cannot, therefore, request changes in drug doses, inclusion criteria, and the like. A demand for such changes is tantamount to disapproving the involvement of the institution in the RCT. The IRB must view the cooperative multicenter RCT as a package deal—take it or leave it. The IRB can, of course, and commonly should ask for changes in the consent procedures and forms.

Because in my writings I usually focus on the problems presented by RCTs and not on their virtues, I am asked often what my position is on the RCT. I have no position on *the* RCT. I can develop positions on particular RCTs, but in order to do this, I almost always require advice from various experts. The closest I have come to developing a position on *the* RCT may be found in a recent publication in which I conclude that the practicing physician has a limited moral obligation to recommend to patients with cancer that they consider referral for participation in a RCT (437). At the time of this writing, the RCT is the *gold standard* for evaluating therapeutic efficacy.

In considering the RCT, the average IRB member must be baffled by its complexity and by the manifold problems it presents. It calls to mind Clarke's third law (138); "Any sufficiently advanced technology is indistinguishable from magic." If you think the RCT is baffling, consider the patient-subject. The National Cancer Institute has developed a very helpful booklet for patients with cancer entitled *What Are Clinical Trials All About?* (522). I recommend its distribution to all patients with cancer for whom enrollment in a RCT might be appropriate.

Chapter 9

Deception

Sissela Bok provides the definition of "deception" as it is used in this book (66, p. 13): "When we undertake to deceive others intentionally, we communicate messages meant to mislead them, meant to make them believe what we ourselves do not believe. We can do so through gesture, through disguise, by means of action or inaction, even through silence." Bok views deception as a large category, of which lying is a subset defined as "any intentional deceptive message which is *stated*."

Research methods dependent upon deceiving subjects are used very commonly by social and behavioral scientists. Deceptive strategies seem particularly prevalent in studies of personality and social psychology (7, 43, 56). Surveys of articles published in major journals in the field of social psychology indicate that deceptive strategies are employed in over one-half of the studies; they are in the majority in some subspecialty areas, e.g., 81% of conformity studies and 72% of cognitive dissonance and balance studies (43). In their study of the *Journal of Personality and Social Psychology*, Adair et al. found that the percentage of published articles using deceptive strategies has increased monotonically during the past three decades, apparently uninfluenced by the promulgation of federal regulations or the publication of professional codes such as the American Psychological Association's *Ethical Principles* (7); moreover, they find no evidence of a decrease in the multiplicity or intensity of the deceptions. Murray (518), Kelman (349), Warwick (713) and others have noted disapprovingly that

213

an essential component in the socialization of good social psychologists is teaching them how to deceive subjects skillfully. The use of deceptive strategies is not uncommon in other types of research including, as we shall see, some types of clinical research.

In this chapter I shall survey some of the major themes in research involving deception with the aim of showing their relevance to clinical research; for a more complete discussion of deception as it is employed by social and behavioral scientists, I recommend Joan Sieber's comprehensive book (646), as well as her series of four fine articles on this subject (644, 645, 647, 648).

Arguments For and Against Deception

Proponents of the use of deception in research claim that it produces beneficial knowledge that cannot be obtained otherwise (40, 56); this point is conceded even by some of the harshest critics of research involving deception (349). There are some types of behavior that simply cannot be observed in naturalistic settings, e.g., studying why bystanders often fail to intervene by offering aid in emergency situations. Some other types of behavior can be studied in naturalistic settings; however, this would be much less efficient than working with a contrived laboratory model. In the field there are many more variables affecting the observed behavior, and this tends to confound interpretation of the 'results.' Additionally, deceptive techniques can be used to mitigate the influence of the Hawthorne effect on the behavior of the subjects (440). The term "Hawthorne effect" refers to the influence that participation in research has on the behavior of the subjects; some aspects of either the research process itself or of the environment in which it is conducted can, to a considerable extent, influence its results. It is seen most commonly when subjects are aware of the fact that they are being studied; they may behave in ways that they think the investigator will approve. Whether or not subjects are aware of being studied, however, psychological and even physiological responses can occur that are due primarily to the fact that they are being treated in a special way rather than to the factors being studied.

Proponents of deceptive strategies also generally claim that critics overestimate the costs and hazards of using such strategies (40, 56). Adair et al. claim that there is a need for empirical research to demonstrate the likelihood and magnitude of harms caused by deception in research (7).

Others claim that it is not necessary to demonstrate harm in order to condemn deception in research; one can be wronged without being harmed (519, 647). Although the notion of being wronged without being harmed has considerable appeal to scholars in the field, some have reported difficulty demonstrating that actual subjects feel wronged (248). In his survey of subjects who had been deceived in research, Smith found none who felt they had been harmed (659); of 36 respondents, all but one reported that his or her overall response to having participated was either positive or neutral; one respondent checked "negative" rather than "very negative." It is further worth noting that in Milgram's studies (*infra*), only 1.3% of the subjects recalled their participation as a negative experience.

Baumrind has provided an extensive account of the various costs of deception (43–45). She classifies these costs as ethical, psychological, scientific, and societal. The chief ethical cost is that it deprives subjects of their right to moral autonomy. Deceptive practices "are not wrong because subjects may be exposed to suffering. Rather, they are unjust because subjects are deprived of their right of informed consent and are thereby deprived of their right to decide freely and rationally how they wish to invest their time and persons."

In her discussion of psychological costs, Baumrind refers to the deceiver as the "aggressor" and the deceived as the "victim."

> The effect on the victim is to: a) Impair his or her ability to endow activities and relationships with meaning; b) reduce trust in legitimate authority; c) raise questions about regularity in cause-and-effect relations; d) reduce respect for a previously valued activity such as science; e) negatively affect an individual's ability to trust his or her own judgment; and/or f) impair the individual's sense of self-esteem and personal integrity.

Baumrind identifies two scientific costs of deception. Firstly, she is concerned that there is already a decreasing pool of naive subjects. Since psychologists are suspected of being tricksters, suspicious subjects may respond by pretending to be naive and by role-playing the part they think the investigator expects. Moreover, "Social support for behavioral science research is jeopardized when investigators promote parochial values that conflict with more universal principles of moral judgment and moral conduct."

The problem of depleting the pool of naive subjects is compounded by the fact that psychologists commonly enroll students in their own universities, often in their own classes, as research subjects (Chapter 4). Sieber quotes one of her subjects as follows (645): "As an undergraduate, I participated in a lot of deception research. It was all harmless; it was also disgusting. Is this the way it has to be? Is this what science is all about? In one study . . . they began to ask . . . more and more intimate questions. I started to think—ah yes, they are using the Guttman scale. And a friend of mine told me yesterday about being in an experiment . . ." which was similar. Cupples and Gochnauer argue "that in those cases where the relationship between the investigator and his subject is reasonably akin to the fiduciary relationship, the obligation to advance scientific understanding is constrained by the moral obligation generated by that relationship'' (170). On this basis they conclude that all research involving deception of students should be forbidden.

Margaret Mead warns that the scientific validity of research involving deception may be undermined even when the subjects are truly naive (484);

> The danger of distortion by cues unconsciously given by the experimenters and stooges and unconsciously picked up by the subject is so great that it can be categorically said that if deception is a necessary and intrinsic part of the . . . design, then human investigators must be removed from the experiment and other means found for giving instructions, faking situations, and making observations. . . . Instructions should be written or at least tape-recorded after careful scrutiny; the observers should be operating with absolutely concealed long-distance TV.

Many social scientists have argued that the data derived from research based upon deception may not permit inferences that are valid in the real world. For example, Baumrind observes (44), ''The rigorous controls that characterize the laboratory setting may prevent generalizations to the free environment.''

Baumrind is also concerned with the societal costs of deception (43, 44). As it becomes known that respected professionals practice deception and that such practices are applauded by society, two consequences may be anticipated. It may be assumed that authority figures are, in general, not to be trusted to tell the truth or to keep promises. Further, persons generally may be expected to emulate their deceptive behavior.

This is a hazard that Sissela Bok takes most seriously in her definitive analysis of lying. She argues that, while all lies always carry a negative value, some may be excused (66, pp. 103–106). However, as we consider excusing particular deceptions, we ''. . . need to be very wary because of the great susceptibility of deception to spread, to be abused, and to give rise to even more undesirable practices.''

An example of even more undesirable consequences is provided by Mead in her assessment of the effects on investigators of practicing deception (484, p. 167):

> Besides the ethical consequences that flow from contempt for other human beings, there are other consequences—such as increased selective insensitivity or delusions of grandeur and omnipotence—that may in time seriously interfere with the very thing which he has been attempting to protect: the integrity of his own scientific work. Encouraging styles of research and intervention that involve lying to other human beings therefore tends to establish a corps of progressively calloused individuals, insulated from self-criticism and increasingly available for clients who can become outspokenly cynical in their manipulating of other human beings, individually and in the mass.

Mead's remarks force us to confront squarely the age-old philosophical question: How do you deal with a person who tells you he or she is a liar? Researchers who will lie to subjects in order to advance what they consider the common good present us with a problem. As we read the results of their research we must wonder whether they have decided that lying to readers, e.g., by faking data, will further advance what they consider the public good. Broad and Wade provide ample examples of scientists who lied for what they thought they knew to be the truth (86).

Illustrative Cases

Now let us consider some of the cases of research involving deception that have evoked sharp criticism by both lay and professional commentators. More comprehensive compilations of cases have been assembled by Baumrind (43, 45), Adair et al. (7), and Sieber (645–648).

The Obedience to Authority Experiments These studies, conducted by Stanley Milgram at Yale University, seem to have generated more published controversy than any other set of experiments based upon deceptive strategies (497). It seems to me that the majority of papers that deal with the ethical problems presented by deception research use the Milgram studies as their prime example. Baumrind has published a harsh and thorough criticism of these experiments (43). Milgram has published a rebuttal to Baumrind's criticism (496), as well as a defense of the ethics of performing these and similar studies (498).

These experiments were conducted in a psychological laboratory (497, pp. 3–4). Subjects were instructed "to carry out a series of acts that come increasingly into conflict with conscience. The main question is how far the participant will comply with the experimenter's instructions before refusing to carry out the actions required of him." The subjects are assigned the role of teacher in these experiments; another individual is assigned to the role of learner. "The experimenter explains that the study is concerned with the effects of punishment on learning. The learner is conducted into a room, seated in a chair, his arms strapped to prevent excessive movement, and an electrode attached to his wrist. He is told that he is to learn a list of word pairs; whenever he makes an error, he will receive electric shocks of increasing intensity."

The teacher is "seated before an impressive shock generator" which has on it a series of switches ranging from 15 volts to 450 volts, in 15-volt increments. Verbal designations are also displayed ranging from "slight shock" to "danger—severe shock." "The teacher is told that he is to administer the learning test to the man in the other room. When the learner responds correctly, the teacher moves on to the next item; when the other man gives an incorrect answer, the teacher is to give him an electric shock." He starts at the lowest shock level (15 volts) and increases the level by 15 volts each time the learner makes an error.

What the teacher-subject does not know is that the learner is not a subject but an actor who actually receives no shock whatever. "Conflict arises when the man receiving the shock begins to indicate that he is experiencing discomfort. At 75 volts, the 'learner' grunts. At 120 volts he complains verbally; at 150 he demands to be released from the experiment. His protests continue as the shocks escalate growing increasingly vehement and emotional. At 285 volts his response can only be described as an agonized scream."

These experiments provide a rather dramatic example of the widespread practice by social psychologists of creating false beliefs in order to manipulate research subjects. Many other examples of staging in the laboratory (as in the Milgram studies) and staging in naturalistic settings (the field) have been published by the American Psychological Association (13), and by Warwick (713), Baumrind (43), and Bok (66). For an example of a debate between two social psychologists over a particular study involving staging in a classroom, see Schmutte (628), who argues from the position of the investigator who performed the study, and Murray (519), who contends that the study was not justifiable. As I shall discuss below, similar problems may be presented in the conduct of clinical research involving the use of placebos.

The Milgram experiments also present a problem characterized by Baumrind as the "*inflicted insight* (in which the subject is given insight into his flaws, although such insight is painful to him and although he has not bargained for such insight)" (43, 44). More specifically, during the debriefing many of these subjects learned that they were capable of egregious cruelty; Milgram found a remarkably high degree of obedience to authority under these laboratory conditions. Upon learning this about themselves, many of the subjects experienced severe and, in some cases, prolonged anxiety reactions. As I shall discuss subsequently, there are some clinical research maneuvers that can produce inflicted insights, e.g., covert monitoring for compliance with therapy with the aim of correcting noncompliant behavior.

The Tearoom Trade Studies   These studies were conducted by Laud Humphreys in the mid 1960s with the aim of developing a sociological description of homosexual practices in public restrooms and of the men who participated in them. ("Tearoom" is the term used by male homosexuals to describe places such as restrooms in which homosexual encounters take place.) Warwick has published an extensive description and analysis of what he considers the ethical improprieties committed by Humphreys (712). In his book, Humphreys provides a rebuttal to various published challenges to the ethical justification of his work (322).

In order to conduct these studies, Humphreys either disguised or misrepresented his identity at several different stages. He gained access to the public restrooms by masquerading as a watchqueen. (A "watchqueen" is an individual who derives pleasure as voyeur in this environment. The watchqueen also plays the role of lookout, warning others in the tearoom of the approach of unfamiliar persons who might be police or blackmailers.) After observing behaviors in the tearoom, he then followed some of the men to their automobiles so that he could record their license plate numbers.

Subsequently, he represented himself to the police as a market researcher. This ploy facilitated enlisting the aid of the police in learning the subject's identities, home addresses, and telephone numbers through their license plate numbers.

In order to interview these men in their homes, it was necessary to change his apparent identity once again. He waited for 1 year after he had made his last observations in the tearoom. In this time he changed his appearance, clothing, and automobile so as to minimize the possibility of being recognized. When he contacted the men in their homes he told them that they had been selected as ". . . part of a random cross-section sample chosen to represent the whole metropolitan area." They were informed further that this was a "Social health survey of men in the community" and that they were anonymous.

Misrepresentation of identity by investigators is a common practice. Among the reasons this particular study became much-discussed was that it presented serious threats to the privacy of a highly vulnerable group of men who had taken great pains to preserve their privacy. Any breach of confidentiality might have engendered serious disruptions in their marriages, their employment, and their freedom. (In the jurisdiction in which these studies were done, the act of fellatio was a felony; Humphreys reported that he had observed and taken notes on "hundreds of acts of fellatio.")

Misrepresentation of the identity of the investigator is commonly done for purposes of developing descriptions of the interactions between health professionals and patients. Some examples follow. Buckingham disguised himself as a patient with incurable pancreatic carcinoma in order to gain access to a "hospital-palliative care unit" (92, 521). Rosenhan lead a team of eight pseudopatients who successfully sought admission to psychiatric in-patient facilities in order to test their hypothesis that the distinction between sane and insane persons is difficult or impossible in psychiatric hospitals (613). Investigators have masqueraded as dental patients in order to gather data on the extent of x-ray use by dentists in Boston (413). Investigators have posed as patients with tension headaches in order to compare the adequacy of treatment provided in government-funded health centers with that provided in private group practices (596). Further examples are provided by Bulmer (93) and Bok (66, pp. 197–202).

Bulmer has reviewed the arguments for and against pseudopatient studies (93). He notes that several investigators who engaged in such studies have reported subsequently that given their revised understanding of what they consider ethical, they would not repeat their studies. He concludes, "The arguments . . . are on the whole stronger against such studies than in their favour." In his view the ethical issues raised by pseudopatient studies are likely to continue to "defy definitive solution." Newton (538) and Weiss (721) argue that pseudopatient studies can almost never be justified. Their arguments are rebutted by Yankauer (744) and Renaud (597), respectively. Levinson has written a particularly insightful article on the problems presented by informers, infiltrators, and entrapment in our society (443).

A Periscope in the Men's Room   Middlemist et al. observed men urinating in a public lavatory in order to test their hypothesis that close proximity to another man produces arousal which, in turn, would influence the rate and volume of micturition through constriction of the external sphincter of the bladder (493). In order to do this, they placed signs over urinals that they did not wish to have the subject use indicating that they were out of order. The urinary stream was observed by a colleague in an adjacent toilet stall with the aid of a prism periscope hidden in a stack of books. To create the condition of crowding, another investigator pretended to use an adjacent urinal. The results of the study indicated that crowding delayed the onset of urination and reduced its volume. This study has been attacked as an unwarranted and undignified invasion of what reasonable men should be able to presume is a private space in the interests of seeking data of trivial consequence (143, 357). Middlemist et al. have published their rebuttal of both charges (494).

This case exemplifies *covert observation*, a strategy that is commonly employed by social scientists. Perhaps the most carefully documented and analyzed case of unseen observer research is the Wichita Jury Recording Case (341, pp. 67–109). Many other examples have been published by the American Psychological Association (13).

Covert observation is also common in medical practice and research. Patients are often observed through one-way mirrors and recorded with the aid of concealed microphones and cameras (303). The use of medical records and left over specimens obtained at biopsy and autopsy may present analogous problems (cf. Chapter 7).

Justification

The Commission made no recommendations concerning research involving deception. However, in the commentary on Recommendation 4F in its report on Institutional Review Boards (IRBs), the Commission stated (527, pp. 26–27):

> In some research there is concern that disclosure to subjects or providing an accurate description of certain information, such as the purpose of the research or the procedures to be used, would affect the data and the validity of the research. The IRB can approve withholding or altering such information provided it determines that the incomplete disclosure or deception is not likely to be harmful in and of itself and that sufficient information will be disclosed to give subjects a fair opportunity to decide whether they want to participate in the research. The IRB should also consider whether the research could be done without incomplete disclosure or deception. If the procedures involved in the study present risk of harm or discomfort, this must always be disclosed to the subjects. In seeking consent, information should not be withheld for the purpose of eliciting the cooperation of subjects, and investigators should always give truthful answers to questions, even if this means that a prospective subject thereby becomes unsuitable for participation. In general, where participants have been deceived in the course of research, it is desirable that they be debriefed after their participation.

Many social scientists have urged officials of the federal government to exempt most social science research activities from coverage by federal regulations (177, 556). They are particularly concerned that the requirement for review and approval by an IRB not be imposed on most of their activities. The thrust of their argument is that social science research is generally harmless; further, because their work generally involves nothing more than talking with people, regulatory oversight constitutes an unwarranted constraint on freedom of speech (cf. Chapter 14). However, they do generally agree that research involving deception should be covered by federal regulations and should be reviewed by an IRB before its initiation.

Although Department of Health and Human Services (DHHS) Regulations permit "a consent procedure which does not include, or which alters, some or all of the elements of informed consent," they offer little guidance on the justification of deceptive research strategies (cf. Chapter 5). Many commentators have argued that DHHS regulations on withholding or altering elements of informed consent do not allow sufficient flexibility to the IRB to approve research protocols that can be ethically justified (cf. Chapter 5). Gray (272) and Levine (415) are among those who have called attention to the discrepancies between the regulations and the Commission's recommendations.

The American Psychological Association (APA) in its *Ethical Principles in the Conduct of Research with Human Participants* (14) provides detailed guidance and commentary on research involving deception:

> *Principle D.* Except in minimal-risk research, the investigator establishes a clear and fair agreement with research participants, prior to their participation, that clarifies the obligations and responsibilities of each. The investigator has the obligation to honor all promises and commitments included in that agreement. The investigator informs the participants of all aspects of the research that might reasonably be expected to influence their willingness to participate and explains all other aspects of the research about which the participants inquire. Failure to make full disclosure prior to obtaining informed consent requires additional safeguards to protect the welfare and dignity of the research participant. Research with children or with participants who have impairments that would limit understanding and/or communication requires special safeguarding procedures.

> *Principle E.* Methodological requirements of a study may make the use of concealment or deception necessary. Before conducting such a study, the investigator has a special responsibility to 1) determine whether the use of such techniques is justified by the study's prospective scientific, educational, or applied value; 2) determine whether alternative procedures are available that do not use concealment or deception; and 3) ensure that the participants are provided with sufficient explanation as soon as possible.

APA distinguishes research in which the participants are uninformed (concealment) from research in which they are misinformed (deception); the former set is seen as ethically less problematic. Note that APA's use of the term "deception" is inconsistent with the definition used in this chapter.

Uninformed Subjects

In its discussion of research involving uninformed participants, APA identified five separate classes of activities (14, pp. 36–40).

1) Covert or unobtrusive observation or recording of public behavior. This includes the use of hidden cameras or microphones as well as the entering of a natural situation by an investigator under an assumed identity. 2) Disguised field experimentation in public situations in which the investigator covertly arranges or manipulates experiences. This class of activities includes the staging of situations that appear to be real in order to observe the reactions of presumably naive persons as they go about their ordinary activities. For example, investigators have feigned injury to learn about helping behaviors; they have entered into false negotiations with automobile dealers in order to investigate the bargaining process. 3) Covert or participant observation in private situations. This is exemplified by the studies of Middlemist et al., as well as some aspects of the Tearoom Trade studies. 4) Adding research manipulation to existing non-

research operations in institutional or action research. For example, evaluation of various educational programs, textbooks, and examinations commonly involves random assignment of pupils to experimental and control conditions. 5) Obtaining information from third parties.

The APA points out that some research activities in which participants are uninformed may offend public sensibilities and exhorts investigators to take this into account. Attention is called to Principle D, which

> reminds the investigator of the responsibility to inform the research participants of every aspect of the research that might affect willingness to participate, including the fact that some information must be withheld. If scientific considerations dictate the withholding of information, ethical responsibility requires the investigator to assume personal responsibility for a careful weighing of the scientific requirements of the situation against the ethical requirements of the situation and for a careful consideration of alternative research questions and methodology that would allow informed consent. Investigators who cannot resolve the conflict about withholding information should seek the advice of others, including persons with different perspectives on values (Principle A) (14, p. 36).

APA's discussion of these issues is generally sensitive to the problems; while it provides much for the ethically concerned investigator to think about, there are few explicit directions.

Misinformed Subjects

APA's discussion of research in which participants are misinformed is presented under the title "Deception," which they define as telling deliberate lies to participants for scientific purposes (14, pp. 40–42):

> Having secured an individual's participation, the investigator may need to use deception to disguise a particular procedure or to conceal the meaning of reactions. Even at the end of the experiment leaving participants misinformed may be better than revealing certain aspects of the study. In each of these instances, Principle D puts the investigator in a serious ethical dilemma: Psychologists cannot pursue research involving deception without some compromise of this guideline.
>
> Since the problem is so difficult, investigators should seek advice before proceeding with a study that involves deception. Moreover, it seems advisable for investigators to consult with people whose values differ on the issue of the scientific and ethical justification for deception. Considerations that may make use of deception more acceptable include the following: (a) the research objective is of great importance and cannot be achieved without the use of deception; (b) on being fully informed later (Principle E) participants are expected to find the procedures reasonable and to suffer no loss of confidence in the integrity of the investigator or others involved; (c) research participants are allowed to withdraw from the study at any time (Principle F), and are free to withdraw their data when the concealment or misrepresentation is revealed (Principle H); and (e) investigators take full responsibility for detecting and removing stressful aftereffects of the experience (Principle I).

For discussions of research strategies that can be used as alternatives to deception, see Geller (259), Sieber (644–648), Adair et al., (7), Levine (428), and Chapter 5. The availability of these alternatives notwithstanding, many psychologists, particularly those of the deterministic school (644), "appear to favor the laboratory experiment, with a measured degree of deception as the most powerful and efficient means for testing hypotheses and advancing our knowledge of behavior" (7).

For an interesting debate over the ethical validity of offering deceived subjects the opportunity to withdraw their data after having been informed that they were deceived, the reader is referred to the exchange between Freedman (249) and Sieber (649). Although both agree that the opportunity to withdraw one's data is a very poor substitute for informed refusal to participate, Sieber rebuts Freedman's argument that withdrawal of data should be forbidden because such withdrawal is detrimental to the scientific validity of the research.

Debriefing Two of APA's ethical principles are concerned with how subjects ought to be treated after they have completed their participation. Although these apply to all types of research, they are especially relevant to subjects of studies involving either concealment or deception.

> *Principle H.* After the data are collected, the investigator provides the participant with information about the nature of the study and attempts to remove any misconceptions that may have arisen. Where scientific or humane values justify delaying or withholding this information, the investigator incurs a special responsibility to monitor the research and to ensure that there are no damaging consequences for the participant.

> *Principle I.* Where research procedures result in undesirable consequences for the individual participant, the investigator has the responsibility to detect and remove or correct these consequences, including long-term effects.

In research involving deception, debriefing is required to compensate the research subject for having been denied at least some powers of self-determination and for having been deliberately misinformed (648); debriefing procedures thus should be designed to

> . . . provide subjects with 1) satisfaction from having contributed to science or society, and 2) new knowledge that will be of educational or therapeutic value to them. In addition, subjects should receive six other benefits or reparations: 3) dehoaxing, 4) desensitizing, 5) restoration of confidence in science, 6) information on ways risk was anticipated and circumvented, 7) an opportunity to ask questions, and 8) an opportunity to withdraw one's own data.

Dehoaxing entails giving the subject whatever explanation or evidence is necessary to convey the truth about the research and to clear up any misunderstandings. Desensitizing involves the detection and amelioration of any undesirable emotional consequences of research participation. Desensitizing should be designed to "alter subjects' feelings concerning the way they behaved or were treated in the study so that they are restored to a state of emotional well-being" (648). Sieber provides good practical advice on how these and the other objectives of debriefing may be accomplished effectively (648).

It is important to note that dehoaxing and desensitizing are not always successful. For example, owing to perseverance phenomena, impressions, once formed, are highly persistent and refractory to new explanations even when the new explanations negate the logic that formed the original grounds for the impressions. In an interesting paper, Ross et al. provide experimental evidence of the existence of perseverance phenomena and show how they undermine attempts at effective debriefing (614). In laboratory conditions, research subjects who were induced to hold false beliefs about their skills continued to hold these beliefs even after careful debriefing. To show the relevance of perseverance to real world situations, the authors suggest several examples. One is, "John may believe that he dislikes abalone because of his first and only exposure and then is assured by an abalone connoisseur that he has sampled an inferior frozen product prepared in a notoriously dreadful restaurant." After this explanation, however, John will probably remain unwilling to give abalone another try. The authors provide evidence that process debriefing, which includes explicit discussion of the perseverance phenomenon, was effective in eliminating erroneous self-perceptions, although not in all cases.

In Baumrind's view, the most serious ethical problems in research based upon deception arise in studies that permit only two possible alternatives: "*Deceptive debriefing* (in which the truth is withheld from the subject because full disclosure would lower the subject's self-esteem or affect the research adversely); or *inflicted insight* (in which the subject is given insight into his flaws, although such insight is painful to him and although he has not bargained for such insight)" (43, 44). Baumrind argues that research that permits only these alternatives must be forbidden. Although APA strongly discourages such practices, their guidelines seem to permit them when other justifications are strong. APA notes that debriefing may, on occasion, provoke strong anger in those who have been deceived. APA further advises that although it is generally important to debrief participants very soon after the completion of the data collection, there are situations in which one might be justified to wait. Examples include studies in which participants are called upon to serve in several sessions separated by intervals of a week or more and studies in which there are multiple participants in which debriefing may be delayed until all participants have provided data in order to avoid premature disclosure to active participants by those who have already completed the project.

In its 1973 *Ethical Principles*, APA branded as "particularly reprehensible" the practice of "double deception" (13). In the 1982 revision, APA's position on this matter seems to have softened somewhat (14, p. 65):

> There have been a few studies that employ "double deception." This practice involves a second deception presented as part of what the participant thinks is the official postinvestigation clarification procedure. Then some further measurement is made, followed by clarification of the deceptions and procedures. In such cases, there is danger that the participant, when finally provided with a full and accurate clarification, will remain unconvinced and possibly resentful. Hence, such procedures should be avoided if possible.

APA also recognizes the problems presented by the possibility of inflicting insights (14, p. 67):

> At times the investigator may discover things about participants that could be damaging to their self-esteem if revealed. . . . This situation raises the question of whether or not full postinvestigation clarification requires the investigator to disclose such uncomplimentary findings even when they might be only incidental to the main purpose of the research and when the participant does not inquire about them or appear to be aware that information of this nature has been obtained.

> In such cases the investigator becomes involved in a basic conflict of values between the obligation to inform the participant fully, on the one hand, and the desire to avoid harming the participant in any way, on the other. If important damaging information that could affect the subject beyond the experiment (e.g., physical or personality problems) is uncovered, the subject should be informed, although not necessarily by the investigator alone, unless the investigator is qualified to handle the resultant distress.

Berkowitz argues that in some sorts of research, debriefing is unnecessary and, in fact, may even be undesirable (56, p. 29). In his opinion no good is done by informing persons that they have been subjects of covert observation. Further, when investigations entail observation of reactions of participants to staged events, inflicting insights is much more likely to produce harm than any good; he contends "that the staged event will probably have only a fleeting impact on the subjects because their ordinary defenses and ways of thinking enable them to adapt readily to the occurrence." In his view, which seems to be shared by the APA, the well-being of the participants takes priority over the obligation to inform them.

Sieber reviews some other circumstances in which debriefing may be either unnecessary or harmful (648). For example, in studies of "behavior that is socially perceived as negative and when the behavior is typical of the subject, desensitizing is usually unnecessary and dehoaxing may be harmful."

Cohen dramatizes the problems of debriefing subjects of covert observation (143). In commenting on the Middlemist study in which men were observed through a concealed periscope as they were urinating, he imagines the debriefing that might be performed as the subjects came out of the lavoratory.

An eager psychologist would bump into the subject. The dialogue might go like this:

"Magnificent, sir, what a flow of urine!"

"Gee, thanks but . . ." subject backs nervously away.

"Your micturitions, sir, have just made their way into the annals of science. We had to take the liberty of observing your lower half, sir, through a periscope . . . would you like to see my periscope, sir?" Psychologist now tries to drag the subject to his periscope. The subject realizes that he has a mad scientist on his hands and, at once, consents to having taken part in the experiment without his consent. He wants to go and have lunch, not listen to gibberings about hypotheses, micturition, the research design and the social benefits this work will confer.

Appeals to the Public Bok provides three general principles that "govern the justification of lies" (66, pp. 105–106). "As we consider different kinds of lies, we must ask, first, whether there are alternative forms of action which will resolve the difficulty without use of a lie; second, what might be the moral reasons brought forward to excuse the lie, and what reasons can be raised as counterarguments." In order to test these two steps, a third question is necessary: "We must ask what a public of reasonable persons might say about such lies." In the most difficult cases of justification of research involving deception or concealment, it seems appropriate to me and to others to take Bok's third suggestion literally, i.e., to put the question directly to a suitable sample of the public: "What do you have to say about such lies or deceptions?"

As discussed in Chapter 4, in some cases in which it is necessary to recruit subjects from vulnerable populations, *community consultation* (discussions of proposed research with assemblies of prospective subjects) might accomplish some important purposes. Proposals to conduct research requiring either deceptive strategies or concealed observers also present problems that might best be resolved through consultation with the community. However, it is not possible to consult a community of actual prospective subjects when the issues at stake are either deception or concealed observation; this would defeat the purpose of the maneuvers that require justification.

When there is cause to consult a community of prospective subjects but consultation with actual prospective subjects is infeasible, surrogate community consultation may serve the needed purposes (398, pp. 27–28). Surrogate consultation involves the assembly of a group of individuals that is in all important respects highly similar to the actual population of prospective subjects. This group may be asked to react to proposals to conduct research involving either deception or concealed observation in terms of the criteria for ethical justification published by the APA. Does the community consider the research problem to be of great importance? In their view, is there sufficient reason for the concealment or misrepresentation so that, on being fully informed later, research participants may be expected to find it reasonable and to suffer no loss of confidence in the integrity of the investigator or of others involved?

As Sieber has observed (644),

> Data may be objective, but interpretations of data are not. The possibility of social Darwinism, blaming the victim rather than seeking to explicate the factors that would give the individual viable choices in life or to understand the problem from the victim's perspective, looms large. Also troublesome is the opposite mindset, the assumption that miseries of the poor are necessarily attributable to the success of the affluent. To help avoid these problems, the investigator may find it useful to involve subjects and others in defining the research problem and interpreting the findings.

In consultation with a community of surrogates, there will often be disagreement as to what vote of the group should be required to authorize proceeding with the proposed research. Veatch has suggested that there should be "substantial agreement (say, 95 percent)" (694, p. 52). In my earlier writings on this issue, I was inclined to accept this suggestion; however, now it seems more consistent to me to call upon the assembly of surrogates to establish the voting rules.

For a description of an actual experience in using a surrogate system, see Fost (232). This system differs from the one described here in that the surrogates did not meet as a group; rather, they met as individuals with the investigator. The purpose of Fost's surrogate system was to determine what sorts of information prospective subjects might find material for informed consent purposes; the research involved no deception.

Soble has rejected the validity of surrogate community consultation on grounds that a unanimous vote would be required to authorize any particular research proposal; unanimity could rarely, if ever, be achieved (661). In his view, if a 95% vote (as suggested by Veatch) were accepted, then 5% would unknowingly be involved in research they disapproved.

Soble proposes instead a method called "prior general consent and proxy consent." Prior general consent, an approach introduced by Milgram (498), involves creating a pool of volunteers to serve in research involving deception; this is a population that states its willingness to be deceived in the interests of science. Then they could be recruited for specific projects without being informed whether the particular project involved deception. Soble's proposal supplements Milgrams's by having each member of the pool designate a friend or relative, a proxy, who is empowered to accept or reject each particular research project on behalf of the prospective subject.

Deception in Clinical Research

Placebos

For thousands of years, physicians have known that persons who feel ill and who are convinced that they will feel better as a consequence of taking some remedy very often do (87, 566, 637). They have capitalized on this knowledge by administering placebos, which are pharmacologically inert substances such as sugar pills and injections of saline solutions. The relief of symptoms engendered by placebo administration is called the placebo effect. The use of placebos in the practice of medicine to induce the placebo effect is a matter of considerable controversy (64; 66, pp. 61–68). It is quite properly charged that the use of placebos is deceptive, that it tends to spread the practice of benevolent lying, and that it tends to undermine public confidence in the medical profession. Proponents of placebo usage argue that placebos are often just as effective as active pharmacologic agents in producing symptom relief and that they are far less toxic. Thus, in this view, their use is justified by the relatively low ratio of harms to benefits. A comprehensive discussion of this issue is beyond the scope of this book.

Unfortunately, some critics have leveled charges against the use of placebos in clinical research that are appropriate only in discussions of medical practice. This generally reflects a lack of careful consideration of the differences between the purpose of placebo administration in the two contexts (428). Moreover, some commentators have defended the use of placebos in clinical research using defenses that are much more appropriate for medical practice than they are for clinical research (450).

In the context of medical practice, physicians who use placebos reinforce their potential efficacy with the aid of various other deceptive maneuvers. They tell *this* patient that *this* pill or injection is very powerful and likely to relieve *this* symptom. Deception is maximized in order to maximize the likelihood and magnitude of the placebo effect.

In the context of clinical research, the situation is much different. For example, in a controlled clinical trial, each prospective subject is told that there is, for example, a 50-50 chance that he or she may receive an *inert* substance. Further, they are told that neither the physician-investigator nor the subject will know whether they are receiving the placebo or the presumably active agent. In the context of clinical trials, we painstakingly minimize deception as well as the likelihood and magnitude of the placebo effect.

The use of placebos in the context of clinical research rarely, if ever, requires the use of deception. Prospective subjects are invited to agree to remain ignorant of which of two or more agents they might receive during the course of a clinical trial (cf. Chapter 5). Unfortunately, some controlled clinical trials have been conducted without adequately informing subjects of the plans to use placebos or of the expected consequence of such usage. For an example of this, see the description of the San Antonio contraceptive study in Chapter 4. For an example of placebo surgery, see Cobb et al. (140). In this study one-half of the patients who thought they were having their internal mammary arteries ligated underwent mock surgery with skin incisions only. Symptomatic relief of angina pectoris was approximately equal in the two groups. In his argument that there ought to be more placebo-controlled trials of surgical procedures, Beecher cited the studies of Cobb et al. as exemplary (50). Placebo-controlled clinical trials are discussed further in Chapter 8.

Monitoring for Compliance

"Compliance" is a term that refers to a behavior of patients characterized by nondeviation from therapeutic plans to which they have agreed with physicians. In the context of research, it refers to nondeviation on the part of subjects with plans agreed upon with investigators. Some commentators find the term infelicitous in that it implies that patients must follow doctors' orders rather than cooperate with mutually agreed upon plans; thus, they prefer the term "adherence." For overviews of the ethical problems presented by the employment of strategies designed to enhance or measure compliance in both medical practice and clinical research, see Jonsen (333) and Levine (419).

Monitoring with the aim of determining compliant or noncompliant behavior commonly requires covert observation and occasionally leads to the development of the necessity to consider either deceptive debriefing or inflicted insights. Among the commonly employed methods are the collection of blood, urine, or saliva for the determination of drug levels (267). In some cases, markers or tracers may be added to drugs or diets, because these are more easily assayed than the materials given for therapeutic or research purposes (61). Some investigators have surreptitiously monitored prescription-filling patterns or counted the number of pills in a patient-subjects' pill container. It is generally agreed that these monitoring strategies must be accomplished without the awareness of patients and subjects; otherwise, they will change their behavior to accord or seem to accord with the investigator's expectations. Thus, at a minimum, monitoring for compliance presents problems characteristic of research involving covert observation.

The ethical problems presented by the actual performance of compliance monitoring strategies can generally be resolved rather easily. Prospective subjects can be told that certain maneuvers will be performed and that for important reasons it is necessary not to inform them of their purpose until the research has been completed; it is not necessary to depart from the Commission's guidelines for incomplete disclosure (*supra*). It is possible to inform them of the amount of risk (usually none) and the amount of inconvenience that will be presented by those maneuvers for which the purpose is not disclosed (cf. Chapter 5).

If the purpose of monitoring is merely to develop data on the general amount of compliance within a population, e.g., subjects in a controlled clinical trial, there is very little difficulty. During the debriefing the subjects can be told that these maneuvers were performed to develop general information on compliance. At times, some subjects will complete their participation in the protocol long before the entire project is completed. Under these circumstances, in order to protect the validity of the clinical trial, some aspects of debriefing may be delayed as advised in the APA guidelines.

Much more difficult problems are presented when the purpose of the monitoring is to enhance compliance. If the information on individual compliance is to be used to identify those subjects for whom special attention is necessary, it may generate a need for either deceptive debriefing (identifying subjects for special attention without letting them know how they were identified) or having to resort to the inflicted insight.

One method for avoiding such problems is to inform prospective subjects that during the course of the clinical trial there will be compliance monitoring and that the purpose of this monitoring is to identify individuals who are complying incompletely. It should be made clear that such identification is in the interests not only of producing high quality generalizable knowledge but also of maximizing each patient's chances of having a satisfactory therapeutic response. It is appropriate to emphasize the mutuality of interest between the investigators and all subjects. Thus, to the extent it is possible to maximize compliance, the interests of all involved will be served. In these circumstances, it is worthwhile pointing out that poor compliers are not considered bad people. Rather, through such identification it is commonly possible to discover reasons for poor compliance and to take remedial action. Finally, it should be pointed out that the nature of the compliance monitoring activities must be kept secret, and good reasons should be given for this.

Following these procedures, one may anticipate that those prospective subjects who do not agree that all of these activities are designed to serve the best interests of all concerned will refuse to participate in the study. Consequently, the subject population may not be representative of the entire population of patients having the condition under examination. For this and other reasons (e.g., introduction of the Hawthorne effect), this approach would not be satisfactory for developing data on general levels of compliance within a population. However, as discussed earlier, such data can be developed by different means and in different studies in which it is not necessary to confront individuals with evidence of their non-compliance.

As an alternative, one might consider employing the device of surrogate community consultation to learn whether such individuals would find it reasonable to conduct compliance monitoring with the aim of detecting and correcting individual noncompliant behavior without informing actual subjects that this is being done. On pragmatic grounds, I favor the first alternative. The first subject upon whom an insight is inflicted is likely to begin the process of debriefing the other subjects prematurely in a manner that is unlikely to enhance their trust in the investigators.

Epilogue

I disapprove of deception. I particularly disapprove of deception when it is practiced by professionals in order to accomplish the goals of their professions. And yet I chair an IRB which occasionally approves research employing deceptive strategies. Does this not appear to be inconsistent?

Deception in research always brings into direct conflict norms arising from two fundamental ethical principles, respect for persons and beneficence; in order to create new knowledge the researcher must violate privacy or autonomy. Parenthetically, there are also often violations of justice-based norms. Researchers who use deceptive strategies tend either to "study down" (i.e., to use as subjects members of less powerful groups such as students) or to study strangers, absolute strangers who they hope will never become aware of the researchers' concealed or disguised activities. It is much more perilous to the investigator overtly to "study up."

What does it mean to decide that, in a particular case, it is justified to override the *prima facie* duty of veracity? Do we simply say that beneficence wins and autonomy loses and that's that? No! The overridden duty does not just vanish like the Cheshire cat leaving behind only a smile. It leaves what may be called "moral traces" (48, p. 48). Those who make a decision to override a *prima facie* duty are expected to feel regret, perhaps even remorse. In this regard it is unseemly for some professional groups to celebrate, or to appear to celebrate, their skills at deception—to convene in order to render homage to the most skillful deceivers. I think it is unseemly and so apparently does the public.

To tend to moral traces we must also do what we can to make up for the damage. We must debrief, dehoax, desensitize, and so on. We must also stand prepared to explain the meaning of our behavior to the public.

Imagine yourself, crouching in a cubicle in a public lavatory, covertly observing men's penises as they urinate (*supra*). You have patiently explained the meaning of your work to your colleagues and your IRB and they have (regret-fully?) approved. Suddenly, the door swings open. Suddenly, you are captured in the suspicious stare of an other. This other will immediately ascribe meaning to your behavior. What do you suppose that meaning will be?

Who is this other? Your son? The loan officer at your bank? Your husband? All those others out there are watching our behavior. We must take care to assure that the meaning of our behavior can be understood and affirmed by those others.

It is inconsistent that I chair an IRB that occasionally approves research involving deception. I regret this inconsistency so much that I experience it as infrequently as circumstances permit.

Children

The Special Populations

When the Department of Health, Education, and Welfare (DHEW) published its first proposals to develop regulations providing additional protections for especially vulnerable populations of research subjects, it designated them as persons having "limited capacities to consent" (181). The choice of this label for the special populations highlights the nature of the fundamental problem in justifying their use as research subjects. Because the Nuremberg Code identifies voluntary consent as absolutely essential, it is clearly problematic to involve subjects who lack free power of choice (e.g., prisoners), the legal capacity to consent (e.g., children), or the ability to comprehend (e.g., the mentally infirm).

In P.L. 93–348, Congress instructed the Commission to ". . . identify the requirements for informed consent to participation in biomedical and behavioral research by children, prisoners, and the institutionalized mentally infirm." The Commission was required further to define the circumstances (if any) under which research involving the living human fetus might be conducted or supported. The Commission went far beyond its obligation to identify requirements for informed consent. Rather, it provided recommendations for nearly all aspects of the ethical conduct of research on children, fetuses, prisoners, and those institutionalized as mentally infirm. The Department of Health and Human Services (DHHS) has, with some modifications, translated the recommendations

235

into final regulations (children, prisoners, and fetuses) and proposed regulations (those institutionalized as mentally infirm).

In general, the Commission concluded that persons having limited capacity to consent are vulnerable or disadvantaged in ways that are morally relevant to their involvement as subjects of research. Therefore, the principle of *justice* is interpreted as requiring that we facilitate activities that are designed to yield direct benefit to the subjects and that we encourage research designed to develop knowledge that will be of benefit to the class of persons of which the subject is a representative. However, we should generally refrain from involving the special populations in research that is irrelevant to their conditions as individuals or at least as a class of persons. *Respect for persons* is interpreted as requiring that we show respect for a potential subject's capacity for self-determination to the extent that it exists. Some who cannot consent can register knowledgeable agreements (assents) or deliberate objections. To the extent that the capacity for self-determination is limited, respect is shown by protection from harm. Thus, the Commission recommends that the authority accorded to members of the special populations or their legally authorized representatives to accept risk be strictly limited; any proposal to exceed the threshold of minimal risk requires special justification.

Many individuals who have limited capacity to consent have not been identified by Congress or DHHS as members of the special populations. The norms established by the Commission for identified special populations may be applied, as appropriate, to others. For example, some norms developed for those institutionalized as mentally infirm may be appropriate for some other prospective subjects with limited capacities for comprehension who are not institutionalized; for further discussion of vulnerable subjects not identified as members of the special populations, see Chapter 4.

Respect for Children

The central problem presented by proposals to do research involving children as subjects is that children, as a class of persons, lack the legal capacity to consent. In addition, many of them, particularly the younger ones, are incapable of sufficient comprehension to meet the high standards of consent to research developed in such documents as the Nuremberg Code. The Declaration of Helsinki and DHHS regulations reflect an awareness of this problem by calling for consent by the parent, guardian, or legally authorized representative; this is commonly referred to as proxy consent. Thus, much of the recent debate about the ethics and law of research involving children centers on the nature of the procedures for which a proxy may consent.

For an excellent survey of the ethical issues, the reader is referred to Chapters 8 and 9 of *Research Involving Children* (525). An extreme position is presented by Ramsey, who bases his argument on a strict interpretation of the principle of *respect for persons*, i.e., that we leave persons alone unless they consent to be "touched" (583). Consequently, he argues that the use of a nonconsenting subject (e.g., a child) is wrong whether or not there is risk, simply because it involves an "unconsented touching." As he puts it, one can be wronged without being harmed. "Wrongful touching" is rectified only when it is for the good of the individual, because then the person is treated as an end as well as a means. Hence, proxy consent may be given for nonconsenting subjects only when the research activity includes therapeutic interventions related to the subject's own recovery. Ramsey acknowledges that, in some cases, there may be powerful moral reasons for doing nontherapeutic research involving children; however, "it is better to leave (this) research imperative in incorrigible conflict with the principle that protects the individual human person from being used for research purposes without either his expressed or correctly construed consent." He continues that it would be immoral either to do or not to do the research, but he maintains that one should "sin bravely" in the face of this dilemma by sinning on the side of avoiding harm rather than attempting to promote welfare. Recently, Ramsey has clarified his position on this issue (585). His use of the term "touching" is literal; if the child is not touched physically, he or she is not wronged. In this interpretation, infants and very young children have no privacy rights to be respected.

When the principle of *respect for persons* is interpreted strictly, the unconsented touching of a competent adult is wrong even if it benefits that person. In that case, why should potential benefit justify such touching for a child? McCormick proposes that the validity of such interventions rests on the presumption that the child, if capable, would consent to therapy (479). This presumption, in turn, derives from a person's obligation to seek therapy, an obligation which people possess simply as human beings. Because people have an obligation to seek their own well-being, we presume that they *would* consent if they *could* and thus presume also that proxy consent for therapeutic interventions will not violate respect for them as persons.

By analogy, people have other obligations, as members of a moral community, to which one would presume their consent; McCormick calls this "their correctly construed consent." One such obligation is to contribute to the general welfare when to do so requires little or no sacrifice. Hence McCormick concludes that nonconsenting subjects may be used in research not directly related to their own benefit so long as the research fulfills an important social need and involves *no discernible risk*. In McCormick's view, respecting persons includes recognizing that they are members of a moral community with its attendant obligations. For a discussion of the philosophical bases for the argument that children bear obligations to a moral community, particularly the community of similarly situated children, see Pence (563).

Ramsey counters this argument by claiming that children are not adults with a full range of duties and obligations (583). Therefore, they have no obligation to contribute to general welfare and respect for them requires that they be protected from harm and from unconsented touching. Ramsey further insists that there is an important difference between *no discernible risk* and *discernibly no risk*; the latter standard presupposes empirical evidence of safety, although the former does not.

Freedman bases his argument on the same premise as Ramsey, i.e., a child is not a moral being in the same sense as an adult (246). However, his analysis yields the same conclusion as McCormick's. Precisely because children are *not* autonomous, they have no right to be left alone. Instead, they have a right to custody. Thus, the only relevant moral issue is the risk involved in research; the child must be protected from harm. Therefore, Freedman agrees with McCormick that children may be used in research unrelated to their therapy provided it presents to them *no discernible risk*.

VanDeVeer argues that McCormick's concept of "correctly construed consent" is indefensible (690). In the case of infants and young children, he contends, we must acknowledge that there is no consent. Thus, he concludes that the overriding obligation regarding children is to do what is in their best interests.

Ackerman argues that we tend to fool ourselves with procedures designed to show respect for the child's very limited autonomy (4). He claims that the child tends to follow "the course of action that is recommended overtly or covertly by the adults who are responsible for the child's well-being." He further contends that, in general, this is as it ought be be. "Once we recognize our duty to guide the child and his inclination to be guided the task becomes that of guiding him in ways which will achieve his well-being and contribute to his becoming the right kind of person."

Gaylin tells the story of a man who acted in accord with Ackerman's position (257). After directing his 10-year-old son to cooperate with venipuncture for research purposes, he explained that his direction arose from his perceived moral obligation to teach his child that there are certain things one does to serve the interests of others even if it does cause a bit of pain: "This is my child. I was less concerned with the research involved than with the kind of boy that I was raising. I'll be damned if I was going to allow my child, because of some idiotic concept of children's rights, to assume that he was entitled to be a selfish, narcissistic little bastard."

In developing his concept of variable competence (*infra*), Gaylin goes even further (257). In considering research procedures presenting low risk to the child, he claims that if a parent has no sense of obligation to the community, it would be good for the child as well as the community for the state to instruct the parent as well as the child on the topic of social responsibility. He uses as his example the collection of urine for epidemiologic research purposes. "Refusal of permission for such an experimental involvement would be trivial, arbitrary, and ungenerous." Therefore, he would place extreme limits on the authority of an individual to refuse to cooperate and be "extremely generous" in according to the child independent authority to choose to participate.

For further discussion of who has the authority or responsibility to make decisions regarding the child's well-being, the reader is referred to two recent and important books; these are *Who Speaks for the Child* (258) and *Children's Competence to Consent* (490). The former is the better of the two for those who are interested in biomedical activities, both research and practice; the latter is more concerned with social, behavioral and educational research and practices. For a recent comprehensive survey of the law of pediatric practice and research, see Holder (313).

As we shall see, the recommendations of the Commission reflect its conclusion that, because infants and very young children have no autonomy, there is no obligation to respond to it through the usual devices of informed consent. Rather, respect for infants and very small children requires that we protect them from harm. No discernible risk seemed to the Commission to be virtually impossible; therefore, they stipulated a definition of "minimal risk" as the amount that would be acceptable without unusual standards for justification. Further, they recommend a mechanism for respecting the developing autonomy of older children and adolescents.

Therapeutic Orphans

As a consequence of the uncertainties about the ethical propriety of and legal authority to do research on children, there has been a great reluctance in the United States to do studies to determine the safety and efficacy of drugs in children. Consequently, as Shirkey observed, "Infants and children are becoming the therapeutic orphans of our expanding pharmacopoeia" (641). Since 1962, nearly all new drugs have been required by the Food and Drug Administration (FDA) to carry on their labels one of the familiar "orphaning" clauses: e.g., "not to be used in children," ". . . is not recommended for use in infants and young children, since few studies have been carried out in this group. . . ," "clinical studies have been insufficient to establish any recommendations for use in infants and children," ". . . should not be given to children. . . ." By

1975, over 80% of all drugs prescribed for children bore such orphaning clauses on their labels (504).

The therapeutic orphan phenomenon is not limited to children. Very similar conditions obtain in the use of drugs in pregnant women (504, 505) and in young women generally (353). Pregnant women are usually excluded from drug trials and from many other types of research because the fetus is seen by some as a nonconsenting subject who might be peculiarly vulnerable to the effects of drugs (cf. Chapter 13). Women who are biologically capable of becoming pregnant are commonly excluded from many types of research, including drug trials. Given suitable plans for contraception, it would be reasonably safe to include them in most studies (353); however, many investigators do not wish to assume the burden of discussing plans for contraception with prospective subjects (Chapter 5). Moreover, many investigators fear the potential legal consequences of a failure to prevent pregnancy during a drug study.

In passing, I wish to point out that the term "therapeutic orphan" has picked up a new meaning. Most current literature that refers to therapeutic orphans concerns orphan drugs which are drugs, that if developed would be useful only in the treatment of relatively uncommon diseases. Because the potential market for these drugs is very small, most drug companies do not wish to invest the huge sums of money required to secure FDA approval for marketing a new drug in the United States; in 1981, this was estimated at approximately $70 million per new chemical entity (150, p. 18). In this book, I use the term "therapeutic orphan" exclusively in the sense intended originally by Shirkey.

The therapeutic orphan phenomenon represents a serious injustice. If we consider the availability of drugs proved safe and effective through the devices of modern clinical pharmacology and clinical trials a benefit, then it is unjust to deprive classes of persons, e.g., children and pregnant women, of this benefit. This injustice is compounded as follows. If we were to do Phase II and III clinical trials in children as we now do in adults, the first administration of various drugs would be done under conditions more controlled and more carefully monitored than is customary in the practice of medicine. It is likely that adverse drug reactions that are peculiar to children would be detected much earlier than they are now; consequently, either we could discontinue administration of the drugs to children or we could issue appropriate warnings to physicians who are using the drugs. The prevailing practice in the United States is to ignore the orphaning clauses on the package labels (456). Consequently, we have a tendency to distribute unsystematically the unknown risks of drugs in children and pregnant women, thus maximizing the frequency of their occurrence and minimizing the probability of their detection. Parenthetically, it should be noted that most drugs proved safe and effective in adults do not produce unexpected adverse reactions in children (398); however, when they do, the numbers of harmed children tend to be much higher than they would be if the drugs had been studied systematically before they were introduced into the practice of medicine. Examples include the development of phocomelia with thalidomide administration to pregnant women and the development of "grey sickness" in infants treated with chloramphenicol. In addition, we have insufficient knowledge of the proper doses of many drugs to be used in children (148).

Capron has reviewed the ambiguous legal status of the FDA-approved package label (112). The FDA's position is that the labeling is not intended "either to preclude the physician from using his best judgment in the interest of the patient, or to impose liability if he does not follow the package insert." However, some courts have held recently that the package insert is admissible in evidence to show that a physician-defendant did not provide reasonable care in the treatment of a child. Some courts have even gone so far as to state that the package insert may be considered *prima facie* evidence of the standard of due care. Because, as noted earlier, over 80% of all drugs prescribed for children carry orphaning clauses in their package labels, there is great potential exposure of pediatricians to malpractice litigation. Thus, whenever a child suffers an adverse reaction to one of these drugs, a complaint might be brought against the prescribing physician that he or she departed from the standard of due care. Of course, such a complaint usually can be rebutted successfully with expert medical testimony.

There is an even more alarming potential for malpractice litigation in this field. Inevitably, someone will bring an action against a pediatrician based upon failure to secure adequate informed consent. It will be alleged that a child was harmed by a drug and that had the parents known that it was not approved by the FDA for use in children, they would not have permitted its use in their child. The court will be called upon to determine whether disclosure of orphaning clauses can be considered material to the consent decision. If a court rules that such information is material, there will be an enormous potential for litigation. A charge that a physician has breached the duty to disclose a material risk usually need not be supported by expert medical testimony unless there is dispute that the risk actually was material (297).

The recommendations of the Commission and the ensuing regulations should go far to reduce the magnitude of the problem associated with the therapeutic orphan phenomenon. In particular, their recommendation that risks "presented by an intervention that holds out the prospect of direct benefit for the individual subject" may be considered differently from the risks presented by procedures designed to serve solely the interests of research should facilitate the ethical conduct of clinical trials in children. It is also noteworthy that the FDA has announced its intention to propose regulations that will require that new drugs with major therapeutic utility in children be tested in children as a condition of approval of such drugs for marketing (218).

Recommendations and Regulations

The Commission's report on research involving children was published in 1977 (525). Public response to this report and its recommendations was generally approving. Moreover, this report was generally well received by researchers and their institutions. In 1982, before DHHS published its final regulations, Kapp's survey of medical schools and children's hospitals indicated that there was already substantial compliance with the spirit of the Commission's assent requirements (339). According to Kapp, these requirements did not present the obstacle to conducting research involving children that some investigators feared they might. DHHS's final regulations providing "additional protection for children involved as subjects in research" were published in March 1983 and became effective on June 5, 1983 (45 CFR 46, Subpart D) (191). The final regulations, which may be found in Appendix I, are largely consistent with the Commission's recommendations. FDA's proposed regulations (224) were substantially identical to DHHS's 1978 proposal (185); it seems reasonable to anticipate that these consistencies will be preserved in the FDA's final regulations. The voluntary cooperation of the research community contrasts strikingly with its reaction to the Commission's recommendations on research involving those institutionalized as mentally infirm (Chapter 11) and prisoners (Chapter 12).

The remainder of this chapter is a survey of the Commission's recommendations for research involving children; it is concerned only with procedures or considerations that differ from those that apply to all research involving human subjects. For example, there is no discussion of the requirements that research involving children be reviewed and approved by an Institutional Review Board (IRB) and that subjects be selected in an equitable manner. Unless otherwise specified, it may be assumed that the DHHS regulations are essentially harmonious with the Commission's recommendations. In connection with the discussion of each recommendation, I will identify any substantial discrepancies that appear in the final regulations.

Recommendations for All Research Involving Children

Recommendations 2 and 7 set forth the norms for all research on children. As we shall see subsequently, certain classes of research call for additional procedural protections. "The IRB is required to determine: 2B: Where appropriate, studies have been conducted first on animals and adult humans, then on older children, prior to involving infants. . . ."

This recommendation reflects the interpretation of the requirement of the principle of *justice* that more vulnerable persons are to be afforded special protection from the burdens of research. Adults are perceived as less vulnerable than older children who, in turn, are less vulnerable than infants. Investigators who propose to do research on children without having first done such research on animals, adults, or both will be obliged to persuade the IRB that this is necessary. The strongest justification is that the disorder or function to be studied has no parallel in animals or adults. In such cases, when the research presents any risk of physical or psychological harm or significant discomfort, investigators are expected to initiate their work on older children who are capable of assent before involving infants.

Recommendation 2D requires that adequate provision be made to protect the privacy not only of the children but also of their parents. Recommendations 2B and 2D are not reflected in the final regulations. However, in the preamble to the publication of the regulations, DHHS states that these recommendations should be followed (191, p. 9816). Also, as we shall see, concern for the privacy and confidentiality of parents and families motivated DHHS to limit the types of research involving children exempted from coverage by the regulations.

Most of the published commentary on the right to privacy and confidentiality of the child centers on adolescents and their legal entitlement to require that professionals withhold certain sorts of information from their parents. It is worth noting, however, that the confidences of young children also should be treated with respect. Children need to be taught how to build relationships based upon trust. They need to be taught that promises (e.g., to keep a secret) are not to be made lightly and that once made they are morally binding, not to be broken without sufficient reason (cf. Chapter 7). Professionals, in their interactions with children, should contribute to their socialization by serving as models of exemplary behavior.

Assent and Permission

The Commission abandoned use of the word "consent" except in situations in which an individual can provide "legally effective consent" on his or her own behalf. As a corollary to this, the term "proxy consent" has been discarded.

Recommendation 7 assigns to the IRB the responsibility for determining that adequate provisions are made for "7A: Soliciting the assent of the children (when capable) and the permission of their parents or guardians. . . ." The transactions involved in negotiating assent and parental permission are essentially the same as those for informed consent (Chapter 5). There are, for example, the familiar basic elements of informed consent set forth in Section 46.116 of the general regulations.

According to the Commission, a child with normal cognitive development becomes capable of meaningful assent at about the age of 7 years, although some may be younger and some older. DHHS did not accept the Commission's suggestion regarding the age of assent. Rather, at the time the proposed regulations were published, DHHS solicited public comment on which of three options it should adopt for nontherapeutic procedures, either age 7, age 12, or leaving the age to the discretion of the IRB. The final regulations specify no age of assent. Rather, the IRB is assigned responsibility for "determining whether children are capable of assenting." In reaching this determination, the IRB shall take into account the age, maturity, and psychological state of the children involved. This judgment may be made for all children to be involved in research under a particular protocol or for each child, as the IRB deems appropriate. For helpful suggestions on how the IRB can accomplish such determinations, the reader is referred to Weithorn's empirical studies (722, 723). Leikin also provides advice in this regard (379); in part, his advice relies on Kohlberg's assessment of the development in children of moral judgment (356), the validity of which has been challenged by Gilligan (261), particularly with regard to moral development in girls and women.

As the assent regulation is written, it seems to reflect the presumption that the capability to assent is an all-or-none phenomenon: the child is either capable or incapable of assent. This presumption is incorrect (722, 723) and, I believe, unintended by the regulation writers. However, a literal reading of the regulations seems to create this requirement: If the child is capable of assent—i.e., passes a certain threshold of capability—he or she must be presented with all of the required elements of information.

In my view the regulations are intended to be interpreted to permit the IRB to determine that prospective child-subjects may be capable of understanding some but not all of the elements of informed consent. Thus, for example, it may be appropriate to provide some children with "a description of any reasonably forseeable risks or discomforts," without providing "an explanation as to whether any compensation . . . (is) . . . available if injury occurs."

When the IRB determines that prospective child-subjects are capable of assent, it "may still waive the assent requirement under circumstances in which consent may be waived in accord with Section 46.116 of Subpart A."

The new regulations make no reference to the child's objection except in the definition of assent (Section 46.402): "Mere failure to object should not, absent affirmative agreement, be construed as assent." However, IRBs should keep in mind that for some protocols in which they have judged the prospective child-subjects incapable of assent, some provision may still be made to respond to the child's "deliberate objection." The term "deliberate objection" is used to recognize that some children who are incapable of meaningful assent are able to communicate their disapproval or refusal of a proposed procedure. A 4-year-old may protest, "No, I don't want to be stuck with a needle." However, an infant who might in certain circumstances cry or withdraw in response to almost any stimulus is not regarded as capable of deliberate objection. In its commentary on Recommendation 7, the Commission suggested that a child's deliberate objection usually should be regarded as a veto to his or her involvement in research. Of course, there are exceptions to this; most importantly, parents and guardians may overrule the young child's objection to interventions and procedures that hold out the prospect of direct benefit to the child and to some research maneuvers designed to evaluate these interventions and procedures (*infra*).

Permission   Unless otherwise specified, the involvement of children as research subjects must be authorized by the permission of the child's parent or guardian. If more than minimal risk is presented by an intervention or procedure that does not hold out the prospect of direct benefit for the individual subject, "permission is to be obtained from . . . both parents . . . unless one parent is deceased, unknown, incompetent, or not reasonably available, or when only one parent has legal responsibility for the care and custody of the child" (Section 46.408). In other classes of research, "the IRB may find that the permission of one parent is sufficient." "Permission by parents or guardians shall be documented in accordance with and to the extent required by Section 46.117 of Subpart A." Whether or not assent must be documented is left to the discretion of the IRB.

The purposes of informed consent, parental permission, and assent are entirely different from those of their documentation (Chapter 5). Because the primary purpose of documentation is to protect the interests of the investigator and the institution against those of the subject, it seems generally unnecessary to document assent except in those cases in which the IRB determines that parental or guardian permission is unnecessary. When children are invited to sign forms, as they often and quite properly are, the principal purpose in my view is to enhance their sense of participation in the process.

In the commentary on Recommendation 7, the Commission indicates that parental or guardian permission should reflect the collective judgment of the family that an infant or child may participate in research. In research projects for which permission of one parent or guardian is sufficient, e.g., research in which the risks or discomforts are related to an intervention that is designed to be therapeutic, diagnostic, or preventive for that particular child, it may be assumed that the person giving formal permission is reflecting a family consensus.

The requirement for parental or guardian permission assumes that the child is living in a reasonably normal family setting. It further assumes a normal loving relationship between the child and his or her parents. In the event there is probable cause to suspect that no such loving relationship exists, different procedures may be required at the discretion of the IRB (*infra*).

The idea that parental or guardian permission should reflect the collective judgment of the family is not reflected in the regulations. In my view this is a very important consideration. This topic is discussed extensively in several chapters in *Who Speaks for the Child* (258) and in two excellent essays by Schoeman (629, 630).

The Commission recommended further: ''7B: The IRB should determine that adequate provisions are made—when appropriate—for monitoring the solicitation of assent and permission, and involving at least one parent or guardian in the conduct of research.'' In contrast to DHEW draft proposals for regulations on the conduct of research on children, this recommendation makes the determination of the need for monitoring and third party participation in the consent process discretionary judgments and assigns them to the IRB (181). Among the factors that are to be considered are the following. In research projects that present to the child more than minimal risk of physical or psychological harm or a substantial burden of discomfort, when these risks and discomforts are not inextricably related to a procedure necessary for the health interests of the child, the IRB may decide that it is necessary to appoint someone to assist in the selection of subjects and to review the quality of interaction between parents or the guardian and the child. The Commission suggested that such persons might be IRB members, the child's pediatrician or psychologist, a social worker, a pediatric nurse practitioner, or any other experienced and perceptive person.

In general, when infants and very small children participate in research that may cause physical discomfort or emotional stress that involves a significant departure from normal routine, a parent or guardian should present. The parent or guardian has the functions of comforting the child and, if necessary, intervening on behalf of the child.

In the commentary under Recommendation 7, the Commission states:

> . . . the IRB may determine that there is a need for an advocate to be present during the decision-making process. The need for third-party involvement . . . will vary according to the risk presented by the research and the autonomy of the subjects. The advocate should be an individual who has the experience and perceptiveness to fulfill such a role and who is not related in any way (except in the role as advocate or member of the IRB) to the research or the investigators.

DHHS regulations do not call for the involvement of persons other than the child and the parent in the consent process unless the research subjects are children who are wards of the state or any other agency, institution, or entity. For further discussion of the optional or mandatory involvement of third parties in the consent process, see Chapter 5.

Requirements Varying with the Degree and Nature of Risk

Research that presents to children no more than minimal risk may be conducted with no substantive or procedural protections other than those specified earlier (*Recommendation 3* and Section 46.404). The following definition was developed by the Commission: "*Minimal risk* is the probability and magnitude of physical or psychological harm that is normally encountered in the daily lives, or in the routine medical or psychological examination, of healthy children." As noted in Chapter 3, DHHS used this definition of minimal risk, developed for research involving children, as a standard applying to all research involving human subjects. Thus, the regulations have no special definitions for minimal risk applying to any specific population of research subject.

The Commission provides examples of procedures presenting no more than minimal risk; these are routine immunization, modest changes in diet or schedule, physical examination, obtaining blood and urine specimens, and developmental assessments. Similarly, many routine tools of behavioral research, such as most questionnaires, observational techniques, noninvasive physiological monitoring, psychological tests, and puzzles, may be considered to present no more than minimal risk. Questions about some topics, however, may generate such anxiety or stress as to involve more than minimal risk. Research in which information is gathered that could be harmful if disclosed should not be considered of minimal risk unless adequate provisions are made to preserve confidentiality.

Minor Increments above Minimal Risk   *Recommendation 5* deals with research proposals that present to the child minor increments above minimal risk. The risks with which this recommendation is concerned are presented by procedures that do not hold out any expectation of direct health-related benefit for that child. In this case, the IRB is charged with the responsibility for determining that: "5A: Such risk represents a minor increase over minimal risk. . . ."

The IRB should consider the degree of risk presented by research procedures from at least the following four perspectives (525, pp. 8–9): a common-sense estimation of the risk, an estimation based upon the investigator's experience with similar interventions or procedures, any statistical information that is available regarding such interventions or procedures, and the situation of the proposed subjects. No definition of minor increment is provided.

"5B: Such intervention or procedure presents experiences to subjects that are reasonably commensurate with those inherent in their actual or expected medical, psychological, or social situations, and is likely to yield generalizable knowledge about the subject's disorder or condition. . . ." The requirement that experiences be reasonably commensurate with those inherent in their actual or expected situations requires some clarification (525, p.9). Firstly, it means that the procedures to be followed are those that they or others with the specific disorder or condition under study will ordinarily experience by virtue of their having or being treated for that disorder or condition. Thus, it might be appropriate to invite a child with leukemia who has had several bone marrow examinations to consider having another for research purposes. It would be more difficult to justify extending a similar invitation to normal children. This requirement will make it difficult to develop normal control data for examinations and other procedures that present more than minimal risk.

The requirement for commensurability reflects the Commission's judgment that children who have had a procedure performed upon them might be more capable than are those who are not so experienced to base their assent on some familiarity with the procedure and its attendant discomforts; thus, their decision to participate will be more knowledgeable.

"5C: The anticipated knowledge is of vital importance for understanding or amelioration of the subject's disorder or condition. . . ." Thus, there is a stronger requirement here than was expressed for research characterized as presenting minimal risk for justification in terms of developing knowledge that will be of use to the class of persons of whom the particular subject is a representative. This requirement thus establishes a higher standard for assessing the importance of the knowledge to be gained. In addition, it strengthens the general requirement to use children as subjects particularly in research that is relevant to their disorder or condition. Thus,it should be very difficult to justify the use of procedures presenting more than minimal risk to develop information irrelevant to disorders or conditions present in the subjects of the research.

DHHS regulations, Section 46.405, are nearly verbatim reproductions of *Recommendations 5A, B, and C*. Thus, the IRB is called upon to make two difficult judgments for this class of research. What constitutes vital importance? What is the upper limit of "minor" in assessing an increment above minimal risk? I think the Commission and DHHS each showed wisdom by resisting demands to define the boundaries of these terms. IRBs and investigators are now challenged to explore these concepts and to develop functionally relevant definitions as they consider problems presented by particular protocols. As they share and debate the fruits of their explorations with one another, we can expect a gradual refinement of our understanding of these concepts and how to use them.

At this point I am aware of only one published case study concerned explicitly with these problems (15). This study was designed to investigate mechanisms of diabetic instability in children. The proposed subject population were juvenile diabetics aged 8 to 18 years; their unaffected siblings were to serve as normal controls. Parenthetically, the IRB disapproved involvement of normal siblings under the age of 12. Justification of the involvement of the older normal siblings was complex; I shall not discuss it further here.

Among the procedures to be done for research purposes were a 48-hour admission to the hospital for each child and placement of two intravenous cannulas, one for sampling of blood and the other for infusion of hormones.

The IRB was satisfied that there was little or no risk of physical injury; however, it was difficult to anticipate the psychological hazards of the 2-day hospitalization with prolonged immobilizations for hormone sensitivity tests.

At the conclusion of the research, Stephanie Amiel, the principal investigator, reported on her experience (15). As anticipated, there were no physical injuries. The main problem seemed to be that the children became bored. The investigator discussed creative maneuvers she introduced to combat boredom. One of the most creative was to borrow a video cassette recorder so that the children might watch movies during the hormone sensitivity tests. Amiel reports,

> Tears disappeared as if by magic. However, if the author has to work through one more showing of *Raiders of the Lost Ark*, she will probably resign and return to clinical practice! But the use of exciting adventure films genuinely has converted the study time into one of enjoyment and pleasure for our youthful volunteers. . . . One word of caution however. After one mother wistfully remarked that "Disney really might have been nicer," we made it a rule to check our films with parents first!

Other procedures that have been approved by Yale's IRB as presenting minor increments above minimal risk include bone marrow aspirations in children with leukemia, single additional spinal taps in adolescents who have already had at least one for a neurologic disorder, and administration of yohimbine in order to gain information about the pathogenesis of a neurologic disorder. The same IRB rejected a proposal to do left heart catheterizations on children at risk for the development of cardiac hemosiderosis.

Recommendation 6 limits the authority of the IRB to approve research that presents more than minor increments above minimal risk when the risk is presented by procedures not expected to provide direct health-related benefit to the particular subjects. In this situation, the IRD is called upon to make the following determination. "The research presents an opportunity to understand, prevent or alleviate a serious problem affecting the health or welfare of children. . . ." In the event the IRB can make such a determination, the proposal must be referred to the federal government. Here it must first be reviewed by a National Ethical Advisory Board (EAB), which should not only conduct its deliberations at meetings open to the public but also publish its conclusions. This is to be followed by an "opportunity for public review and comment." After this, the Secretary of the responsible federal department (or highest official of the responsible federal agency) must determine either 1) that the research can be approved by the local IRB (this would represent a judgment made at the federal level that either the risk is only a minor increment above minimal or is justified by the direct health-related benefit that will accrue to the individual subjects) or 2) that the conduct of the research is approved on the following grounds:

I. The research presents an opportunity to understand, prevent or alleviate a serious problem affecting the health or welfare of children; and

II. The conduct of the research would not violate the principles of respect for persons, beneficence and justice. . . .

In its comment under this recommendation, the Commission advises that if the local IRB is in doubt, this procedure should be followed. The National EAB may find that there is no need for the full procedure, in which case it may refer the decision back to the local IRB.

The class of research discussed by the Commission in Recommendation 6 is covered by Section 46.407 of DHHS regulations. No provision is made for referral to an EAB; as discussed in Chapter 13, there has been no EAB since 1980. Instead, the DHHS Secretary is called upon to make judgments about the permissibility of conducting or funding such research ". . . after consultation with a panel of experts in pertinent disciplines (for example; science, medicine, education, ethics, law) and following opportunity for public review and comment. . . ." Criteria for approval of such research are those recommended by the Commission.

Interventions Presenting the Prospect of Direct Benefit   Research protocols that present more than minimal risk of physical or psychological harm or discomfort to children but in which the risk "is presented by an intervention that holds out the prospect of direct benefit for the individual subjects, or by monitoring procedures required for the well-being of the subjects" may be considered differently (*Recommendation 4*). In this case, the IRB must determine that:

4A: such risk is justified by the anticipated benefits to the subjects; and

4B: The relation of anticipated benefit to such risk is at least as favorable to the subjects as that presented by available alternative approaches. . . .

In this recommendation, the Commission calls for an analysis of the various components of the research protocol. Procedures that are designed solely to benefit society or the class of children of which the particular child-subject is representative are to be considered as the research component. Judgments about the justification of the risks imposed by such procedures are to be made in accord with other recommendations. For example, if the risk is minimal, the research may be conducted as described in Recommendations 3 and 7, no matter what the risks are of the therapeutic components.

The components of the protocol "that hold out the prospect of direct benefit for the individual subjects" are to be considered precisely as they are in the practice of medicine. Risks are justified by anticipated benefits to the individual subjects and, further, by the assent when appropriate of the child and the permission of the parents or guardians. Recommendations 4A and B became Section 46.405 of the regulations.

Commonly, in research designed to test the safety and efficacy of nonvalidated practices or investigational therapies, investigators use placebo-controlled designs. Several commentators on DHHS's proposed regulations raised the question of whether such studies could be approved. DHHS's response was (191, p. 9816): ". . . research activities involving placebos may be conducted in accord with Subpart D, depending upon the individual activity. An IRB may find that a particular activity is approvable under Sections . . . 46.404, 46.405, or 46.406, or in some cases, a combination of these sections." Thus, the IRB may find that placebo administration either does or does not hold out the prospect of direct benefit to the child; if it does not, the IRB may find that it presents either minimal risk or a minor increment above minimal risk. DHHS's advice reflects succinctly the controversies on placebo controls (Chapters 8 and 9).

In Recommendation 4B, the Commission made one further statement relevant to risk presented by potentially therapeutic procedures: "A child's objection . . . should be binding unless the intervention holds out a prospect of direct benefit that is important to the health or well-being of the child and is available only in the context of the research."

Section 46.408 (a) reflects this concept in a somewhat altered form: "If the IRB determines . . . that the intervention or procedure holds out a prospect of direct benefit that is important to the health or well-being of the children and is available only in the context of the research, the assent of the children is not a necessary condition. . . ."

The general presumption is that parents may make decisions to override the objections of school-age children in such cases. However, in some circumstances the objection of teenagers to decisions made on their behalf by parents may prevail. For a recent analysis of the law governing the minor's authority to consent to treatment, see Holder's Chapter 5 (313). Capron also reviews the law on this subject and provides a table of statutes defining the legal ability of minors to consent to medical care in the various states (113).

In the practical world of decision-making about who can authorize a therapeutic procedure, whether it be investigational or accepted, it rarely suffices to point to the law and thereby identify the person who has the legal right to make the decision. Many factors must be taken into account in reaching judgments about the capability of various persons to participate in and, in the event of irreconcilable disputes, to prevail in such choices. In general, these judgments become more complicated as the child gets older or as the stakes get higher. Gaylin provides a good discussion of our arbitrary legal standards regarding the competence of children (257). He notes the absurdity of the events attending the 18th birthday. In 1 day an essentially disenfranchised member of the state is "miraculously transformed . . . welcomed into the decision-making apparatus of his country. No longer would 'they' decide his future; today he was a man!" He suggests that this arbitrary standard should be replaced by a concept he calls "variable competence." He presents a detailed discussion of the various factors that should be considered in deciding how "generous" we ought to be in according decision-making authority to a maturing minor. I believe that many of these factors are taken into account by sophisticated pediatrician-investigators in their negotiations with teenagers and their parents.

Thomasma and Mauer illustrate that the recruitment of an adolescent to participate in a protocol designed to test a nonvalidated therapy can be a very complicated process (682). They emphasize that both the ethical analyses of such cases and the eventual decisions to which they lead require great sensitivity to the life-situations of the patients. They argue cogently that it is particularly inappropriate to apply general ethical concepts to all cases without attention to factors that may permit a choice of several ethical alternatives for resolution of the case.

Their illuminating case history describes a 15-year-old boy who required an above-the-knee amputation of his right leg for osteosarcoma. Subsequently, he was invited to participate in a randomized clinical trial of adjuvant immunotherapy.

Four months after agreeing reluctantly to participate in this clinical trial, he expressed his wish to withdraw. The treatments were affecting adversely his capacities to perform in his school work, as well as in other social contexts. Although his mother supported this decision, his father seemed reticent about getting involved. Within this context, the authors discuss the ethical propriety of the physician-investigator's use of his powers of persuasion to continue the boy's involvement in the protocol. Before making judgments about recruiting adolescents to participate in clinical trials, one should read and reflect on this important article.

Children in Special Situations

Recommendation 8 states that when:

> . . . a research protocol is designed either for conditions or for a subject population for which parental or guardian permission is not a reasonable requirement . . . (the IRB) may waive such requirements provided an appropriate mechanism for protecting the . . . subjects . . . is substituted.

DHHS Regulations (Section 46.408(c)) essentially repeat the language of Recommendation 8, inserting examples of the sorts of subjects to whom they intend to apply the permission waiver—"(for example, neglected or abused children)." One should also consider the additional examples provided by the Commission in its commentary on Recommendation 8:

> Research designed to identify factors related to the incidence or treatment of certain conditions in adolescents for which, in certain jurisdictions, they may legally receive treatment without parental consent; research in which the subjects are "mature minors" and the procedures involved entail essentially no more than minimal risk that such individuals might reasonably assume on their own; research designed to meet the needs of children designated by their parents as "in need of supervision," and research involving children whose parents are legally or functionally incompetent."

The regulations provide a new definition of "children" as "persons who have not attained the legal age for consent to treatment or procedures involved in the research, under the applicable law of the jurisdiction in which the research will be conducted" (Section 46.402). I think this should be interpreted to mean that those who have attained the legal age for consent to treatment or procedures involved in the research should also be recognized as having the authority to consent to other procedures such as blood tests or questionnaires that might be used to identify factors related to the incidence or treatment of those conditions.

Herceg-Baron presents a case study in which she details some of the special problems involved in research in the field of family planning involving minors as subjects (289). Many adolescents wish to seek advice about such matters as contraception without the awareness of such others as their parents. She details her institution's policy for protecting the minor's confidentiality. For example, with regard to follow-up, investigators are required to offer various options, e.g., telephone calls during certain hours when the minor knows she will be at home alone; contacts through school personnel such as nurses, teachers, or counselors; contacts by mail containing no agency letterhead or other identifying information; or leaving messages with friends.

In commenting on this case study, Carol Levine raises several concerns (381). First, she suggests that many adolescents are ambivalent about clandestine sex and would, with some encouragement, welcome open discussion with their parents. The IRB's great concern with privacy seems to undermine the possibility for what could be valuable communication within the family. Levine is further concerned about the fact that the institution not only approves deception, it collaborates with the adolescent in deceiving her parents. This, she argues, sets a very poor example for the adolescent. The authors continue their discussion in the letters section of the February 1982 issue of *IRB: A Review of Human Subjects Research*.

What if we are recruiting adolescents to participate in research presenting little or no risk that does not involve any procedure relevant to the "legal age for consent to treatments?" Is parental permission unnecessary to authorize such recruitment? According to Holder (with whom I agree), the answer is yes when dealing with teenagers living away from home (e.g., 16-year-old college students) and the research procedures present minimal risk (302, 308). She would extend even to preteenagers the authority to assent without parental permission to participate in such innocent studies as public opinion polls. She cautions, however, that there are practical reasons to avoid providing therapy to minors without parental permission, except in emergencies or for conditions specified by state law (e.g., venereal disease or abortion). Among other reasons, parents are not obliged to pay for therapy (other than emergency treatment) they have not authorized. Veatch, on the other hand, contends that the requirement for parental permission ought not be waived unless the waiver is necessary to serve the health interests of the minor, e.g., to secure treatment for venereal disease in cases in which the minor would remain untreated rather than inform the parents of the infection (698). He concludes that parental permission should almost always be required to involve teenagers as research subjects even when the studies are of such problems as venereal disease, pregnancy, and abortion; on this point he disagrees with both Holder and the Commission.

The foregoing discussion is confined to considerations of mature minors and those who are approaching maturity. It is not applicable to emancipated minors. As a matter of law, emancipated minors have full decision-making authority on all matters; thus, there is no need to consult their parents for permission. The most common ways in which minors become legally emancipated are through marriage or military service. For further discussion of the concepts of emancipated and mature minors, the reader is referred to Capron (113) or Holder (313).

For further elaboration of the issues arising in considering consent and confidentiality in providing health care for adolescents, the reader is referred to a recent American Academy of Pediatrics Conference (512). This report identifies "problem areas that are currently unmet in the law" and provides a discussion of some possible solutions to these problems.

The regulations state that the IRB may "waive the requirement for parental permission . . . provided an appropriate mechanism for protecting the children is substituted." The Commission provided some suggestions for substitute protective mechanisms (525, pp. 18–19):

> There is no single mechanism that can be substituted for parental permission in every instance. In some cases the consent of mature minors should be sufficient. In other cases court approval may be required. The mechanism invoked will vary with the research and the age, status and condition of the prospective subject. . . .
>
> Assent of . . . mature minors should be considered sufficient with respect to research about conditions for which they have legal authority to consent on their own to treatment. An appropriate mechanism for protecting such subjects might be to require that a clinic nurse or physician, unrelated to the research, explain the nature and the

purpose of the research . . . emphasizing that participation is unrelated to provision of care.

Another alternative might be to appoint a social worker, pediatric nurse, or physician to act as surrogate parent when the research is designed, for example, to study neglected or battered children. Such surrogate parents would be expected to participate not only in the process of soliciting the children's cooperation but also in the conduct of the research, in order to provide reassurance for the subject and to intervene or support their desires to withdraw if participation becomes too stressful.

Recommendation 9 limits the circumstances under which children who are wards of the state may be exposed to more than minimal risk by research (procedures performed without the intent of producing direct benefits to the child-subject). In the language of the corresponding regulation (Section 46.409):

> Children who are wards of the state or any other agency, institution, or entity can be included in research approved under Sections 46.406 or 46.407 only if such research is: (1) Related to their status as wards; or (2) Conducted in schools, camps, hospitals, institutions, or similar settings in which the majority of children involved as subjects are not wards.

The regulations further direct the IRB to

> . . . require appointment of an advocate for each child who is a ward, in addition to any other individual acting on behalf of the child as guardian or *in loco parentis* . . . The advocate shall be an individual . . . who is not associated in any way (except in the role as advocate or member of the IRB) with the research, the investigator(s), or the guardian organization.

Recommendation 10 requires:

> Children who reside in institutions for the mentally infirm or who are confined in correctional facilities should participate . . . only if the conditions regarding research on the institutionalized mentally infirm or on prisoners (as applicable) are fulfilled in addition to the conditions set forth herein.

This recommendation is not reflected in any form in the regulations.

Exemptions

Subpart A lists six categories of research that are exempt from coverage by the regulations (Section 46.101). Exemptions numbered 1, 2, 5, and 6 are equally applicable to research involving children. Exemption 4, research involving the observation of public behavior, is applicable to research involving children only when the investigator does not participate in the activities being observed. Exemption 3, research involving survey or interview procedures, does not apply to research involving children.

In DHHS's view (191, p. 9815),

> children being surveyed or interviewed . . . may not be capable of recognizing that their responses to questions on sensitive issues could be potentially damaging to themselves or others. Therefore, it is appropriate that the IRB at least review such research to determine whether the rights and welfare of children participating as subjects are adequately protected and when the requirements of permission or assent can be waived. Such waivers shall be in accordance with the requirements of Sections 46.116 and 46.117 of Subpart A.

Similarly, "children involved in observation research, where the investigator is also participating in the activities being observed, may not have the capability to determine whether or not to participate and therefore IRB review of such research is appropriate" (191, p. 9815). This seems to reflect the view that the children may not be aware of their privacy interests and their implications and, further, may have different perspectives on the privacy interests of their families than those held by their parents. As I have noted earlier, there is considerable controversy over whether children should have the authority to assent to certain sorts of research without parental permission. This will be particularly problematic to IRBs as they are assessing the necessity for parental permission to involve teenagers in survey and interview research.

Chapter 11

Those Institutionalized as Mentally Infirm

Proposals to do research involving as subjects those institutionalized as mentally infirm present a very complicated array of ethical issues. Much of the literature on the ethics of research involving those institutionalized as mentally infirm tends to see the problems as extremely similar to, if not identical with, those presented by proposals to do research involving children (371, 373). Consequently, as we shall see, the Commission's recommendations for these two classes of subjects are similar in many respects. There are some important differences. 1) There is a problem in defining a class of persons as "those institutionalized as mentally infirm." 2) There are often disputes over what constitutes a therapeutic benefit. 3) The fact of institutionalization presents important problems. 4) Some of these persons are capable of consent; for those who are not, finding a replacement for the parent who is presumed to stand in a loving relationship to the prospective child-subject may be problematic.

Definition

The Commission was charged to report on research involving the "institutionalized mentally infirm," defined by Congress as "individuals who are mentally ill, mentally retarded, emotionally disturbed, psychotic, or senile, or who have other impairments of a similar nature and who reside as patients in an institution"

(PL 93-348). The Commission declined to use this term and presented several important reasons for doing so (529). 1) The term "mentally infirm" is not a diagnostic term; thus, it does not adequately define a class of persons. 2) There is considerable debate about whether symptoms that may result in institutionalization are properly characterized as diseases or illnesses in the conventional sense or whether they represent problems in social adaptation. 3) Labeling may lead to stereotyped conceptions of people and their problems. 4) Some persons are institutionalized as mentally infirm because of misdiagnosis or error. Therefore, the Commission chose to use the term "those institutionalized as mentally infirm."

The Commission (529, pp. xviii–xix) went on to stipulate that their report is applicable to persons who are residents either by voluntary admission or by involuntary commitment in public or private mental hospitals, psychiatric wards of general hospitals, community mental health centers, half-way houses or nursing homes for the mentally disabled, and similar institutions. Its recommendations are applicable to research involving all residents of such institutions, including those not mentioned by Congress, e.g., alcoholics and drug abusers. Further, the recommendations are intended to apply to persons who are in the transitional phases of deinstitutionalization, including those who reside outside traditional institutions in foster homes, group homes, or other facilities and those who are on leave or furlough. The defining attribute is that they remain listed in the census of the institution and, therefore, remain under its administrative responsibility.

The Commission further states its expectation that the investigator and the Institutional Review Board (IRB) will apply, as appropriate, the norms expressed in its recommendations to subjects who, although they are not institutionalized, have disorders either specified in the Commission's definition or presenting similar problems.

Risk and Benefit

As in the report *Research Involving Children,* the Commission developed separate recommendations for research characterized as presenting minimal risk, minor increments above minimal risk, and more than minor increments above minimal risk. The definition of minimal risk for those institutionalized as mentally infirm is presented in the commentary under Recommendation 2; it differs slightly from that provided for children:

> For the purposes of this report, minimal risk means the risk (probability and magnitude of physical or psychological harm or discomfort) that is normally encountered in the

daily lives, or in the routine medical or psychological examination, of normal persons. Thus, for subjects who are institutionalized as mentally infirm, routine examination procedures present no more than minimal risk if the likely impact of such procedures on them is similar to what would be experienced by normal persons undergoing the procedure.

Also, as in *Research Involving Children,* there is a separate recommendation (Recommendation 3) for research activities in which risk is presented by procedures that hold out the prospect of direct benefit for the subjects. The commentary under this recommendation reflects an understanding that in those institutionalized as mentally infirm there is often controversy over what constitutes a benefit and who the beneficiary is (134, 336, 340, 447). The problems discussed in Chapter 1 in defining a class of activities as practice for the benefit of others are often encountered. For example, a decision to administer large doses of phenothiazines to disruptive patients in a mental institution may be challenged on several grounds. Tranquilization may, at times, be done more in the interest of maintaining order in the institution than in the interest of fostering the patient's recovery. Tranquilization may at times be chosen as a more convenient and less expensive alternative to individualized psychotherapy which, in some cases, may be more advantageous to the patient. The disruptive behavior itself might be interpreted by some observers as an appropriate protest to illegitimate institutionalization.

In research designed to test the safety and efficacy of many types of therapy other than drugs, it is often difficult to develop adequate control data (cf. Chapter 8). In studies of the utility of psychotherapy it is, of course, impossible to utilize either double-blind or placebo-control designs. Richman et al. argue that the strategies employed by most investigators in this field for developing control data are deficient on both scientific and ethical grounds (599). Such strategies include assignment of controls to the waiting list, to receive nonspecific attention from individuals who are not trained in psychotherapy, and to receive low or minimal contact with psychotherapists. Richman et al. propose instead that control should be assigned to a "nonscheduled treatment" group.

> With this method, patients are seen periodically for research assessment and are given no active scheduled treatment. However, they are explicitly told that, should their condition worsen, they may request immediate treatment. Methodologically, this method provides a control for the effect of the availability of support. It also provides greater safety for acutely depressed patients. This was the control utilized in the Boston-New Haven Collaborative Study of acutely depressed patients (their article describes this study).

Institutionalization

Many commentators have expressed concern that institutionalized persons in general are used disproportionately and unfairly as research subjects because they are administratively convenient to the researchers (cf. Chapter 4). Consequently, there are those who argue that an individual who is institutionalized as mentally infirm should not participate, or be asked to participate, in research for which noninstitutionalized persons would be suitable subjects (529, pp. 115–117). This argument is grounded in the requirement of the principle of *justice* that we protect vulnerable persons from harm and that we minimize the burdens imposed upon those who already bear more than their fair share. It is argued that those institutionalized as mentally infirm are particularly vulnerable to exploitation; this vulnerability arises from their relative incapacity to protect themselves through the usual negotiations for informed consent. In addition, some of them tend to be totally dependent upon the institution in which they reside. Moreover, persons who are institutionalized as mentally infirm already carry burdens—often heavy burdens—imposed by their disabilities; therefore, it is unjust to ask them to assume additional burdens. Those outside the institution may be burdened by similar incapacities (diseases and social maladjustments); however, they are relatively less vulnerable because they are more likely to have caring persons to assist and protect them.

Opponents to these arguments claim that it is incorrect to assume that participation in research is always a burden or that being in an institution is always a damaging experience. At times, participation in research may have beneficial effects; e.g., research subjects may receive additional attention and may be afforded opportunities for interaction with people from outside the institution. The tasks associated with research may be interesting and a welcome change from the boredom of institutional life. In addition, deinstitutionalization of mental patients is not always associated with satisfactory integration into a community of caring persons; in some cases, it may be perceived as abandonment to ghettos where they have no one to look after their personal, health, and social needs.

As we shall see, the Commission addressed this argument by recommending two requirements. Firstly, investigators proposing to recruit subjects from an institution assume the burden of justifying their involvement. Secondly, institutionalized individuals are permitted to participate in research that is not relevant to their conditions only if they are capable of consenting and the research presents no more than minimal risk. In Chapter 12, there is more extensive discussion of the problems presented by the fact of institutionalization.

Competence and Comprehension

Although the Nuremberg Code requires both "legal capacity" to consent and "sufficient understanding" to reach an "enlightened" decision, contemporary interpretations tend to link the two by defining competence in terms of comprehension. In discussion of incompetence, it is useful to distinguish two different types, *de jure* and *de facto* incompetency (486). *De jure* incompetence means that individuals are deprived of the authority to make decisions as a matter of law; they are legally incompetent. For example, all persons under the legal age of majority are *de jure* incompetent. (The legal authority of minors to make decisions varies from state to state and within states depending on the nature of the decision.) Persons above the age of majority may be determined to be incompetent by a court. Upon adjudicating an individual to be incompetent, the court ordinarily appoints a guardian to make those decisions that the individual has been found unable to make. The adjudication of incompetence may be plenary (the individual is deprived of the authority to make all decisions having legal significance) or partial (deprivation of authority is confined to a narrow and specified area, e.g., to sign contracts or manage one's financial affairs).

I shall not discuss legal incompetence further, as it is an enormously complicated topic. For a comprehensive survey of this concept as it relates to consent to become a research subject, see Chapter 6 in the Commission's Report on those institutionalized as mentally infirm. In this rapidly evolving field, this reference, although still important conceptually, is becoming dated. More recently, the President's Commission has provided an authoritative overview of the problem of incompetence, which it calls incapacity (571, Chapter 3 and 8); explicitly excluded from their considerations are "the distinctive issues posed by consent to mental health care or consent to health care by the mentally ill, whether or not institutionalized." Moreover, this report does not address the problems peculiar to consent to research. The important book by Lidz et al. provides an adequate portal of entry into the recent literature of legal incompetence as it relates to consent to participate in research (447). For a recent thorough survey of the widely varying standards for determinations of legal incompetence in each state, see Brakel (79).

Persons are said to be incompetent *de facto* if in the judgment of a qualified clinician they have those attributes that ordinarily provide the grounds for adjudicating incompetence. Thus, while the standards for considering a person incompetent *de facto* are essentially the same as those for determining *de jure* incompetence, it is important to keep in mind that a person is not legally incompetent unless he or she has been so adjudicated.

There are two classes of standards for the identification of persons as incompetent: status and functional (571). That is, persons may be labeled incompetent either because they occupy a certain status or because they cannot perform a certain function. The status that defines incompetence may be permanent (e.g., severe mental retardation), temporary (e.g., inebriation), or transitory or subjective (e.g., peculiar behavior or symptoms of psychosis). Childhood is a status standard for defining incompetence. Status standards presuppose that persons occupying that status have relevant functional incapacities; however, they do not require that their presence be verified in any particular candidate for the label incompetent.

The President's Commission recognizes the validity of status standards for incompetence for individuals "in certain basic categories (such as under the age of 14, grossly retarded, or comatose)." For others, the President's Commission "recommends that determinations of incapacity be guided largely by the functional approach. . . . The fact that a patient belongs to a category of people who are often unable to make general decisions for their own well-being or that an individual makes a highly idiosyncratic decision should alert health care professionals to the greater possibility of decisional incapacity. But it does not conclusively resolve the matter (571, pp. 171–172).

Functional standards require that individuals be tested for lack of capacity to perform specified functions that are deemed by the standard-setters to be relevant to determining their competence. The literature on competence to consent to research tends to be both confused and confounding because many authors discuss it without stating their criteria for defining it or articulating the standard by which they measure it.

Stanley and Stanley have reviewed the literature on functional standards for determinations of competency; this review is helpful in that it brings order and clarity to this field (669). Functional standards for determinations of competence are portrayed as variations of five basic themes:

1. *Evidencing a choice.* According to this standard, if a person makes a decision, any decision, he or she is considered competent. Of the five standards, this is most respectful of the individual's autonomy in that there is no assessment of the quality of the decision or of the means used to reach it. Although this test is virtually never used in the real world in determining that individuals are competent to consent to participation in research, there are those who argue that any other standard should be rejected as excessively paternalistic (266). On the other hand, it is generally agreed that a negative choice expressed by a person who is incapable of assent must be respected; thus, the deliberate objection of such a person usually constitutes a veto to his or her participation in research.

2. *Reasonable outcome of choice.* According to this test, the individual who fails to make a reasonable decision is considered incompetent. A decision is

judged reasonable if it accords with what the competency reviewer either considers reasonable or presumes a reasonable person might make. This is a highly paternalistic standard in which the individual's autonomy is respected only if he or she makes the "right" choice. It is for this reason that the President's Commission explicitly repudiated this standard (571, p. 61).

3. *Factual comprehension.* According to this standard, the individual is expected to understand, or at least to be able to understand, the information disclosed during the consent negotiations. Prospective subjects may be called upon to display their ability to master this information through such devices as the two-part consent form (cf. Chapter 5).

4. *Choice based on rational reasons.* According to this standard, it is essential to test the person's capacity for rational manipulation of information. For example, persons might be called upon to demonstrate that they not only understand the risks and hoped-for benefits, but also that they have weighed them in relation to their personal situations. In common with the reasonable outcome standard, this one is paternalistic. Competency reviewers may express their biases for particular styles of reasoning. Moreover, it is difficult to provide objective criteria for the identification of rational reasons and to assess reasons that, although they are important to the decision-maker, are not verbalized.

5. *Appreciation of the nature of the situation.* This standard is closely aligned to the type of assessment that is performed when determining whether an individual was responsible for a criminal act. The person must display not only comprehension of the consent information but also the ability to use that information in a rational manner. In addition, they must appreciate the fact that they are being invited to become research subjects and what that implies. As the Stanleys point out, even nonvulnerable populations have a difficult time appreciating that fact. This, too, is a highly paternalistic standard.

The President's Commission did not identify any of these standards as the preferred standard. Rather, it identified three elements of capacity (571, p. 57): "Decisionmaking capacity requires, to greater or lesser degree; (1) possession of a set of values and goals; (2) the ability to communicate and to understand information; and (3) the ability to reason and deliberate about one's choices."

Most importantly, and, in my view, wisely, the President's Commission recognized that particular individuals may have sufficient capacity to make some decisions but not others; in short, they recognized the importance of context (571, p. 60):

> How deficient must a decisionmaking process be to justify the assessment that a patient lacks the capacity to make a particular decision? Since the assessment must balance possibly competing considerations of well-being and self-determination, the prudent course is to take into account the potential consequences of the patient's decision. When the consequences for well-being are substantial, there is a greater

need to be certain that the patient possesses the necessary level of capacity. When little turns on the decision, the level of decisionmaking capacity required may be appropriately reduced (even though the constituent elements remain the same) and less scrutiny may be required about whether the patient possesses even the reduced level of capacity. Thus a particular patient may be capable of deciding about a relatively inconsequential medication, but not about the amputation of a gangrenous limb.

For a philosophical analysis of the importance of taking into account the nature of the choices to be made (consent or refusal to research or therapy), as well as the consequences of the choice (how much may be gained or lost), see Drane (196). The recently introduced National Institutes of Health (NIH) durable power of attorney policy (*infra*) requires consideration of such factors.

There is a widely held presumption that persons who are mentally ill, particularly schizophrenic and severely depressed patients, generally are relatively incompetent to consent. This presumption, which is reflected implicitly in Department of Health, Education, and Welfare (DHEW) proposed regulations, has the effect of creating a status standard for incompetence. In reviewing the literature on this subject, Stanley found that most studies showed that psychiatric patients do not have a very high level of understanding of consent information (667). However, when compared directly with medical patients, they seem to be no less understanding. For example, Soskis found that schizophrenic patients were more aware than were medical patients of the risks and side effects of their medications; medical patients, by contrast, were more knowledgeable about the names and doses of their medications, as well as the relationship of drug treatment to their diagnoses (664). In another study designed to examine patients' rationale for deciding about participation in research, Stanley et al. compared inpatients on a general medical service with psychiatric patients on a locked unit who were mostly schizophrenic and psychotically depressed patients (670). These patients were asked whether they would consent to a series of six hypothetical research studies. They found no significant differences between the two groups in the frequency with which they would agree to participate in these studies. Moreover, there was no difference in their willingness to accept a high degree of risk or in the frequency of their refusal to accept highly favorable risk/benefit ratios.

Stanley summarizes her review of the literature by concluding that, although empirical research shows that psychiatric patients do have some impairment in their abilities, some studies also show that in some important respects they do not differ from medical patients (667). She calls for further studies in which psychiatric patients are compared directly with other types of patients. Most importantly, investigators should state precisely the standards they use for determining competency so that comparisons between studies can be made more readily. Conclusions such as only a "quarter of the patients could give true consent'' are of value only if the criteria for true consent are identified and articulated.

Stanley and Stanley conclude that, because we have overestimated the prevalence of incompetence among the mentally ill, we have adopted a paternalistic stance as we develop policies for their protection (669). Such policies have consequences that are detrimental to the interests of the overprotected. Unwarranted deprivation of their autonomy and decision-making authority is an affront to their dignity; moreover, it lowers their self-esteem, fosters childlike behavior, and often retards recovery. Stigmatization of this highly heterogeneous population may give tacit societal approval to discrimination against members of this group. This, in turn, may inhibit our goal of reintegrating psychiatric patients into the larger community. Last, but not least, excessive bureaucratic procedural requirements designed to protect the mentally ill are likely to turn investigators away from their efforts to develop more efficacious therapy for this group and focus on goals that may be accomplished with fewer bureaucratic encumbrances.

Consent and Assent

Some of those institutionalized as mentally infirm are for various reasons incapable of consent. To some of these the Commission assigns the authority to *assent*. In recognition of the differences between these persons and children, the Commission stipulates a different definition for the term "assent" (529, pp. 9–10) than it did for children. Those institutionalized as mentally infirm may assent to participate in some sorts of research if they are capable of understanding what they are being asked to do, of making reasonably free choices, and of communicating their choices clearly and unambiguously. The ability to assent is unrelated to judicial determinations of incompetency or to involuntary commitment (307). An individual may be involuntarily committed to an institution, and even have been adjudicated incompetent, yet still able to make a knowledgeable choice to participate in research. On the other hand, living for several years in an institutional setting may render some individuals quite incapable of making an autonomous choice even though they may have entered the institution voluntarily and have never undergone incompetency proceedings. One must be concerned in general that 1) potential subjects may agree to participate in research out of fear that necessary services or attention will be withheld if such permission is denied. Additionally, 2) when research involves participation over an extended period of time, one cannot presume from initial assent that there will be a continuing willingness to participate; the capacity to exercise the right to withdraw may fluctuate (447).

The Commission further recognizes that some of those institutionalized as mentally infirm may be incapable of meaningful assent. As with children, the deliberate objection of a prospective subject constitutes a veto to participation in all research except that designed to evaluate a therapeutic intervention; such involvement must be authorized by a court of competent jurisdiction (Recommendation 3).

As with children, the Commission recommended that, to the extent one's capacity to consent or assent is limited, there is an increasing requirement to minimize risk and to scrutinize carefully the consent process to secure the rights of prospective subjects to refuse to participate or to withdraw from the research.

In reviewing proposals to do research involving children, the IRB is to determine the need for individuals to assist in the selection of subjects or to supervise the negotiations for assent and parental permission. In the recommendations for those institutionalized as mentally infirm, similar functions are assigned to an individual identified as the consent auditor (Commentary under Recommendations 3 and 4).

The *consent auditor* is to be appointed by the IRB and should not be involved except in the capacity of consent auditor with the research for which subjects are being sought. The auditor should be a person who is familiar with the physical, psychological, and social needs of the class of prospective subjects as well as with their legal status. The auditor should determine whether subjects consent, assent, or object to participation in research. In some instances it may be appropriate for the auditor to observe the conduct of the research in order to determine whether the subject's willingness to be involved continues.

In some sorts of research presenting more than minimal risk, appointment of a consent auditor is mandatory (Recommendation 4). For other types of research, the need for a consent auditor is a discretionary judgment of the IRB; in general, they should be employed when there are substantial questions about the ability of the subjects to assent.

The Recommendations of the Commission

This survey is concerned only with those recommendations calling for different procedures or considerations than those required for research involving children. The reader may assume that unless mention is made of a difference the recommendations for the two classes of subjects are substantially identical.

Recommendation 1 For all research involving those institutionalized as mentally infirm, the IRB must determine that: "1B: The competence of the investigator(s) and the quality of the research facility are sufficient for the conduct of the research. . . ." Although the ethical codes require that researchers should be competent, the Commission assigns to the IRB responsibility for determining that they are only for this class of subjects (cf. Chapter 2).

"1D: There are good reasons to involve institutionalized persons in the conduct of research. . . ." The investigator must satisfy the IRB that it is appropriate to involve those institutionalized as mentally infirm in the research project. The IRB should consider whether the research is relevant to the subjects emotional or cognitive disability, whether individuals with the same disability are reasonably accessible to the investigator outside the institutional setting, or whether the research is designed to study the nature of the institutional process or the effect of some aspect of institutionalization on persons with a particular disability. For an example of a well-reasoned justification involving institutionalized persons exclusively as subjects of research not designed to evaluate therapeutic interventions, see Marini (466). Following the Commission's guidelines, he argues that some types of research on some types of behavior (in this case, aggression) should only be done on institutionalized persons.

The IRB is instructed to review with special care any proposal to institutionalize subjects or to extend their stay in an institution solely for research purposes; this is to be permitted only if subjects "knowledgeably agree." In addition, careful consideration should be given to the possibility of using alternate means to solve the research problem.

> 1H: Adequate provisions are made to assure that no prospective subject will be approached to participate in the research unless a person who is responsible for the health care of the subject has determined that the invitation to participate in the research and such participation itself will not interfere with the health care of the subject. . . .

In general, one should not invite patients to become research subjects without authorization of the physicians responsible for their care (cf. Chapters 4 and 5). However, this is the only recommendation for a regulation that would require such consultation.

In the commentary, the Commission further elaborates that when the potential subject's physician or other therapist is involved in the proposed research, independent clinical judgment should be obtained regarding the appropriateness of including that patient in the research. This is intended to reduce conflicts of interest between the objectives of health care and those of research, while still permitting clinicians, who may be especially knowledgeable regarding promising avenues of research, to apply their expertise in both enterprises. This recommendation is addressed to the same problem as Principle I.10 of the Declaration of Helsinki; however, it does not require that a third party obtain informed consent. Further discussion of separating the roles of clinician and investigator can be found in Chapter 5.

Recommendation 2   For research that presents no more than minimal risk, the IRB must determine that:

2B: Adequate provisions are made to assure that no subject will participate in the research unless: (I) the subject consents to participation; (II) if the subject is incapable of consenting, the research is relevant to the subject's condition and the subject assents or does not object to participation. . . .

Persons who object to participation may not participate in any research except as specified in Recommendation 3. For research activities covered by both Recommendations 2 and 3, consent auditors should be appointed at the discretion of the IRB.

Recommendation 3   For research in which more than minimal risk is presented by an intervention that holds out the prospect of direct benefit for the individual subjects or by a monitoring procedure required for the well-being of the subjects, the IRB must determine that:

3D: . . . (N)o adult subject will participate . . . unless: (I) the subject consents to participation; (II) if the subject is incapable of consenting, the subject assents to participation (if there has been an adjudication of incompetency, the permission of a guardian may also be required by state law); (III) if the subject is incapable of assenting a guardian of the person gives permission (if a guardian of the person has not been appointed, such appointment should be requested at a court of competent jurisdiction) or the subject's participation is specifically authorized by a court of competent jurisdiction; or (IV) if the subject objects to participation, the intervention holding out the prospect of direct benefit for the subject is available only in the context of research and the subject's participation is specifically authorized by a court of competent jurisdiction. . . .

These provisions are similar to those authorizing similar activities for research involving children. In addition, they reflect a concern that the guardian from whom permission is sought is truly a "legally authorized representative." Moreover, they recognize that there will at times be differences of opinion as to what constitutes a benefit for a particular patient. In jurisdictions that grant institutionalized individuals an unqualified right to refuse therapy, their objection to participation in research will be binding.

Recommendation 4   For research in which minor increments above minimal risk are presented by interventions that do not hold out the prospect of direct benefit for individual subjects, the recommendations are very similar to those for research involving children. The only substantive difference is that there is no requirement that the procedures be reasonably commensurate with those inherent in the subject's life situation. The requirements for consent and assent are the same as those specified in Recommendation 3. For this class of research, the appointment of a consent auditor by the IRB is mandatory.

Recommendation 5 For research in which more than minor increments above minimal risk are presented by interventions that do not hold out the prospect of direct benefit to the subjects, review by a National Ethical Advisory Board is required; the criteria for its approval are substantially identical to those prescribed for children.

DHEW Proposed Regulations

In its publication of proposed regulations, DHEW changed not only the name of the class of person to "those institutionalized as mentally disabled," but also its definition (188). DHEW accepted the Commission's recommendation that consent auditors be appointed by the IRB; however, although the Commission recommended that the need for such persons should be determined as a discretionary judgment of the IRB, except that they are mandatory when procedures presenting more than minimal risk are to be performed for research purposes, DHEW "is giving consideration to requiring that a consent auditor monitor all research covered by these Regulations, including research involving no more than minimal risk" (188, p. 53952). In addition, DHEW proposes to create the position of advocate, an individual appointed by the IRB but having no other tie to any institution or agency with an interest in the research. The advocate will "be construed to carry the fiduciary responsibilities of a guardian *ad litem*" (Section 46.503k). "Consideration is being given to mandating that . . . whenever the consent auditor determines that a subject is incapable of consenting, the subject may not participate without the authorization of the advocate" (Section 46.505b). The Commission did not recommend advocates.

Thus, in order to involve some of those institutionalized as mentally infirm (or disabled) in research, it may be necessary for an investigator to first seek the approval of 1) DHHS, 2) the IRB, 3) the subject, 4) the subject's "legally authorized representative" (who, in many cases will be a guardian *ad litem*, 5) the consent auditor, 6) the advocate (who carries the fiduciary responsibility of a guardian *ad litem*) and 7) a court of competent jurisdiction. It is not surprising that this proposal evoked from "the public" a massive and generally disapproving response (591).

In 1981 (569, pp. 74–76) and again in 1983 (574, pp. 23–29), the President's Commission insisted that it was necessary to promulgate final regulations in this field. The Secretary of the Department of Health and Human Services (DHHS) "responded that, while continuing to consider specific issues regarding protections for institutionalized mental patients, the Department is not intending to issue additional regulations in the near future." He provided two justifications (574, p. 26): "first, that the rules proposed by the Department in November

1978 had produced a 'lack of consensus' and, second, that the basic regulations on human subjects research adequately respond to the recommendations made by the National Commission to protect persons institutionalized as mentally disabled.''

I share the opinion of the President's Commission that this response is unacceptable and, further, that the regulations do not provide adequate guidance for the conduct of research involving those institutionalized as mentally infirm. Because it appears unlikely that there will be final regulations in the foreseeable future, I shall not detail the proposed regulations' several other deviations from the Commission's recommendations.

Reatig suggests that the tendency to excessive bureaucracy expressed in the proposed regulations reflects their authors' preoccupation with the most seriously impaired "back ward" population, the relatively few for whom such procedural protections might be necessary (591). She argues that there are many persons for whom modest levels of consent auditing could serve some useful purpose; these are generally those described as vulnerable in Chapter 4. Thus, in order to avoid stigmatization of the mentally infirm, she calls upon DHHS to issue guidelines (not regulations) suggesting to the IRB when and how it should consider at their discretion the application of additional procedural protections.

Reatig further calls for a second set of guidelines that would require special procedural protections for institutionalized persons, but only when there is reason to believe that specific individuals or groups of persons have impaired rationality or comprehension. Although this second set of guidelines would be more directive than the first, they should allow the IRB sufficient flexibility to meet the needs of particular individuals or groups.

Those who make judgments of the ethicality of conducting specific research projects in specific institutions should have a thorough knowledge of the empirical realities of those institutions. For an excellent study of decisionmaking in a university mental hospital and the extent to which it conforms to legal and ethical informed consent standards, I recommend a recent book by Lidz et al (447).

Anticipated Permanent Incompetence

Physicians can predict with a high degree of confidence that persons having certain diseases sooner or later will lose their capacities to make rational choices. Most prominent among these diseases is Alzheimer's disease or senile dementia of the Alzheimer's type (SDAT) (489). If we are to learn how to treat these diseases, research must be done on persons afflicted with them; some of this research can be done only during the stage of irrevocable incompetence. How can such research be justified? In many cases, particularly those in which there is no more than minimal risk, it seems reasonable to proceed to accord with the Commission's recommendations. One never knows, however, whether one is doing what any particular incompetent subject might have chosen. This can and should be a source of misgivings on the part of investigators, relatives, and IRB members.

In making decisions on behalf of incompetent persons, there are two approaches available to us. In the language of the law, we may use either the best interests standard, or the substituted judgment standard (279). The former entails deciding what is best for the incompetent individual. What course of action is most likely to yield the most favorable balance of harms and benefits? The latter is concerned with discovering what course of action the incompetent individual would choose if he or she were capable of making and communicating choices. For persons who never were competent, such as infants, the best interests standard seems most rational; although we may, and all too often do, attempt to guess what any particular infant might choose, such attempts seem to me to be merely embarrassing efforts to show respect for nonexistent autonomy.

On the other hand, formerly competent persons may have made statements, spoken or written, that may be used later as evidence to help vicarious decision-makers to discern what they would choose in their present circumstances. To the extent that these statements are increasingly clear and relevant to present circumstances, use of the substituted judgment standard may be increasingly warranted ethically (550, 728).

Bruce Miller, in several important articles (499–501), has explored in detail ways in which persons can assure that they will be treated in accord with their autonomous wishes when they are no longer capable of making or expressing their choices. His early work was designed to facilitate the development of living wills, documents that express the choices of individuals concerning the medical therapy they would accept or reject if they become incapable of speaking for themselves (499). Subsequently, he applied similar reasoning to the development of what I shall call research living wills, which are instruments designed to express persons' choices about being research subjects when they become incompetent (500, 501).

Miller's recommendations are grounded in his analysis of the concept of autonomy. As this concept is used in medical ethics, there are at least four senses of autonomy: 1) as free action, 2) as authenticity, 3) as effective deliberation, and 4) as moral reflection.

1. Autonomy as free action means that an action is both voluntary and intentional. It is voluntary in the sense that it is not the result of coercion or undue influence. Similarly, it is intentional if it is chosen consciously by the actor.

2. Autonomy as authenticity means that an action is consistent with the person's attitudes, values, dispositions, and life plans. Roughly, the person is acting in character. . . . For an action to be labeled "inauthentic" it has to be unusual or unexpected, relatively important in itself or its consequences, and have no apparent or proffered explanation.

3. Autonomy as effective deliberation means action taken where a person believes that he or she was in a situation calling for a decision, was aware of the alternatives and the consequences of the alternatives, evaluated both, and chose an action based upon that evaluation.

4. Autonomy as moral reflection means acceptance of the moral values one acts on. The values can be those one was dealt in the socialization process, or they can differ in small or large measure. In any case, one has reflected on these values and now accepts them as one's own (499).

Having thus analyzed the four senses of autonomy, Miller then provides a taxonomy of proxy consent consisting of six models: 1) specific authorization, 2) general authorization with instructions, 3) general authorization without instructions, 4) instructions without authorization, 5) substitute judgment, and 6) deputy judgment. In discussing these models, he uses the terms "principal" and "agent" in ways that do not conform to standard legal usage. "Principal" refers to "the person who authorizes another to act, or on whose behalf another acts; and 'agent' is the person authorized by the principal to act or who acts on behalf of the principal without the principal's authorization, but with the authority of some law or social custom" (499).

1. Specific authorization entails the appointment by a principal of "an agent to give or to deny permission to some specific procedure at some future time when the principal anticipates that he or she will not be competent."

2. General authorization with instructions entails the appointment by the principal of "an agent to either give or deny permission to any procedures at some time in the future when the principal anticipates that he or she will not be competent, and the principal gives the agent instructions to follow in making the decisions."

3. General authorization without instructions entails the appointment of an agent to make a specified type of decision without instructions for making such decisions. For example, a patient with Alzheimer's disease could authorize an agent to consent to research procedures involving no more than minimal risk and designed to develop therapy for memory loss.

4. Instructions without authorization is exemplified by the living will, which specifies circumstances under which life-sustaining procedures should be withdrawn but identifies no agent; it is addressed to "my family, my physician, my clergyman, any medical facility in whose care I happen to be and to any individual who may become responsible for my health, welfare, or affairs."

5. Substitute judgment is the model in which the principal has not provided explicit instructions and has not identified an agent. In such circumstances a close relative or friend of the principal may be able to determine what the principal would decide given a certain set of facts.

6. Deputy judgment refers to "the appointment of an agent by someone other than the principal;" it differs from substitute judgment in that the agent cannot say that the decision he or she will make is the decision the principal would make if competent. This may be "because the agent does not know the principal; does not know the principal's values and attitudes well enough to be confident about what they imply in the situation at hand; or, though the agent is familiar with the principal's values and attitudes, there is nothing in them that points to any specific decision for the matter at hand" (500). In such circumstances the agent must appeal to what appears to be in the best interests of the principal or to what a reasonable person would want.

Now let us consider the extent to which each of these models of proxy consent can be respectful of the principal's autonomy in each of its senses. It is necessary to acknowledge that none of these can be responsive to the principal's autonomy in the sense of moral reflection. This should not be a matter of grave concern for, as Miller points out, autonomy in this sense requires a self-analysis that is not of interest to most persons. It requires "awareness of alternative sets of values, commitment to a method for assessing them, and the ability to put them in place. . . . Autonomy as moral reflection is the sense of autonomy most discussed by philosophers, but it is the sense of autonomy least relevant to questions about the ethics of research" (501).

The extent to which the various models of proxy consent satisfy the relevant aspects of autonomous action is compared in the following table. For each type of proxy consent, the listing of one of the aspects of autonomy means that it is *possible* to respect it. Whether it is respected in any particular case depends upon the specifics of the situation.

1. Specific Authorization
   Free action
   Authenticity
   Effective deliberation
2. General authorization with instructions
   Free action (secondary)
   Authenticity
   Effective deliberation (secondary)
3. General authorization without instructions
   Free action (secondary)
   Authenticity

4. Instructions without authorization
   Authenticity
   Effective deliberation (secondary)
5. Substitute judgment
   Authenticity
6. Deputy judgment
   None

In the foregoing table, the expression "(secondary)" appears several times. This reflects Miller's distinction between primary and secondary senses of free action (499):

> One feature of any act of autonomous authorization is of great importance. When a principal authorizes another as an agent, then the principal makes the act of the agent an act of the principal. When the agent acts, that action is a free action of the principal in a secondary sense. . . . When one person voluntarily intends that another person act for him or her in a given situation, then the former person voluntarily intends the free action of the second person.

In the case of "specific authorization," in which the agent merely serves as the principal's messenger, autonomy is in all senses primary. By contrast, in "general authorization with instructions," autonomy as free action is secondary in that it is accomplished by the agent in accord with the principal's instructions on how to act without specifying particular actions to be chosen.

Miller argues, and I agree, that in circumstances in which research is to be done during periods of anticipated permanent incompetence, plans for proxy consent should be worked out very carefully in the early stages of the prospective subjects' diseases, at times when they are most capable of expressing their choices. Moreover, one should in general be trying to develop plans that are most respectful of prospective subjects' autonomy in its relevant senses; i.e., "specific authorization" and "general authorization with instructions." Specific authorization offers the advantage of being responsive to the prospective subject as the primary actor. However, it has the disadvantage of requiring anticipation of all procedures that might be done to serve research purposes. In my view it is best to employ a blend of the two models in which the prospective subject first makes clear statements about all procedures that can be anticipated and then provides general instructions about how the agent should go about accepting or rejecting those which have not been anticipated.

Durable Power of Attorney

The Warren Grant Magnuson Clinical Center of the NIH has established a policy for proxy consent for "research with subjects who are impaired by severe physical or mental illness'' (220). According to Miller's taxonomy, this policy is one of general authorization without instruction in which the agent is assigned durable power of attorney. The power of attorney form authorizes the agent

> . . . in the event that I become disabled or incompetent, to exercise Power of Attorney over my person for the sole purpose of providing informed consent on my behalf for participation in research protocols and routine clinical care. I understand that, unless revoked by me, the appointee shall hold the power of attorney for this purpose until completion of my participation in research at the National Institutes of Health.

This policy is on trial at the Clinical Center for 1 year beginning September 15, 1985, after which it is to be reevaluated.

The durable power of attorney is designed to accommodate the needs of patients in whom anticipated incapacity may be either permanent or temporary. Among the diagnoses specified by its authors are Alzheimer's disease, schizophrenia, manias with depression and suicidal behavior, types of aphasia, and states of partial or total coma. Also mentioned are Korsakoff's psychosis, acquired immune deficiency syndrome, and septic shock.

All protocols in which use of the durable power of attorney is intended must have such use approved by the IRB. In addition, the IRB as well as the Institute Director must be notified of each appointment of any individual to hold a durable power of attorney and the Clinical Center bioethicist must be notified before informed consent is negotiated with any holder of a durable power of attorney. When feasible, assent is required in addition to the consent of the durable power of attorney holder. It is recognized that some persons may be capable of authorizing a durable power of attorney at times when they are too impaired to negotiate a valid informed consent.

Requirements of the policy vary according to the degree of impairment of the subject as well as the level of the risk of either the research or the nonvalidated practice. If the subject is or is expected to become incapable of consent but is capable of executing a durable power of attorney and the research presents no greater than minimal risk, there are no requirements other than the aforementioned notifications. However, if the risks are more than minimal, there is an additional requirement for prior consultation with the bioethicist who must determine the appropriateness of the individual designated as the holder of the durable power of attorney.

If the subject is incapable not only of consent but also of executing a durable power of attorney and there is no greater than minimal risk, there is to be no durable power of attorney. Rather, after prior consultation with the Clinical Center bioethicist, the next-of-kin may be designated as surrogate. If such subjects are to be involved in protocols presenting greater than minimal risk, the risk must be justified by a prospect of direct benefit to the subject; in such cases, after prior consultation with the Clinical Center bioethicist, the next-of-kin must take the initiative to obtain court-appointed guardianship to provide consent.

No research or investigational therapy involving greater than minimal risk is permitted if the subjects are unable to give informed consent or execute a durable power of attorney *and* have no next-of-kin available. Exceptions to this rule are permitted in medical emergencies in which the physician, in accord with federal regulations, may employ investigational therapies (cf. Chapter 5).

If the subject is capable of executing a durable power of attorney but is unable to designate an individual to hold this power of attorney, this function may be served by the Clinical Center patient representative. In the absence of the patient representative, a Clinical Center chaplain or social worker may serve, but only those who are not assigned to the unit in which the subject is taking part in research.

The design of the Clinical Center's durable power of attorney policy reflects its authors' interpretation of Maryland State Law. In this regard, it is worth noting that other procedures may be required in other jurisdictions. At the time of this writing, many states have not taken action to establish the legal standing of the living will. Very few state statutes authorizing the living will mention the use of the durable power of attorney as a means for implementing the living will. To the best of my knowledge, the research living will with or without durable power of attorney has not been given legal recognition in any jurisdiction. Thus, although the research living will is grounded in sound ethical reasoning, it is impossible to predict the outcome of any future challenges to its legal validity.

Chapter 12

Prisoners

"Ladies and Gentlemen: You are in a place where death at random is a way of life. We have noticed that the only place in this prison that people don't die is in the research unit. Just what is it that you think you are protecting us from?" This is the statement of an inmate to representatives of the Commission during their site visit to the Jackson State Prison in Michigan on November 14, 1975.

The use of prisoners as research subjects presents two special and, perhaps, intractable ethical problems (524, pp. 5–13). 1) Are prisoners "so situated as to be able to exercise free power of choice"; that is, are they capable of a truly voluntary consent? 2) Do prisoners bear a fair share of the burdens and receive a fair share of the benefits of research?

Voluntariness

Erving Goffman (263, p. xiii) characterized the typical American prison as a " 'total institution,' a place of residence and work where a large number of like-situated individuals, cut off from the wider society for an appreciable period of time, together lead an enclosed, formally administered round of life." Activities within the total institution are designed to make the prisoner submissive, dependent, and conforming.

277

> First, all aspects of life are conducted in the same place and under the same single authority. Second, each phase of the member's daily activity is carried on in the immediate company of a large batch of others, all of whom are treated alike and required to do the same thing together. Third, all phases of the day's activities are tightly scheduled (p. 6).

The prison is designed deliberately to surround the inmate with symbols of constraint, isolation, and intimidation; these include the massive gates, thick walls, and armed guards. The inmate of a total institution is subjected to "a series of abasements, degradations, humiliations and profanations of self," designed to bring about total conformity to the environment. Eventually, "The inmate appears to take over the official or staff view of himself and tries to act out the role of the perfect inmate" (263, p. 63).

It is in this context that many commentators have argued that prisoners are not "so situated as to be able to exercise free power of choice" (28, 84). Individuals who are trying to "act out the role of the perfect inmate" are, so the argument goes, likely to do anything they presume that the prison authorities will regard as good behavior. They are not likely to refuse to cooperate with investigators who seem to them to be in close contact with prison authorities. Thus, they may agree to participate in research for the wrong reasons. Once having enrolled in a research project, they might be afraid to displease authorities by exercising their rights to withdraw, notwithstanding assurances that they may do so without prejudice.

Coercion

Although all commentators agree that coerced consent is not valid, they differ in their understandings of the concept of coercion. Cohen (142) and West (725) provide philosophical analyses of the concept of coercion. In his analysis, West distinguishes coercion from bribery (725, p. 5):

> Coercion always involves a threat, which is understood by the person coerced and is intended to alter that person's behavior. A threat makes the consequences of one's action *worse* than they would have been in the normal, expected course of events.

> Bribery is ". . . defined as the manipulation of incentives to get persons to perform a certain action. Bribery is not coercion; it does not involve a threat. A person is not coerced into performing an action if he performs it because someone has offered him something."

Cohen and West agree that bribery, unlike coercion, does not invalidate consent. Accordingly, Cohen concludes that prisoners should be permitted to consent to participate in research. West, on the other hand, concludes that the conditions that prevail in prison are such as to undermine the rationality (not the voluntariness) of the prisoner's consent; thus, until these conditions are rectified, he would restrict the involvement of prisoners in research. Among other things, he cites the limited range of options available to the prisoners as restricting their capacities to make rational choices.

Goldiamond (264, p. 14.39) analyzes the ideas of coercion and freedom in terms of the number of true options available to an individual to secure things (consequences) that he or she considers critical to his or her well-being. To the extent that the number of options that will secure the critical consequence is increased, the choice of any one of the options may be considered increasingly free. Coercion is described as the inverse of freedom; if an individual, must perform a specified function in order to secure a critical consequence, there are no meaningful options. In this condition, coercion is maximal and freedom is minimal. In this view, it seems illogical to respond to the prisoners' condition by saying that because they are insufficiently free to provide a voluntary consent we will deprive them of yet another option, that of participation in research.

Goldiamond distinguishes two types of institutional coercion. "Institutionally instigated coercion" exists when the "institution which delivers a critical consequence has set up the very conditions which make the consequence critical." This is exemplified by identifying liberty as the critical consequence and informing prisoners that if they participate in research they will become eligible for early parole. In this situation, the institution (the criminal justice system) has deprived the prisoner of liberty, establishing it as a critical consequence, and then offered to return it in exchange for a specified behavior.

"Institutionally opportune coercion" is that in which "the institution which delivers a critical consequence has not made it so. It is merely capitalizing, so to speak, on an opportunity provided by a state of nature (actual or manmade)." For example, persons having serious illnesses commonly identify recovery as a critical consequence. The means through which they can secure recovery may be totally controlled by an institution (a hospital) and its agents (health care professionals).

In both types of institutional coercion, consent must be considered suspect. However, Goldiamond (264, pp. 14.57–14.58) identifies as especially problematic consent obtained for participation in some program (research or treatment) when the consequence of participation is diminution of "institutionally-instigated coercion."

Rothman argues that it is ethically permissible to conduct a "study in nature," which he defines as one in which the researcher is a passive observer of the course of some natural process, such as a disease, that he or she is powerless to change (618). However, he distinguishes from that studies in which subjects live under conditions of overwhelming social deprivation. In his view, researchers who attempt to take advantage of the social predicament in which subjects are found become accomplices to the problem, not observers of it. Using as his examples the Tuskegee and Willowbrook studies (cf. Chapter 4), he claims that the investigators, instead of exploiting their subjects, could have invested their energy in improving their conditions. He acknowledges that the investigators had neither the immediate authority nor the funds to carry out such ameliorative efforts. However, ". . . had they brought the same zeal to providing treatment as to conducting their experiment, had they pursued foundation and government grants for treatment, they might well have succeeded."

In a very different context, Curran implicitly endorses capitalizing on institutionally opportune coercion (175). He called attention to the plans developed by the National Heart, Lung and Blood Institute to do epidemiological research on acquired immune deficiency syndrome (AIDS) by screening samples of blood donated by apparently healthy blood donors at blood banks. For various reasons, it was possible to accomplish the research objectives before the development and widespread use of a reliable screening test for AIDS. Now that such a test is available, such research would be much more difficult to justify ethically. In the language of the space program, he referred to this as a one-time-only "window of opportunity." He supported the researchers' plans to exploit this window of opportunity without exploiting the research subjects.

Quite often persons are offered a choice between the sick role and the criminal role. After their convictions for such crimes as driving while under the influence of alcohol or various illegal sexual behaviors, they may be told by the judge that they may avoid spending time in jail by agreeing to enroll in some therapeutic program. In some cases they may be referred to a treatment program in which a nonvalidated approach to treatment of their "sickness" is being evaluated. Murphy and Thomasma discuss such a case involving court-ordered therapy for exhibitionism (516). They argue that under certain conditions, which they specify, it is ethically acceptable to enroll such persons in nonvalidated therapeutic programs. First, the court mandate must not itself be coercive. In my view, their argument that in this case the court mandate was not coercive can be read in part as an endorsement of Goldiamond's position on institutionally opportune coercion. They further argue that the range of choices for therapy must not be limited to the nonvalidated therapy; prospective subjects must be assured access to some other program in which only validated therapies are employed. Thus, withdrawal from the nonvalidated therapy would not necessarily result in immediate imprisonment; rather, subjects could enroll in another program. Among their additional recommendations is that there should be a patient advocate to ensure subjects' rights. In passing it is worth noting that many persons in methadone maintenance programs in the days when they were considered experimental chose to be enrolled in these programs as an alternative to jail sentences.

Legal Status

In their review of the legal status of prisoners' informed consent, Annas et al. identified no cases in which consent to participate in research was ruled invalid on grounds that conditions of imprisonment rendered voluntary consent unattainable (23). One possible exception was *Kaimowitz v. Department of Mental Health,* a case that is probably not germane to this issue. In this case, an involuntarily detained inmate in a state mental hospital was found incapable of giving informed consent to "experimental psychosurgery" because (among other reasons) "the very nature of his incarceration diminishes the capacity to consent to psychosurgery" (95). Subsequently, the issue has been tested definitively in *Bailey v. Lally* (33). In this case, the prisoner-plaintiffs claimed "that poor prison conditions, idleness, and high level of pay relative to other prison jobs rendered their participation in medical studies coerced and in violation of their constitutional rights to due process, privacy, and protection against cruel and unusual punishment." Judge Kaufman found that:

> Plaintiffs have not proven any violations of their constitutional rights. Some persons may prefer that if society's needs require that human beings be subjects of non-dangerous, temporarily disabling, unpleasant medical experiments, such subjects should either be chosen by a lot or at least not come solely from the ranks of the socially or economically underprivileged, including prison inmates. Such preference, however, even if valid, does not add up to a presently established constitutional absolute. Accordingly, judgments will be entered for defendants (p. 225).

The Commission concluded that, owing to the closed and coercive nature of prisons, the capability of prisoners to consent is seriously curtailed; however, it is not vitiated totally. As we shall see, the Commission's recommendations are designed, in part, to mitigate the impact of the total institution on the prisoner's capacity to consent.

Motivation

Many commentators on the law and ethics of using prisoners as research subjects have expressed great concern with the motivation of the prisoners. Legal scholars are concerned that if the prisoner is motivated ". . . for improper reasons, or by forces that are coercive or that unduly influence him, his consent may be involuntary and therefore invalid" (23, p. 106). Others are concerned that if the volunteer is motivated by improper rewards, involvement of that volunteer constitutes unethical behavior on the part of both subject and investigator. For example, Wartofsky contends that putting one's health or life at risk as a research subject is ethically commendable behavior when it is done freely (711, p. 3.20). However, when this is done in exchange for money (or presumably any other material reward), the act is akin to prostitution. In this case the researcher and subject are, in effect, dehumanizing each other just as the "John" and the prostitute. He sees this as a relationship of mutual exploitation.

Wells et al. studied the reasons given by prisoners for volunteering to participate in drug research (724). The reasons given in rank order of importance are 1) do something worthwhile that will benefit other people; 2) do something important for medical science; 3) just thought is was something to try; 4) improve living conditions; 5) do something requiring courage; 6) need the money; 7) relieve the boredom of prison life; 8) receive better treatment from the officers and other correctional staff; 9) be part of a group and meet others; 10) greater chance of getting a job or position when I get out; 11) might help keep me out of trouble; 12) feel better when taking drugs (parenthetically, it should be noted that this was a mock drug study comparing the effects of two placebos, one of which was an active placebo made up of a small dose of lithium carbonate; this fact notwithstanding, the subjects assigned nearly the same importance to this reason after the study that they had before the study); 13) greater chance of getting an earlier parole; 14) get out of having to do other things; 15) close buddy had joined up; 16) get away from feared others; 17) be close to people like the doctor who will take good care of me (some prisoners who testified before the Commission asserted that volunteering for drug research was in their view the only sure way to get an adequate physical examination in prison); 18) feel more important; 19) have fewer restrictions on me for a while now or in the future; and 20) be looked up to by the other prisoners.

These reasons are very similar to those identified in prisoners by Arnold et al. (28) and Martin et al. (470). Ayd (30) observes that the motives attributed most commonly to prisoners are, with the exception of the hope for a reduction in sentence, identical to those listed for normal nonprisoner volunteers by Lasagna and von Felsinger (369), Beecher (53), and Pollin and Perlin (567).

Of the various incentives that are or might be offered to prisoners, two have been the subject of special attention; these are cash payments and early parole.

Cash Payments As noted earlier, there are those who contend that any cash payments for participation in research activities that present a risk of injury are unethical (711). On the other hand, there are those who recognize that cash payments for participation in such research activities are both customary and ethically acceptable. However, some commentators are concerned that the customary practice of paying prisoners far less than free-living volunteers represents exploitation. For example, Mitford reported that the amount of pay offered to prisoners to participate in drug trials may be three to fifteen times as great as the rate of pay for other jobs available in the prison (506). In her view, this constitutes an undue inducement to participate in research. At the same time, the amount of money drug firms pay to prisoner volunteers is approximately one-tenth that customarily paid to free-living volunteers. In her view, this constitutes an incentive to drug companies to do most of their drug testing on prisoners. It should be noted that the pay for all jobs in prisons is remarkably low. At the prisons visited by the Commission, for example, most jobs paid between $1 and $3 per day, and payments to research subjects were approximately in this range.

In the case *Bailey v. Lally*, some details of the economic arrangements were documented. For most jobs within the prison, the payment ranged from $0.63 to $1.46 per day, with the exception of those prisoners working in the laundry, who earned $2.22 per day. Prisoners in the educational program were paid $0.70 per day but were forbidden to simultaneously hold a job. About 85% of the prisoners who worked earned less than $1.10 per day. Payment for participation in medical research projects was $2.00 per day (23, pp. 205–206). The sponsor of the research paid $10.00 per prisoner per day; the $8.00 that the prisoner did not receive went to a special fund for use by the prison hospital (23, pp. 113–114).

Although the Commission made no recommendation on the matter of cash payments to the prisoner-subject, it published this comment (524, pp. 10–11):

> There are at least two considerations that must be balanced in the determination of appropriate rates for participation in research not related to the subjects' health or well-being. On the one hand, the pay offered to prisoners should not be so high, compared to other opportunities for employment within the facility, as to constitute undue inducement to participate. On the other hand, those who sponsor the research should not take economic advantage of captive populations by paying significantly less than would be necessary if nonprisoner volunteers were recruited. Fair solutions to this problem are difficult to achieve. One suggestion is that those who sponsor research pay the same rate for prisoners as they pay other volunteers, but that the amount actually going to the research subject be comparable to the rates of pay otherwise available within the facility. The difference between the two amounts could be paid into a general fund, either to subsidize the wages for all inmates within the prison, or for other purposes that benefit the prisoners or their families. Prisoners should participate in managing such a fund and in determining allocation of the monies. Another suggestion is that the difference be held in escrow and paid to each participant at the time of release or, alternatively, that it be paid directly to the prisoner's family.
>
> A requirement related to the question of appropriate remuneration for participation in research is that prisoners should be able to obtain an adequate diet, the necessities of personal hygiene, medical attention and income without recourse to participation in research.

For further discussion of due and undue inducements, see Chapter 4.

Parole   In the 1950s and earlier, prisoners were often rewarded for their services as research subjects by early parole or commutation of sentence (239). Beecher defended this practice as follows (53, pp. 70–71):

> Among the five usually accepted purposes of imprisonment—punitive, expiative, deterrent, protective of society from criminals, and reformative—the last is especially important in the present consideration. Under the parole system, a reduction of prison sentence is recognized as encouraging and rewarding good conduct and industry, and it is also allowed for exceptional bravery or fidelity in a good cause. The purpose of the use of prisoners in medical research is reformative to the prisoner and constructive in terms of the advancement of medical knowledge. It is assumed that service in a medical experiment is consonant with the parole system's statutory "good time," "merit time," "industrial credits."

Hodges and Bean, two clinical investigators, reflect their sympathy with Beecher's opinion in the face of contrary prison policy (295):

> . . . for their participation in research activities, they receive no reduction of their sentences nor any favoritism regarding paroles. We do, however, send a letter to the warden at the termination of each experiment expressing our appreciation for the inmate's participation. . . . It is possible that this letter in the prisoner's file may favorably influence the parole board.

In 1952, the House of Delegates of the American Medical Association adopted a resolution in which they took note of the fact that some prisoners who participated in research ". . . have not only received citations, but have in some instances been granted parole much sooner than would otherwise have occurred, including several individuals convicted of murder and sentenced to life imprisonment." Because they disapproved of such rewards for "persons convicted of vicious crimes," they resolved to (341, p. 1025):

> express . . . disapproval of the participation in scientific experiments of persons convicted of murder, rape, arson, kidnapping, treason, or other heinous crimes, and also urge that individuals who have lost their citizenship by due process of law be considered ineligible for meritorious or commendatory citation . . .

Freund provided a concise statement of the opinion that currently prevails (251):

> . . . experiments should not involve any promise of parole or commutation of sentence; this would be what is called in the law of confessions undue influence or duress through promise of reward, which can be as effective in overbearing the will as threats of harm. Nor should there be a pressure to conform within the prison generated by the pattern of rejecting parole applications of those who do not participate. It should not be made informally a condition of parole that one go along, be a good prisoner, and subject himself to medical experimentation.

The Commission in the commentary under Recommendation 4 stated:

> There should be effective procedures assuring that parole boards cannot take into account prisoner's participation in research, and that prisoners are made certain that

there is absolutely no relationship between research participation and determinations by their parole boards.

Nature of Research Involving Prisoners

A comprehensive survey of the nature of research involving prisoners was prepared by the Commission Staff (524). By far the most common class of research was drug studies done with no intention of providing direct health-related benefit to the subjects. Most of these studies were Phase I drug trials. Also common were drug metabolism and bioavailability studies. Research conducted at the Addiction Research Center in Lexington, Kentucky, was designed to determine the potentiality of new narcotic derivatives and analogues to produce addiction. This entailed the administration of new drugs to prisoners who are addicts who have withdrawn successfully from narcotics.

In other studies, prisoners are deliberately infected with various diseases ranging in severity from the common cold to malaria. In general, these are diseases that occur in no animals other than humans. Over the years the purposes of these studies have been to elucidate the natural history of these diseases and to develop vaccines and various therapies for them.

Another class of activities, innovative prison practices, involves manipulations of either the prison environment or the prisoner with the aim of fostering the prisoner's rehabilitation. To some extent, this reflects a highly controversial tendency to identify certain types of criminal behaviors as sicknesses for which cures might be developed. Included in this class of activities are behavior modification programs (732). As Rothman, in his history of the development of behavior modification programs in prisons, observed (617): "What we might be reluctant or unwilling to do in the name of retribution, deterrence or incapacitation, we do eagerly and enthusiastically in the name of rehabilitation."

Some examples of this class of activities are provided by the Commission (524, p. 25).

> Examples range from use of "therapeutic community" and reinforcement techniques in prison, to use of aversive conditioning (employing electric shock or drugs with unpleasant effects) in treating sex offenders or uncontrollably violent prisoners, to use of a structured tier system (token economy) in which a prisoner progresses from living conditions of severe deprivation to relative freedom and comfort as a reward for socially accepted behavior. At the extreme of research or treatment designed to change behavior were castration for sexual offenders and psychosurgery for uncontrollable violence.

Finally, in the language of the Commission's Recommendation 1, there are "Studies of the possible causes, effects, and processes of incarceration and studies of prisons as institutional structures or of prisoners as incarcerated persons. . . ." Although these studies draw largely on the techniques of the social sciences, there are also studies designed to identify cytogenetic abnormalities, to identify the ways in which infections are disseminated in crowded environments, to develop new methods of treating drug addiction, and so on.

Martin et al. investigated the proclivities of various classes of persons to volunteer as research subjects (470). They examined the likelihood of securing consent to participate in four different protocols presenting various levels of risk and discomfort. To the extent that a research proposal was perceived as presenting increasing amounts of either risk or discomfort, prisoners were most likely to volunteer, followed in descending order of probability by persons with low income, fire and police personnel, and professionals. For the project presenting the lowest degree of risk or discomfort, all four groups were highly likely to volunteer.

Considerations of Justice

In its deliberations on research involving prisoners, the Commission found particularly onerous the task of interpreting the requirements of *respect for persons* and *justice* (524, pp. 6–10). The Commission observed that the choices of prisoners in all matters except those explicitly withdrawn by law should be respected, a principle that is increasingly affirmed by the courts. Thus, it appears that the principle of *respect for persons* requires that prisoners not be deprived of the opportunity to volunteer for research. Any systematic deprivation of this freedom would be unjust as an arbitrary exclusion of one class of persons from access to benefits available to others, viz. the benefits of participation in research.

However, the Commission concluded that, although prisoners who participate in research claim that they do so freely, their freedom is compromised by the conditions of social and economic deprivation in which they live. Further, it appears that prisoners are much more likely than free-living individuals to volunteer for participation in certain types of research. Thus, the Commission concluded that the most appropriate expression of respect for prisoners consists in protection from exploitation.

The approach taken by the Commission to responding to the requirements of *justice*—the fair distribution of burdens and benefits—resembles that adopted in its recommendations on other vulnerable populations. Research designed to develop knowledge that is likely to benefit prisoners as a class of persons (Recommendation 1) is relatively easy to justify. Not only are prisoners likely to share in the benefits, there are no other populations suitable to serve as subjects for such research. Similarly, research designed to evaluate nonvalidated practices is relatively easy to justify (Recommendation 2). In such activities, the prisoner-subjects receive the benefits of the nonvalidated practices.

Recommendation 3 is concerned with research activities not covered by the first two recommendations; this class of activities is exemplified by Phase I drug studies. The Commission expresses its concern that prisoners as a group bear a disproportionate share of the burdens of research activities in this class. For example, most Phase I drug studies were done on prisoners; some studies show that most free-living individuals would decline participation in such protocols. In addition, prisoners are seldom full beneficiaries of improvements in medical care and other benefits accruing to society from such research activities.

In its concern to distribute the burdens of participation in research equitably, the Commission points out (524, p. 7) that it

> is not primarily intending to protect prisoners from the risks of research; indeed, the Commission notes that the risks of research, as compared with other kinds of occupations, may be rather small. The Commission's concern, rather, is to ensure the equitable distribution of the burdens of research no matter how large or small those burdens may be.

Recommendation 3 calls for stringent requirements that must be met in order to justify the involvement of prisoners in this class of research activity. Among other things, these requirements are designed to alleviate those conditions in prison that are detrimental to the prisoners' capacity to make free choices.

The Commission also considered issues of discriminatory treatment based upon race and sex; it made no recommendations on these matters. Many individuals told the Commission of their concern that minorities bear a disproportionate share of the burdens of research conducted in prisons. In part, this concern was based on evidence that prison populations are disproportionately nonwhite. However, evidence presented to the Commission indicated that those prisoners who participate in research are predominantly white even in those institutions in which whites are a minority in the prison population. In fact, some black prisoners protested that their disproportionately low representation reflected racial discrimination. In their view, the whites who control selection of subjects discriminated against blacks. The Commission further found that those who participated in research tend to be better educated and more frequently employed at better jobs than the prison population as a whole. This evidence suggested to the Commission that nonwhites and poor or less educated prisoners do not carry a disproportionately large share of the burdens of research.

The Commission further noted that less research is conducted in women's prisons than in men's. Owing to the possibility that they might become pregnant, women in general are employed less frequently than men as research subjects (cf. Chapter 4). However, the Commission opined that questions of distributive justice may still need to be addressed with respect to participation in research by women prisoners.

The Recommendations of the Commission

Recommendation 4 applies to all research involving prisoners.

> 4A: The head of the responsible federal department or agency should determine that the competence of the investigators and the adequacy of the research facilities involved are sufficient for the conduct of any research project. . . .

This reflects the Commission's concern that closed institutions might provide havens for incompetent or inadequately equipped investigators.

> 4B: All research involving prisoners should be reviewed by at least one . . . institutional review board comprised of men and women of diverse racial and cultural backgrounds that includes among its members prisoners or prisoner advocates and such persons as community representatives, clergy, behavioral scientists and medical personnel not associated with the conduct of the research or the penal institution; . . . the board should consider at least the following: the risks involved, provisions for obtaining informed consent, safeguards to protect individual dignity and confidentiality, procedures for the selection of subjects, and provisions for providing compensation for research-related injury.

It should be noted that the recommendations for composition of the Institutional Review Board (IRB) are substantially different from those provided in other Commission reports.

In the commentary under this recommendation, the Commission elaborates its intent as well as its concerns with issues of *justice*. The risks should be commensurate with those that would be accepted by nonprisoner volunteers. In the event the IRB is uncertain on this point, it ". . . might require that nonprisoners be included in the same project."

The Commission further advises that procedures for the selection of subjects should be fair and immune from arbitrary intervention by authorities or prisoners.

Compensation for research-related injury is discussed in Chapter 6.

> *Recommendation 1:* Studies of the possible causes, effects, and processes of incarceration and studies of prisons as institutional structures or of prisoners as incarcerated persons may be conducted or supported, provided that they present minimal or no risk and no more than mere inconvenience to the subjects. . . .

The Commission found this class of research relatively easy to justify in that it is designed to develop information that will improve the lot of prisoners as a class of persons. Although most of this research employs the devices of the social sciences (e.g., questionnaires or surveys), some may require the collection of samples of blood or urine or the performance of various physiologic measurements. In this report, the Commission did not define "minimal risk;" the definition of "mere inconvenience" is discussed in Chapter 3.

Recommendation 2: Research on practices, both innovative and accepted, which have the intent and reasonable probability of improving the health or well-being of the individual prisoner may be conducted or supported. . . .

In the commentary, the Commission expresses its concern that prisoners not be discriminated against with respect to research protocols in which a therapeutic result might be realized for the individual subject, e.g., Phase II and Phase III drug studies. However, IRBs are cautioned to analyze carefully any claim that activities are designed to improve the health or well-being of subjects, particularly when the purpose of the practice under study is to induce or enforce conformity with behavioral norms established by prison officials or even by society. We cannot assume that such conformity will improve the condition of the prisoner. If the IRB finds that such claims are not substantiated sufficiently, the research may not be conducted unless it conforms to the requirements of Recommendation 3.

Recommendation 3: Except as provided in Recommendations (1) and (2) research involving prisoners should not be conducted or supported . . . unless . . . the head of the responsible Federal department or agency has certified, after consultation with a national ethical review body, that the following three requirements are satisfied: (A) The type of research fulfills an important social and scientific need, and the reasons for involving prisoners in the type of research are compelling. . . .

The Commission provides no guidance as to what would be considered a compelling reason. It does, however, suggest that alternative populations should be utilized more fully than they are (524, p. 11); alternative populations are discussed in the next section. The fact that prisoners are easily available or administratively convenient for use in certain types of research is not considered a compelling reason.

3B: The involvement of prisoners in the type of research satisfies conditions of equity; and 3C: A high degree of voluntariness on the part of the prospective participants and openness on the part of the institution(s) to be involved would characterize the conduct of the research; minimum requirements for such voluntariness and openness include adequate living conditions, provisions for effective redress of grievances, separation of research participation from parole conditions, and public scrutiny.

In the commentary, the Commission indicates that prisoners should be able to communicate, without censorship, with persons outside the prison and, on a privileged basis, with attorneys, the accrediting office that assists the national ethical review body, the grievance committee, and the IRB. Moreover, the latter three agencies should be allowed free access to the prison.

There should be a grievance committee made up of elected prisoner representatives, prisoner advocates, and representatives of the community. This committee, in addition to enabling prisoners to obtain effective redress of their grievances, should facilitate inspections and monitoring by the accrediting office to assure compliance with the provisions of Recommendation 3C.

A list of conditions is provided to improve the living standard of the prisoners; the Commission recommends that compliance with *all* of these conditions be required:

> 1) The prison population does not exceed designed capacity, and each prisoner has an adequate amount of living space; 2) There are single occupancy cells available for those who desire them; 3) There is segregation of offenders by age, degree of violence, prior criminal record, and physical and mental health requirements; 4) There are operable cell doors, emergency exits and fire extinguishers, and compliance with state and local fire and safety codes is certified; 5) There are operable toilets and wash basins in cells; 6) There is regular access to clean and working showers; 7) Articles of personal care and clean linen are regularly issued; 8) There are adequate recreation facilities and each prisoner is allowed an adequate amount of recreation; 9) There are good quality medical facilities in the prison, adequately staffed and equipped, and approved by an outside medical accrediting organization such as the Joint Commission on Accreditation of Hospitals or a state medical society; 10) There are adequate mental health services and professional staff; 11) There is adequate opportunity for prisoners who so desire to work for remuneration comparable to that received for participation in research; 12) There is adequate opportunity for prisoners who so desire to receive education and vocational training; 13) Prisoners are afforded opportunity to communicate privately with their visitors and are permitted frequent visits; 14) There is a sufficiently large and well-trained staff to provide assurance of prisoners' safety; 15) The racial composition of the staff is reasonably concordant with that of the prisoners; 16) To the extent that it is consistent with the security needs of the prison, there should be an opportunity for inmates to lock their own cells; and 17) Conditions in the prison satisfy basic institutional environmental health, food service and nutritional standards.

Alternative Populations

As we have seen, the Commission stated its general intention to encourage the involvement as research subjects of noncaptive populations as alternatives to prisoners. In particular, it recommended that the types of research activity that comprise the majority of research involving prisoners, e.g., Phase I drug testing, should not be permitted in prisons unless there are compelling reasons to do so. Further, the Commission suggested that administrative convenience should not be considered a compelling reason. As we shall see, the regulations promulgated subsequently by the Department of Health and Human Services (DHHS) and the Food and Drug Administration (FDA) are even more restrictive with respect to this class of research activities than the Commission recommended. Thus, if these research activities are to continue, increased use of alternative populations will be necessary.

Those who speak in favor of using prisoners as subjects in Phase I drug studies point to several characteristics of the prisoner population and prison conditions that make them highly suitable for such use (511). These characteristics and conditions are virtually all in the class that the Commission would call administrative conveniences. Lasagna, a strong proponent of continuing Phase I drug studies in prisoners, concedes, "It would be silly to assert that non-prison environments could not substitute for much of the drug research now being done in prisons" (368). However, he points out that there are some types of research that it would be nearly impossible to do in alternative populations. He cited as an example the work done at the Addiction Research Center in Lexington, Kentucky, where new narcotic analogues and derivatives are tested for addiction potential in recidivist prisoners (cf. Shubin, 643). It remains to be seen whether a national ethical review body would find the reasons to continue such research compelling.

The Commission Staff provided a description of some research programs that employ alternative populations (524). These include the normal volunteer program at the National Institutes of Health (NIH) Clinical Center, which draws upon college students and members of various religious sects (182); the University of Maryland; and the Quincy Research Center in Kansas City, Missouri, under the directorship of John Arnold. Arnold discussed in detail the advantages of using alternative populations as compared with prisoners in similar research protocols (26).

In a recent interview, Arnold stated some important reasons to avoid using prisoners as subjects in Phase I drug studies (643): ". . . much of the testing . . . is not scientifically valid for free world people because of the physiological changes that take place in prisoners. . . . Prisoners are the greatest con artists in the world. . . . We have documentation of some tests at Connecticut State Prison. We've got affidavits from prisoners that, with 24 prisoners in the test, only 2 were taking the pills each week, the others were hiding them or getting rid of them. And the 2 shared their urine specimens for the week with the other 22."

Some drug companies regularly use their employees as subjects for drug studies; Meyers has surveyed the ethical and administrative problems encountered by the industry in establishing and conducting such programs (491).

Political Considerations

During its deliberations, the Commission took note of the American public's increasingly strong sentiment against the use of prisoners as research subjects (524, pp. 2–5). It recalled that the Nuremberg Code had been developed in 1949 largely as a reaction to the Nazis' outrageous treatment of prisoners in the name of research. It observed that little, if any, drug testing is conducted in foreign prisons. Indeed, an early draft of the Declaration of Helsinki included the following provision (524, p. 16.4):

> Persons retained in prisons, penitentiaries, or reformatories—being "captive groups"—should not be used as subjects of experiment; nor persons incapable of giving consent because of age, mental incapacity, or being in a position in which they are incapable of exercising the power of free choice.

Deletion of this provision from the final version of the Declaration is said to have occurred as a consequence of pressure from the United States (524, p. 16.4).

Several popular books and articles expressed strong disapproval of research involving prisoners. Perhaps the most influential was Jessica Mitford's book, *Kind and Usual Punishment,* which was published shortly before the Commission was convened (507). Mitford portrayed the involvement of prisoners as subjects in drug studies as exploitation by the drug industry, as well as by investigators and prison authorities, based largely upon economic considerations. In addition, she leveled heavy criticism against the use of various medical, surgical, and behavioral techniques designed to "cure" some criminal behaviors, particularly violent behaviors. This is exemplified by the 1973 case, *Kaimowitz v. Department of Mental Health,* in which the court found that "the very nature of his incarceration diminishes the capacity to consent to psychosurgery."

By the time the Commission began its deliberations on research involving prisoners, eight states had already outlawed their involvement as research subjects. Further, in March 1976, the Federal Bureau of Prisons forbade the use of Federal prisoners as subjects in "medical experimentation." The Department of Health, Education and Welfare's (DHEW) proposed regulations had specified very strict limitations on the use of prisoners as research subjects (181).

This political climate had its influence on the outcome of the Commission's deliberations. Indubitably, the Commission would have found it difficult to defend a permissive stance on the issue of research involving prisoners, much more difficult than defending the restrictive position it adopted.

There is yet another factor that seems to have had an influence on the Commission's recommendations (524, p. 5):

> The Commission, although acknowledging that it has neither the expertise nor the mandate for prison reform, nevertheless urges that unjust and inhumane conditions be eliminated from all prisons, whether or not research activities are conducted or contemplated.

Lasagna (368) and Levine (424) have suggested that the Commission may have confused the agenda of prison reform and protection of prisoners as research subjects. I think the Commissioners believed that if they set very high standards for prison conditions as a prerequisite for research, then researchers would develop a vested interest in improving prison conditions so that they could conduct research. There are those who agreed that this strategy might prove effective. For example, Dubler, who opposes research involving prisoners, asks us to "simply imagine the strength of a lobby that would unite medicine, correctional officers, administrators, and drug companies. It would be in the overwhelming interest of all of these groups to continue a system that would provide an unlimited supply of available, trackable, and willing subjects" (198).

I think the Commission overestimated the economic and other incentives of researchers and the drug industry to do research in prisons (424, 427). If DHHS and the FDA had promulgated regulations specifying the Commission's standards for prison conditions, I think that researchers and the drug industry would have made some minor investments in improving conditions in the best of the prisons— those that already nearly met the Commission's standards—and they would have abandoned those others that are most in need of reform, thus accomplishing nothing but removing one of the options available to prisoners for pursuing their own ends. Because the DHHS and FDA regulations did not follow the Commission's recommendation, my speculations were not and may never be tested.

Federal Regulations

DHHS published final regulations on research involving prisoners in 1978 as 45 CFR 46, subpart C (187). Section 46.306 specifies the types of research that may be conducted or supported; those types of research that are not specified in this section are forbidden. Most importantly, the regulations preclude the types of research that would have been authorized under the Commission's Recommendation 3; no provision is made for demonstrating a compelling reason to the Secretary as recommended by the Commission.

The types of research described in the Commission's Recommendation 1 are permitted (46.306a). This class of research is extended and clarified as follows (46.306a):

> Research on conditions particularly affecting prisoners as a class (for example, vaccine trials and other research on hepatitis which is much more prevalent in prisons than elsewhere; and research on social and psychological problems such as alcoholism, drug addiction and sexual assaults) provided that the study may proceed only after the Secretary has consulted with appropriate experts including experts in penology, medicine, and ethics, and published notice, in the *Federal Register*, of his intent to approve such research. . . .

"Minimal risk," one of the standards for justification of this class of research, is defined as ". . . the probability and magnitude of physical or psychological harm that is normally encountered in the daily lives, or in the routine medical, dental, or psychological examination of healthy persons." This is essentially the definition that the Commission proposed for children.

In accord with the Commission's Recommendation 2, research on practices, both innovative and accepted, may be conducted (46.306a). However, "In cases in which those studies require the assignment of prisoners . . . to control groups which may not benefit . . ." the Secretary must secure consultations and publish a notice of intent as specified in the preceding paragraph.

IRBs that review research involving prisoners must have at least one member who is either a prisoner or a prisoner representative; further, a majority of IRB members (exclusive of prisoner members) must have no association with the involved prison apart from IRB membership (46.304).

Section 46.305a prescribes duties for the IRB that are in addition to those required by general DHHS regulations. The IRB must determine that:

> (2) Any possible advantages accruing to the prisoner through . . . participation in the research . . . are not of such a magnitude that his or her ability to weigh the risks of the research against the value of such advantages in the limited choice environment of the prison is impaired;

> (3) The risks . . . are commensurate with risks that would be accepted by nonprisoner volunteers;

> (4) Procedures for the selection of subjects . . . are fair . . . and immune from arbitrary intervention by prison authorities or prisoners. Unless the principal investigator provides . . . justification in writing for following some other procedures, control subjects must be selected randomly from the group of available prisoners who meet the characteristics needed for that particular research project.

> (7) Where the Board finds there may be a need for follow-up examination or care of participants after the end of their participation, adequate provision has been made . . . taking into account the varying lengths of individual prisoners' sentences, and for informing participants of this fact.

Section 46.305 further requires that parole boards are not to consider participation in research as they make decisions regarding parole; each prisoner is to be informed in advance that research participation will have no effect on parole.

On May 30, 1980, the FDA promulgated substantially identical regulations (228). In response to FDA regulations, prisoners at the Jackson State Prison filed a lawsuit claiming that they had been deprived unconstitutionally of their liberties (211). Rather than argue the case, the FDA "stayed indefinitely the effective date" of the regulations and in December 1981 proposed new regulations (229). The new proposed regulations (which the FDA calls a reproposal) omit most, but not all, of the objectionable provisions of their predecessors. However, the reproposal would permit research not relevant to the well-being of prisoners either as individuals or as a class of persons only if the "reasons for involving prisoners are compelling." According to the FDA, the compelling reasons standard recommended by the Commission had been omitted from the original regulations because in the view of the FDA "sponsors of research could never establish a compelling need to use prisoners" (229, p. 61668). Given this opinion, it appears that the FDA intends to abolish from prisons all research other than that permitted by DHHS regulations. Mechanisms for approving other types of research including Phase I drug testing are described in the reproposal; however, it appears that the FDA does not expect any sponsor or investigator to meet the standards prescribed for justification of such types of research.

Chapter 13

The Fetus and the Embryo

In the past two decades, one of the most controversial issues in the American political arena has been the ethical and legal permissibility of abortion. Although a discussion of these matters is beyond the scope of this book, I shall comment briefly on its relevance to considerations of research involving the fetus. It is worth noting that the Commission was established by Congress in the year following the *Roe v. Wade* decision in which the United States Supreme Court established the right of women to secure abortions. Subsequently, there have been a variety of bills and Constitutional Amendments introduced in Congress the purpose of which is either to overrule or to limit the scope of the Supreme Court's decision (631). Clendinen has written a dramatic account of the impact this controversy has had on the lives of a physician who performs abortions and his patients (139).

Two provisions in the Congressional mandate (P.L. 93–348) to the Commission signal the high priority assigned by Congress to addressing the ethical dilemmas presented by proposals to do research on the fetus (221, 390). Firstly, in an act which allotted 2 years to a comprehensive investigation of all other research involving human subjects, Congress directed the Commission to report on research on the fetus within 4 months. Secondly, pending receipt of this report, Congress imposed a moratorium (its only moratorium) on the conduct or support by the Department of Health, Education, and Welfare (DHEW) of all ". . . research . . . on a living human fetus, before or after the induced abortion of such fetus, unless such research is done for purposes of assuring the survival of such fetus."

297

As a consequence of these time constraints, the Commission completed its report, *Research on the Fetus* (523), before it had an opportunity to address the general conceptual issues in its mandate. If the conceptual clarifications discussed in Chapter 1 had preceded this report, it is likely that the Commission would have developed substantially different recommendations (394).

For example, in this report the Commission used the terms therapeutic and nontherapeutic research; these were defined as follows (523, p. 6):

> *Research* refers to the systematic collection of data or observations in accordance with a designed protocol.

> *Therapeutic research* refers to research designed to improve the health condition of the research subject by prophylactic, diagnostic or treatment methods that depart from standard medical practice but hold out a reasonable expectation of success.

> *Nontherapeutic research* refers to research not designed to improve the health condition of the research subject by prophylactic, diagnostic, or treatment methods.

Because virtually all literature on the ethics and regulation of research on the fetus uses the terms "therapeutic" and "nontherapeutic" research, it will be necessary to use them in this chapter.

As noted in Chapter 1, all ethical codes, regulations, and commentaries relying on the distinction between therapeutic and nontherapeutic research contain serious errors. One of the more serious errors is displayed by inserting into the Commission's definition of research its definition of therapeutic research. There is, of course, no such thing as a "systematic collection of data or observations . . . designed to improve the health condition of a research subject . . . that departs from standard medical practice." Thus, the Commission developed recommendations for the conduct of a nonexistent set of activities. The reader should keep in mind that those who use these terms are, in general, thinking of specific procedures as they discuss more complex activities such as those described in research protocols.

There is one important problem created by the Commission's definition of therapeutic research not resolved by the Department of Health and Human Service's (DHHS) interpretation—viz. the inclusion of "diagnostic . . . methods." Investigational techniques such as chorionic villus biopsy may be performed with the purpose of meeting "the health needs of the mother and fetus." However, such techniques often are used to identify impaired fetuses for selective abortion. In such cases, the health needs of the mother and fetus may be, or may appear to be, in conflict (*infra*).

DHHS regulations on research on the fetus (Section 46, Subpart B) were promulgated shortly after the Commission published its report. The regulations reflect the concept of therapeutic research as follows: ". . . activities, the purpose of which is to meet the health needs of the mother or the fetus."

The Commission contracted with eight ethicists to prepare analyses of the complex ethical issues presented by proposals to do research on the human fetus. Their papers were published in the Appendix to the Commission's Report; excerpts of them were published in the June 1975 issue of the *Hastings Center Report*. Since the publication of the recommendations and the subsequent promulgation of the regulations, the controversy over the ethical and legal issues has continued; interested readers are referred to a symposium (678) in which various commentators assessed the Commission's recommendations shortly after their publication and to a recent survey of political activities relevant to regulation of fetal research at the state level (39).

Questions of Personhood

The fetus is, of course, totally incapable of consent. Because the requirement for consent is derived from the principle of *respect for persons*, there is considerable debate as to whether the fetus is to be considered a person. Extreme positions are reflected in the following quotations.

> An actual person, as distinguished from a potential one, is . . . both legally and ethically a human being who has left the maternal/fetal unit, is born alive, and lives entirely outside the mother's body with an independent cardiovascular system. Only the pregnant patient is a "human subject" to be protected . . . the fetus is an object, not a subject—a nonpersonal organism (219).

Louisell, on the other hand, believed that personhood begins at the moment of conception; in his dissent to the Report of the Commission (523, p. 80), he argued:

> I would, therefore, turn aside any approval, even in science's name, that would by euphemism or other verbal device, subject any unconsenting human being, born or unborn, to harmful research, even that intended to be good for society.

In Bok's view, the question of personhood had been settled by the United States Supreme Court in its two 1973 decisions, *Roe v. Wade* and *Doe v. Bolton*, in which it ruled that no state could intervene in the plans for abortion worked out between a woman and her physician (65). However, she took note of the continuing debate over whether the fetus possessed humanity. In Bok's opinion, in discussions of the fetus, the notion of humanity is a "premature ultimate," citing I.A. Richards' concept:

> The temptation to introduce premature ultimates—Beauty in Aesthetics, the Mind and its faculties in Psychology, Life in Physiology, are representative examples—is especially great for believers in Abstract Entities. The objection to such ultimates is that they bring an investigation to a dead end too suddenly.

Wasserstrom reviewed the arguments for and against four differing perceptions of the status of the human fetus; he also surveyed the consequences of adopting each of these perceptions (715). The fetus is in most if not all morally relevant respects like 1) a fully developed, adult human being, 2) a piece of tissue or a discrete human organ, or 3) an animal, such as a dog or a monkey. Alternatively, one might view the fetus as 4) being in a distinctive, relatively unique moral category, in which its status is close to but not identical with that of a typical adult. It is the last of these four perceptions that seems to have been adopted by the Commission.

What are the consequences of adopting the view that a fetus is neither a person nor an object but rather a potential person? We are to show respect for it by refraining from actions that would violate its dignity and integrity. Our actions should be directed toward fostering the well-being of each individual fetus and minimizing harm. Although we must respect the authority of a pregnant woman to have an abortion, we are not to encourage abortions in the interests of doing research. Research on fetuses should be designed to develop knowledge that would be of benefit to this class of potential persons.

Before we proceed, I wish to voice my agreement with Bok that the question of personhood, as this concept is commonly understood, is not particularly helpful. Even less relevant is the question so often raised in proposed legislation and constitutional amendments: When does life begin? There is life in a sperm, in a white blood cell, and in a fingernail. As we consider how we ought to treat the human fetus or embryo, the most constructive questions are: When does a developing human begin to acquire the entitlements of membership in the moral (human) community? When does it begin to count as one of us? When should it become enfranchised by the Fourteenth Amendment to the United States Constitution? These are metaphysical questions and, thus, are not susceptible to resolution using the devices of ethics (461). Practical answers, if any, will issue from the political process (534). An excellent collection of papers on the problems of personhood may be found in a recent special issue of the *Milbank Memorial Fund Quarterly* (676). The reader should begin with the last essay in this volume by Fox and Willis, entitled *Personhood, Medicine, and American Society* (242). These problems will be discussed further in this chapter in the section on the moral status of the embryo.

Relevance of Abortion

The Commission's Report reflects the fact that proposals to conduct nontherapeutic research either on the pregnant woman or on the fetus generated the most heated controversies (523). Proposals to do such research before, during, or after induced abortions produced irreconcilable conflicts.

There was little opposition to proposals to conduct "nontherapeutic research directed toward the pregnant woman" which would ". . . impose minimal or no risk . . . to the fetus . . ." if abortion was not being considered (Recommendation 3). However, when an abortion is anticipated, the issue is much more problematic. Many commentators expressed concern that a woman, or a pair of parents, who have chosen abortion have abandoned the fetus. Ordinarily, we call for maternal or parental consent based on the assumption that the parents love and care for the fetus and therefore will tend to protect its interests. However, if they have signaled their intention to abandon the fetus, they might too readily do things that might harm it. Therefore, it was argued, they should be disqualified from consenting to any nontherapeutic research. Opponents to this position contended that a decision to abort should not be construed as a complete loss of interest in the well-being of the fetus. It was suggested that some decisions to abort were made "in the best interests of the fetus." Moreover, for this class of research, the standard of minimal or no risk to the fetus should afford sufficient protection even to those carried by nonprotective mothers.

Proposals to conduct nontherapeutic research directed at the fetus produced still more complex ethical dilemmas. One particularly controversial issue was whether fetuses-going-to-term differed in any morally relevant manner from fetuses-about-to-be-aborted. If so, would this difference justify proposals to do research on the fetuses-about-to-be-aborted that one would not consider doing also on those going to term? Firstly, let us consider the reasons that scientists might wish to treat these two classes of fetuses differently.

There are two types of research which, if they are to be done at all, are most appropriately initiated shortly before an abortion procedure. The first is research in which the proposed objective cannot be achieved unless the fetus can be examined shortly after it is aborted. This most commonly entails administration of a drug to the mother before the abortion is begun and the examination, shortly after the abortion, of the tissues of the fetus to determine whether the drug has gotten into the fetal tissues and, if so, what effects, either good or bad, it might have had. Development of rational drug therapy for the pregnant woman and for the fetus is dependent upon the performance of such research. For example, it would be absurd to attempt to treat an infection in the fetus with a drug that did not even penetrate the fetal tissue. It is highly uncommon for drugs to produce any harm to the fetus that could not be anticipated from the results of research done before such research on the fetus is performed (394). For an elaboration of the consequences of not proceeding with research of this sort, the reader is referred to Mirkin's review (505); briefly, it tends to exacerbate and perpetuate the therapeutic orphan problem (cf. Chapter 10).

The second type of research that it seems most appropriate to do in anticipation of abortion is that in which there is some cause to suspect that some harm might befall the fetus as a consequence of the research. The sort of harm with which one is most commonly concerned is that a maneuver might induce labor prematurely. If this is done at a time when the fetus is previable, the delivered fetus will be nonviable and will die. Usually, in the early stages of such research, it is not known whether premature labor will be induced. However, during the early stages of development of diagnostic instruments that penetrate the uterine cavity, one may ordinarily assume that labor will be induced prematurely in some cases. Thus, in the early stages of development of such techniques as amniocentesis, fetoscopy, chorionic villus biopsy, and intrauterine umbilical cord blood sampling, researchers have preferred trials performed shortly before initiation of a procedure to terminate pregnancy. In the view of researchers, the greatest harm that might befall such a fetus is that its abortion might be induced an hour or perhaps a day earlier than it would otherwise. For a discussion of the consequences of not proceeding with this sort of research, the reader is referred to Mahoney's review (463).

Lebacqz has analyzed the ethical dilemmas presented by proposals to treat the fetus-to-be-aborted differently than the fetus-going-to-term (372). As she sees it, the two fetuses are similar in that they are both vulnerable subjects deserving of protection from harm; the fact that one is scheduled to die does not make it any less deserving of respect or protection. However, the locus of the disagreement is over what equal protection or similar treatment means. Lebacqz argues that similar treatment does not mean subjecting both fetuses to the same procedure, but rather putting both to equal risk.

Consider, for example, research designed to determine whether a drug administered to a pregnant woman crosses the placental barrier. If it does, it could conceivably damage the fetus. If the fetus is aborted within a few days, it is not likely that any harm will have been done. On the other hand, if the fetus is brought to term, there might be life-long disability. Therefore, Lebacqz reasons that the risk to the fetus should be calculated by multiplying the risk of harm were it to be carried to term by the probability that the woman will change her mind about the abortion and carry it to term. Because less than 1 in 100 women change their minds after they have contacted an abortion clinic the risk to a fetus-about-to-be-aborted is about 0.01 times that presented by the same procedure to a fetus-going-to-term (75–77). On this basis, Lebacqz concludes: ". . . justice requires that fetuses to be carried to term not be subjected to some experiments which might be done on fetuses scheduled for abortion."

Parenthetically, opponents to exposing the fetus-to-be-aborted to risky research procedures grounded their argument, in part, in the possibility that such exposure might deprive some women of the option of changing their minds; if they did, there would be a relatively high probability of delivery of a damaged infant. To obviate this possibility, Toulmin suggested that possibly harmful nontherapeutic interventions upon the fetus be permitted only as part of or during a single operative procedure designed to terminate in abortion (684).

Proposals to do nontherapeutic research during the abortion procedure and on the nonviable fetus *ex utero* generated another heated controversy. For the prospective subjects of these types of research, death is imminent. This fact is assured through the Commission's Recommendation 6 that no such research be performed on fetuses unless they are less than 20 weeks of gestational age; survival of such fetuses *ex utero* was unprecedented (391). Because there is no possibility of providing for these fetuses anything that we ordinarily construe as a benefit, how can we justify imposing any burden on them?

Some commentators argued that we should show respect for the wishes one might presume a dying fetus might express if it had the capacity to do so. For example, Lebacqz proposed that it would not be unreasonable to presume that a dying fetus might have an "interest" in the cause of its dying or in the development of technology that would allow others like it to survive (523, p. 87). Her proposal is similar to those offered to justify nontherapeutic research on children (McCormick (479)), on dying persons (Jonas (329)) and on vulnerable subjects generally (Chapters 3 and 4).

Assessment of the burdens presented by research to nonviable fetuses was similarly problematic. The Commission was presented with strong evidence that a fetus of less than 20 weeks of gestational age is incapable of experiencing anything at a conscious level (389); in particular, the tracts in the central nervous system that transmit experiences of pain to the cerebral cortex have not developed (708). However, for this class of research, the Commission recommended that no intrusion be made into the fetus that alters the duration of life.

Other Arguments Against Research on the Fetus   Many opponents to nontherapeutic research on the fetus expressed concern that such activities would tend to brutalize the physicians who performed such research. For example, Walters expressed concern with ". . . the possible dehumanizing effects on investigators of their performing highly invasive procedures on still living fetuses" (704, p. 8). Similarly, Toulmin referred to the ". . . fear that any relaxation in the general feelings of reverence and concern towards the tissues and remains of the dead and dying could give the color of extenuation to other forms of callousness, violence and human indifference" (684, p. 11).

There were those who argued that if nontherapeutic research on the fetus were permitted, this would motivate physician-researchers to encourage pregnant women to have abortions so that they might have a more abundant supply of research material (678). Such concerns were clearly groundless because even before the *Roe v. Wade* decision, there was no shortage of aborted fetuses on which to do research (394).

Another type of argument advanced against research on the fetus is that commonly termed "the edge of the wedge." Edge of the wedge arguments are employed commonly in ethical debates, particularly when the debator wishes to oppose a proposal that seems to threaten an important norm. Consider the norm: "One should not do research on persons without informed consent." This norm is correctly portrayed by our debator as a barrier against immoral acts. A proposal to do research on the fetus, for example, may be countered by our debator using a wedge argument. The proposal is described as the leading edge of the wedge, a sharply honed point which penetrates the normative barrier. The driving force behind the wedge—the mallet, if you will—is the demand for consistency. (Recall, for example, the formal statement of the principle of justice in Chapter 1: "Equals ought to be treated equally and unequals unequally.") The demand for consistency could cause the entire barrier to collapse; if not the entire barrier, more than intended. LeRoy Walters, for example, in his discussion of proposals to do fetal research (704, p. 9), observed that "One would also wish to inquire whether such research would set a precedent for the performance of similar procedures on other classes of human organisms—for example, on newborns who are mortally ill or comatose elderly persons."

Edge of the wedge arguments are, in my view, not good reasons to reject proposals to do research, or, for that matter, anything else. Instead, they call attention to the need to identify grounds for morally relevant distinctions, reasons to say that the subjects under discussion are not equals and that they may or ought to be treated unequally. If no morally relevant distinctions can be found, the barrier probably will collapse; then the proposal ought to be rejected. This is among the reasons that controversies over the moral status of the fetus and the embryo are so important.

Edge of the wedge arguments are also commonly called "slippery slope," "camel's nose under the tent," or "domino effect" arguments.

Therapeutic Research

"In the judgment of the Commission, therapeutic research directed toward the health care of the pregnant woman or the fetus raises little concern, provided it meets the essential requirements for research involving the fetus, and is conducted under appropriate medical and legal safeguards" (523, p. 72).

In order to reach this judgment, the Commission reasoned (523, pp. 65–66):

> In therapeutic research directed toward the fetus, the fetal subject is selected on the basis of its health condition, benefits and risks accrued to that fetus, and proxy consent is directed toward that subject's own welfare. Hence, with adequate review to assess scientific merit, prior research, the balance of risks and benefits, and the sufficiency of the consent process, such research conforms with all relevant principles and is both ethically acceptable and laudable. In view of the necessary involvement

of the woman in such research, her consent is considered mandatory; in view of the father's possible on-going responsibility, his objection is considered sufficient to veto.

Therapeutic research directed toward the pregnant woman may expose the fetus to risk for the benefit of another subject and thus is at first glance more problematic. Recognizing the woman's priority regarding her own health care, however, the Commission concludes that such research is ethically acceptable provided that the woman has been fully informed of the possible impact on the fetus and that other general requirements have been met. Protection for the fetus is further provided by requiring that research put the fetus at minimum risk consistent with the provision of health care for the woman. Moreover, therapeutic research directed toward the pregnant woman frequently benefits the fetus, though it need not necessarily do so. In view of the woman's right to privacy regarding her own health care, the Commission concludes that the informed consent of the woman is both necessary and sufficient.

In general, the Commission concludes that therapeutic research directed toward the health condition of either the fetus or the pregnant woman is, in principle, ethical. Such research benefits not only the individual woman or fetus but also women and fetuses as a class, and should therefore be encouraged actively.

Subsequent developments suggest that such a benign view of therapeutic research, particularly that involving the fetus, may no longer be warranted. As Walters has observed, recent developments in the field of fetal diagnosis and therapy have made the human fetus seem much more like a distinguishable individual patient, particularly during the second and third trimesters of pregnancy (706).

One of the most dramatic of these developments was the advent in 1981 of fetal surgery, which is the performance of surgical procedures on the fetus *in utero*. At several medical centers, surgery has been done on the fetus to ameliorate such problems as obstructive uropathy and hydrocephalus (38, 221). In July 1982, a conference entitled "Unborn: Management of the Fetus with a Correctable Congenital Defect," was convened at Santa Ynez Valley, California (281). This resulted in a report reviewing the state of the art of fetal therapeutics, including not only fetal surgery but also blood transfusions and the replacement of vitamins, nutrients, and hormones. This report announced the establishment of an international fetal treatment registry in Winnipeg, Canada and the authors requested that all experiences with fetal surgery be reported to it. Interested individuals may contact this registry for up-to-date information on the results of various procedures.

Therapeutic research directed at the fetus may be conducted with the consent of the parents (*infra*). But what if one or both parents object? Robertson has written an excellent review of the law relevant to this question (609). He points to two relevant precedents in case law. In one case, the New Jersey Supreme Court ordered blood transfusions be given to a pregnant Jehovah's Witness because, in the view of her physicians, transfusions might be necessary to save the fetus's life. According to the Court, "The unborn child is entitled to the law's protection and . . . an order should be made to ensure blood transfusions to the mother in the event that they are necessary in the opinion of the physician in charge at the time." In another case, the Georgia Supreme Court ordered that a major surgical procedure, a cesarean section, be performed on a woman against her will. Robertson concludes, "Doctors and parents acting in good faith may use an experimental *in utero* intervention on a fetus that is at great risk without it, but they are not legally or morally obligated to do so. When the therapy is established, a duty to use it may arise, and a mother's refusal of the intervention may be penalized or overridden by the courts."

Hallisey's review substantially agrees with Robertson (280). She recommends standards to be adopted by the courts, one of which is germane to our question:

> The court will be authorized to intervene against the wishes of the mother, however, if and only if (a) the recommended fetal therapy is established, proven, and would be a clear benefit to the fetus, (b) the use of the particular fetal therapy would prevent significant irreversible physical or mental impairment, (c) no less intrusive medical alternative is available to prevent the impairment, and (d) the procedure would not result in serious harm to the mother.

From these analyses, it appears that the crucial judgment lies in determining when the proposed fetal therapy should be considered established, rather than innovative or investigational.

Recently, there have been several cases in which pregnant women whose fetus's are approaching the gestational age at which they could be viable *ex utero* have been pronounced dead by brain death criteria. Dillon et al. reported two such cases (195). In their view, it is important to maintain the mother's biological function in order to enhance the chances of the fetus for survival. According to them, a fetus delivered during Week 24 of gestation has a 36% chance of survival; by Week 27, the chance of survival increases to 76%. Thus, they recommend maintenance of the woman's biological function until at least Week 28 of gestation.

In commenting on this proposal, Siegler and Wikler conclude that generally one should proceed to maintain the mother's biological function in order to maximize the chances of the fetus to survive (651). For the mother, therapy can be neither harmful nor directly beneficial. One can and should presume that the mother's only relevant interest is to give birth to a healthy infant. Veatch's analysis of the case is somewhat more complicated (701). In his view, if the physicians and the child's father agree to maintain biological function, then this should be done. However, if the father refuses, whether or not one may proceed turns on a technical point—i.e., whether the mother has been pronounced dead. In his view, if she has not been pronounced dead and the father refuses, it is likely that one could secure a court order to proceed with treatment. On the other hand, if she has been pronounced dead, the decision will be made according to the provisions of the Uniform Anatomical Gift Act; thus, further treatment would require consent from the individual having legal authority to consent. Consequently, the father, if he is the woman's husband, could veto further treatment.

Kerenyi and Chitkara reported on selective abortion of one of a pair of twins (351). In their case, one twin was normal and the other was afflicted with trisomy 21 (Down's syndrome). The latter fetus was killed by exsanguination. At 40 weeks of gestation, the woman was delivered of one normal infant and the remnants (*fetus papyraceus*) of the "terminated" fetus. In this case, the physicians went to court to secure confirmation of the parents' right to consent on behalf of the normal fetus. In commenting on this case, Somerville protests that destroying a fetus without removing it from the uterus is inconsistent with the privacy grounding used by the United States Supreme Court to establish a woman's right to abortion (662). For further discussion of the importance of the distinction between fetal removal and fetal destruction, the reader is referred to an interesting essay by Fost, Chudwin, and Wikler (235).

Placebo-controlled randomized clinical trials are, in some cases, especially difficult to justify ethically (cf. Chapter 8). Lipsett and Fletcher presented a case study illustrating how such justification may be particularly difficult in research involving the fetus (453). This case entails an assessment of the effects of folic acid, a B-complex vitamin, in preventing serious neural tube defects such as spina bifida. Preliminary studies indicate that folic acid in doses substantially higher than the estimated daily requirement for pregnant women may be effective in this regard. The evidence, however, is not conclusive; moreover, there is little information available on the adverse effects of high doses of folic acid during pregnancy.

Fletcher and Schulman point out that in addition to the usual ethical problems in justifying such studies, there are technical problems in meeting the requirements of the regulations (221):

> First an IRB would have to determine that the fetuses on the higher dosage vitamin arm of the trials meet the requirements of 46.208(a)(1) if that arm was seen to present greater than minimal risk: that is, would inclusion in this arm meet the health needs of the particular fetus? Second, an IRB would have to determine that the fetuses in the placebo arm met the requirements of 46.208(a)(2), that is, that no greater than minimal risk was involved and that important biomedical knowledge was a likely outcome which could not be obtained in another way. Exact knowledge to answer these objections is unavailable. Indeed, the trial is designed to seek (such) information. . . . Literally applied, the regulations appear to prevent a trial that seeks scientific information in the context of a *possibly* therapeutic trial. (Emphasis is in original.)

The authors recommend an interpretation of the regulations

that allows some attention to early, favorable results in uncontrolled trials but includes true uncertainty as to whether vitamin-taking around the time of conception will prove . . . (safe and effective) . . . when rigorously tested. On this view, since the trial is at least partly designed to "meet the health needs of fetuses at higher risk for neural tube defects . . . and since the trial will distribute the risks (if any) equally to all fetuses in the trial (by random chance), each particular fetus will be placed at risk only to the minimum extent necessary to meet such needs.

DHHS Regulations

DHHS Regulations (Section 46, Subpart B) on research on the fetus contain a few substantive departures from the language and intent of the Commission's Recommendations upon which they are based (201). These will be pointed out in the following discussion. Section 46.203 provides the following definitions:

Pregnancy encompasses the period of time from confirmation of implantation (through any of the presumptive signs of pregnancy, such as missed menses, or by a medically acceptable pregnancy test) until expulsion or extraction of the fetus.

Fetus means the product of conception from the time of implantation . . . until a determination is made, following expulsion or extraction of the fetus, that it is viable.

Following delivery, if a fetus is found to be either viable or possibly viable, it is referred to as an infant; the regulations require that it be treated as such. Wikler published an excellent discussion of how one ought to treat a fetus which, after an induced abortion designed to terminate its life, is found to be viable *ex utero* (730). According to the regulations, those that are found nonviable continue to be called fetuses. The confusing terminology used to describe fetuses was reviewed in detail by Levine (391). Fost et al. published a careful analysis of the concept of fetal viability and the need to change its definition from time to time to comport with the capabilities of advancing technology (235).

Section 46.204 indicates that one Ethical Advisory Board (EAB) or more will be established by the Secretary, DHHS. The EAB ". . . may establish . . . classes of applications or proposals which . . ." either must be or need not be submitted to the EAB.

The Commission recommended that certain classes of activities be conducted only after approval by a National Ethical Review Body, e.g., nontherapeutic research directed toward the fetus during the abortion procedure and that directed toward the nonviable fetus *ex utero*. This recommendation is not reflected as such in the regulations. Rather, Section 46.211 states that any applicant, with the approval of the Institutional Review Board (IRB), may request a waiver or modification of specific requirements in the regulations. The Secretary may grant such a request with the approval of the EAB. Such requests must be published in the *Federal Register* along with an announcement that public comment is invited.

The first application for DHHS support to do nontherapeutic research directed at the fetus to be submitted to the EAB was a proposal to test the safety of fetoscopy as a technique for diagnosis of sickle cell anemia and other hemoglobinopathies. The EAB approved the requested waivers of DHHS regulations and further recommended that similar waivers be granted without EAB review for subsequent applications for support of research involving fetoscopy provided certain conditions specified by the EAB are met (221).

Relying on this precedent, an IRB approved a protocol designed to develop diagnostic tests for prenatal detection of several diseases caused by inborn errors of metabolism (304). This protocol entailed performing amniocentesis on women known to be at risk for having children with such diseases; however, because the validity of these tests remained to be established by the proposed studies, the amniocentesis could not be labeled therapeutic. Among the justifications for IRB approval of this protocol articulated by Holder, the author, was: "Since she (the pregnant prospective subject) had a clear right to impose a 100% risk of death on the fetus with abortion, she must also have the authority to impose the 0.25% risk of death from miscarriage following amniocentesis."

Section 46.205 requires the IRB to perform several functions in addition to those it performs in reviewing research not involving those having limited capacities to consent. These include determinations:

> . . . that adequate consideration has been given to the manner in which potential subjects will be selected, and adequate provision has been made . . . for monitoring the actual informed consent process (e.g., through such mechanisms, when appropriate, as participation by the IRB or subject advocates in: (i) Overseeing the actual process by which individual consents . . . are secured either by approving induction of each individual . . . or verifying, perhaps through sampling, that approved procedures . . . are being followed, and (ii) monitoring the progress of the activity and intervening as necessary . . .).

The requirement that adequate provision be made for monitoring has elicited considerable discussion among IRB members. Some have voiced the opinion that this demands that some provision be made. The majority, however, seem to rely on the expression "when appropriate." They contend, and I concur, that an IRB may determine that no provision for monitoring may be appropriate for any particular protocol (cf. Chapter 5).

Section 46.206 sets forth general limitations on research involving fetuses and pregnant women. No such research may be undertaken unless:

> (1) Appropriate studies on animals and nonpregnant individuals have been completed:
> (2) except where the purpose . . . is to meet the health needs of the particular fetus,

the risk of the fetus is minimal and, in all cases, is the least possible risk for achieving the objectives of the activity; (3) Individuals engaged in the activity will have no part in: (i) Any decisions as to the timing, method, and procedures used to terminate the pregnancy, and (ii) determining the viability of the fetus at the termination of the pregnancy; and (4) No procedural changes which may cause greater than minimal risk to the fetus or the pregnant woman will be introduced into the procedure for terminating the pregnancy solely in the interest of the activity. (b) No inducements, monetary or otherwise, may be offered to terminate pregnancy for purposes of the activity.

Section 46.207 presents the requirements for research ". . . activities directed toward pregnant women as subjects.'' They may not be involved unless:

(1) The purpose of the activity is to meet the health needs of the mother and the fetus will be placed at risk only to the minimum extent necessary to meet such needs, or (2) the risk to the fetus is minimal.

Further, both . . . the mother and father (must be) legally competent and have given their informed consent. . . . The father's informed consent need not be secured if: (1) The purpose of the activity is to meet the health needs of the mother; (2) his identity or whereabouts cannot reasonably be ascertained; (3) he is not reasonably available; or (4) the pregnancy resulted from rape.

The requirements for consent of both the father and the mother are repeated verbatim in all subsequent sections governing research on the living fetus. Section 46.208 requires that:

No fetus *in utero* may be involved as a subject . . . unless: 1) The purpose of the activity is to meet the health needs of the particular fetus and the fetus will be placed at risk only to the minimum extent necessary to meet such needs, or (2) the risk to the fetus . . . is minimal and the purpose of the activity is the development of important biomedical knowledge which cannot be obtained by other means.

The requirement that the purpose of the activity be the development of important biomedical knowledge which cannot be obtained by other means applies to all research on fetuses *ex utero* as well.

Section 46.209 provides additional requirements for research directed toward fetuses *ex utero*:

(a) Until it has been ascertained whether or not a fetus *ex utero* is viable, (it) may not be involved . . . unless: (1) There will be no added risk to the fetus . . . or (2) the purpose . . . is to enhance the possibility of survival of the particular fetus to the point of viability. No nonviable fetus may be involved . . . unless: (1) Vital functions . . . will not be artifically maintained (and) (2) experimental activities which . . . would terminate the heartbeat or respiration . . . will not be employed. . . .

According to Fletcher and Schulman, a strict interpretation of Section 46.209 could be construed as forbidding the development of methods to sustain the previable fetus *ex utero* until it develops to the point of sufficient maturity for independent survival—i.e., the development of an artificial uterine environment or artificial placenta (221). In their view, ". . . the regulations go too far in an effort to prevent a recurrence of nontherapeutic research with nonviable fetuses that had no relevance to potential therapy." They recommend that in studies designed to "develop important biomedical knowledge which cannot be obtained by any other means; and which is related clearly to the development of life-saving therapy, the nonviable fetus-subject should be anesthetized; they further recommend, that the research procedures should be terminated at a specific predetermined point.

Section 46.210 specifies that research activities involving the dead fetus, fetal material, or the placenta are to be conducted in accord with applicable state or local laws. For a comprehensive discussion of the law relevant to fetal research, I recommend Holder (313, Chapter 3). Baron published a survey of state laws governing fetal research (39). This article includes an interesting case study demonstrating how such rules are made in a pluralistic society. Copies of such state statutes and regulations are provided in the 1984 volume of *Bioethics Reporter*, Legislation Section, pp. 204–239.

As we come to the close of this discussion of fetal research, it is worth noting that almost all of the efforts of the Commission and the regulation writers were devoted to protecting fetuses from nontherapeutic research presenting serious risk of injury. One wonders, then, how much research of this sort is done. A recent survey provides some perspective on this matter (221); Fletcher and Schulman reviewed 183 research projects on high risk pregnancies and fetal pathophysiology supported by the National Institute for Child Health and Development, the primary source of federal support for fetal research. Among these there were only three projects that even "approached the threshold of minimal risk;" two employed ultrasound (which theoretically may present some risk of injury, although none has ever been demonstrated) and the other involved a placebo-controlled trial of antibiotics in the treatment of genitourinary infections in pregnant women. At first glance, this seems to provide reassurance that very little research of the sort the Commission considered problematic is being done. However, this may be an underestimate. In the absence of a clearly defined federal authority to review and approve nontherapeutic research presenting more than minimal risk, it is likely that most research in this class is supported by nonfederal funds (*infra*).

In Vitro Fertilization

Subpart B of 45 CFR 46 is also concerned with *in vitro* fertilization (IVF), a subject that was not discussed by the Commission. The Regulations state only that:

> No application or proposal involving human *in vitro* fertilization may be funded . . . until [it] . . . has been reviewed by the Ethical Advisory Board and the Board has rendered advice as to its acceptability from an ethical standpoint (46.204d).

The first application for DHEW support of work on human IVF was reviewed and found ethically acceptable by the EAB in 1979 (221, 671). In its report, the EAB set forth a list of requirements for such activities (207):

> A. if the research involves human *in vitro* fertilization without embryo transfer, the following conditions [must be] satisfied: 1) the research complies with all appropriate provisions of . . . 45 CFR 46; 2) the research is designed primarily: (A) to establish the safety and efficacy of embryo transfer and (B) to obtain important scientific information toward that end not reasonably attainable by other means; 3) [requires informed consent]; 4) no embryos will be sustained *in vitro* beyond the stage normally associated with completion of implantation (14 days after fertilization); and 5) all interested parties and the general public will be advised if evidence begins to show that the procedure entails risks of abnormal offspring higher than those associated with natural human reproduction.

> B. in addition, if the research involves embryo transfer following human *in vitro* fertilization, embryo transfer will be attempted only with gametes obtained from lawfully married couples.

Following the convention established by workers in the field, I shall refer to IVF followed by embryo transfer, i.e., placing the embryo in a woman's uterus with the aim of allowing the embryo to develop normally, as IVF/ET. This convention further designates the woman from whom the ova (eggs) are obtained as the donor, and, unlike standard dictionaries, the earliest product of conception, as the embryo.

It is not difficult to understand why the Commission and the regulation writers paid so little attention to IVF. The first birth of a human infant following IVF/ET occurred in England in 1978, fully 3 years after the regulations were published. A thorough examination of IVF is beyond the scope of this book. For an excellent review of the literature through 1978, the reader is referred to Walters (705). More recently there have been several very useful and comprehensive publications. One is the proceedings of the Third World Congress on *In Vitro* Fertilization and Embryo Transfer held in Helsinki in May 1984 (633); however, many of the papers in this volume are probably beyond the scope of those who do not understand the highly technical language and concepts of this field.

Another excellent and comprehensive publication written in language accessible to the lay person was published in 1984 by the Council for Science and Society (159); this volume covers not only IVF, but also artificial insemination and gene therapy. Good recent reviews of the legal issues in this field are provided by Quigley and Andrews (579) and by Holder (313, Chapter 1). In the next sections, I shall refer to several articles which, taken together, cover the major themes in the current ethical literature.

The Moral Status of the Embryo

Crucial to an understanding of how we ought to treat the human embryo is an assessment of its moral status. In particular, at what point does the embryo become a member of the moral community entitled to the privileges and protections society accords to such members?

The arguments in this field are quite reminiscent of those concerning the moral status of the human fetus (*supra*). These arguments are surveyed by Walters (705), Macklin (461), Zaner (747), and Ozar (553). There are those who argue that full human status for all moral purposes begins at the moment of conception. It is at this point that the conceptus should "count as one of us." The next milestone that seems to have considerable support is the time at which the embryo normally implants in the endometrium (the inner wall of the uterus); implantation begins about 5½ days after fertilization and is considered complete by the end of the second week. Proponents of this milestone point out that in nature many embryos do not implant and, consequently, do not develop further. Moreover, until implantation the embryo can divide; such division usually results in the development of identical twins. Thus, Jones argues (330): "It is difficult to conceive of ensoulment as occurring prior to the establishment of the definitive number of persons resulting from the fertilization of a single egg by a single sperm." Some authors refer to the early embryo as being in a stage of preindividuality. The 14-day milestone is compatible with the position adopted by the EAB.

Recently, many commentators have focused on the time at which the embryo develops a human brain (747). They argue that we have recognized brain death as the point at which persons cease to be members of the human community. Thus, in order to be consistent, we cannot claim that the embryo which has not yet developed a human brain is entitled to such membership. Similarly, but in a different context, Chervenak et al. conclude that third trimester abortions can be justified if the fetus "is afflicted with a condition that is . . . characterized by the total or virtual absence of cognitive function . . ." (131). As an example of such a condition, they offer anencephaly. Singer and Wells argue (656):

The internationally recognized criterion for the permissibility of transplants . . . is brain death. . . . If the medical profession (and indeed the Churches) recognize, as they do, a body's lack of a functional brain as a sufficient condition for utilizing transplantable material, then this condition is clearly met by the early embryo. That is to say that the medical profession's own criterion, logically applied, should legitimate the surgical use of fetal material up to the point of brain development.

Of course, it may be objected that a brain dead individual does not have the potential to have a functional brain, whereas an early embryo does. But so do the egg and sperm, yet nobody feels guilt about failing to bring them together.

The reader cannot have failed to recognize this as the sort of argument that most concerns those who employ "edge of the wedge" defenses (*supra*).

Even more extreme positions on this point are voiced by Michael Tooley (683) and by William Walters (708). Their arguments strictly construed would lead one to conclude that the normal full-term infant is not yet eligible for membership in the human community. As Walters put it:

If rights are to be attributed on morally defensible grounds, we must base them on some morally relevant characteristics of those beings to whom the rights are attributed. Examples of such moral characteristics would be consciousness, rationality, autonomy and the other criteria of personhood. Consequently, as the embryo does not possess these criteria it is difficult to see how moral rights can be attributed to it.

But what if we agree that the decisive criterion is the presence of a human brain? According to William Walters:

The first development of the central nervous system in the embryo occurs with the formation of the neural plate . . . at 18 days of embryonic age. The neural plate becomes the neural tube at 22–23 days. . . . Three primary vesicles which will form the fore-, mid- and hind-brains develop at 28 days and at 35 days the secondary brain vesicles appear. . . . At 12 weeks of age the cerebral lobes are small and maturation of nerve cells is just beginning. In the second trimester of pregnancy the cerebral lobes increase appreciably in size but the complex folding of the cerebral cortex associated with higher cerebral functions does not begin until the last trimester of pregnancy.

In general, before nerve fiber tracts can become fully functional they must receive coverings in the form of myelin sheaths. This process of myelination begins in the spinal cord at 20 weeks gestation and continues during the first post-natal year of life. . . .

In fact, there is a good deal of evidence to suggest that the higher brain centers that are necessary for the cerebral functions responsible for manifestation of the criteria of personhood have not developed until late in the pregnancy or in the first few months of post-natal life.

Thus, if the development of a human brain is the decisive criterion, there is a large range of options available from which to choose.

Another criterion that has received widespread support is viability. This seems to be the position adopted by the United States Supreme Court in *Roe v. Wade*.

Those who claim that the embryo is fully human from the moment of conception contend that the embryo's informed consent is required to justify IVF (747). This, of course, is impossible. Consequently, these commentators argue that IVF can never be justified ethically. Parenthetically, it is worth noting that the same sort of argument has been used to oppose research involving fetuses, infants, young children, and those with severe mental incapacities. Zaner (747) endorses Edwards' response to this claim as "unrealistic in practice because it leads to total negation—even to denying a mother a sleeping pill, a cesarean section, or an amniocentesis for fear of disturbing the child." Moreover, "fetuses are not asked beforehand about their own (natural) conception or even their abortion." Regardless of the specific mode of procreation, "natural" or otherwise, a requirement for consent would forbid any pregnancy whatever. It is not only pointless to insist upon consent from embryos, fetuses, and newborns, it is not expected. I agree with Zaner in his concluding remark: "What we do expect is that others (parents, health professionals, teachers, etc.) will act responsibly on behalf of their best interests."

Similarly, I concur in Zaner's dismissal of the argument that one should not proceed with IVF/ET until the risks to the embryo are known. Such arguments are commonly used to obstruct the movement of biomedical science into new areas. In this particular case, experience over many years with IVF/ET in animals has revealed no cause for alarm. At this point, experience with IVF/ET in humans does not carry a higher than normal risk of fetal abnormality (632).

The Problem of "Spare" Embryos

By "spare" embryos I mean those resulting from IVF that for various reasons will not be reinserted into the uterus of the donor. What should be done with them? What may be done with them? At this time, these are apparently the most controversial questions in the field of IVF among those who agree that IVF may be condoned.

By definition, I have excluded from this discussion embryos produced *in vitro* for purposes of doing research on them with no intention of proceeding with ET. The Ethics Advisory Board has already decided that, under carefully circumscribed conditions, such activities are permissible. This is not to say that this decision is not controversial (274, 671). Rather, I do not now wish to discuss this issue further. Walters has provided a thorough account of the types of research that might be contemplated and the ethical problems raised by each of these types (705).

The recent introduction of cryopreservation, which is the freezing of embryos at the 8- to 16-cell stage so that they may be reinserted into the uterus at a later date, has exacerbated the problem (276). Now there are more spare embryos the fates of which must be decided.

What can be done with these spare embryos? The alternatives are these (274, 276): they can be used for research purposes, they can be implanted in the uterus of a woman other than the donor, or they can be destroyed. Workers in the field of IVF refer to destruction of frozen embryos as thawing. Judgments regarding the proper disposition of spare embryos will be highly dependent upon one's view of their moral status.

Grobstein et al. have surveyed the advantages and disadvantages of cryopreservation (276). In their view, we can be reassured by the results of extensive animal studies that freezing does not increase the risk of abnormalities in the embryo. They caution, however, that the judgment of embryo damage in human beings is complicated by the special concern about effects on the human brain that could not be detected in animal studies. At the time of this writing, there is little experience with cryopreservation of human embryos from which one can derive estimates of risk.

Embryos apparently can be stored indefinitely in the frozen state. This presents a variety of novel problems that are not susceptible to easy resolution. I shall offer just a few examples. Suppose two eggs from the same donor are fertilized simultaneously *in vitro*. Suppose further they are born as infants 30 years apart. Which one is older? Which one has priority in inheritance disputes? Grobstein et al. inquire further (276):

> To whom do such embryos "belong"—to living parents, to the estate of deceased parents, to the storage facility that maintains them, to the state? If they are not "property," what are their social relationships? Who is responsible for their welfare? Do they, under some circumstances, become wards of the state? For the first time, we are directly confronting such questions about embryos as immature entities that are capable of maturing into adults that are physically independent of their parents.

These questions may seem fantastic. Recently, however, the issue of control over the fate of frozen embryos became the subject of a dramatic public dispute.

According to *The New York Times* (June 23, 1984) an American couple named Rios went to Melbourne, Australia, where three eggs were removed from Mrs. Rios and then fertilized with sperm from an anonymous donor. The initial implantation of one of the resulting embryos failed; the other two were frozen. The couple left shortly thereafter, saying they might return later.

About 2 years later, both Mr. and Mrs. Rios died in an airplane crash. There ensued considerable public debate over the disposition of the remaining two frozen embryos. A committee was established by the Victoria State Government, headed by professor Louis Waller of the Victorian Law Reform Commission. This committee recommended destruction of the two frozen embryos.

According to *The New York Times* (October 24, 1984), this decision precipitated such a public outcry that the matter was taken up by the Upper House of Victoria's Parliament. They passed an amendment to a bill calling for an attempt to have the embryos implanted in a surrogate mother and then placed for adoption. The Attorney General of Victoria further recommended that in the future couples should be required to determine what should be done with frozen embryos in the event they die before the embryos could be implanted.

Grobstein et al. have issued a strong call for a national policy on IVF, embryo freezing and gene transfer (276). Similar appeals have been advanced by the President's Commission (576), by Fletcher and Schulman (221), and by several others.

Grobstein et al. have offered recommendations for dealing with frozen embryos while awaiting the formulation of the national policy (276). The key points in their recommendations include the following. The clinical community involved with IVF should voluntarily limit use of embryo freezing to its initial purpose, which is to circumvent infertility in patients. Freezing should be carried out only with surplus embryos obtained from a clinically justifiable laparoscopy; on thawing, embryos should be returned to the uterus of the donor, usually after an unsuccessful first attempt to transfer unfrozen embryos. Frozen embryos should be transferred to a nondonor only with the consent of the donor and the approval of an IRB or a hospital ethics committee.

The development of public policy on IVF is substantially more advanced in the United Kingdom (152) and in Australia (151, 655) than in the United States. Abramowitz (2) and Grobstein et al. (276) provide good surveys of policy developments in other countries as well as by various professional societies.

The American Fertility Society has published ethical guidelines for IVF with and without ET (12). Among the interesting features of these guidelines are that gametes and concepti are the property of the donors, who therefore have the sole right to decide their disposition, providing such disposition is within the Society's medical and ethical guidelines. Spare embryos may be used for research purposes within 14 days of fertilization. Frozen embryos may be retained no longer than the reproductive life of the female donor.

The Human Investigation Committee (HIC) (the IRB at Yale University School of Medicine) has reviewed and approved a protocol involving IVF/ET and cryopreservation of embryos. Adopting a highly conservative position, the HIC required the investigators to agree that the frozen embryos would be used for no purpose other than that originally intended, which was to circumvent infertility. Explicitly precluded were implantation of embryos in women other than the donors and the use of sperm donors other than men expected to be the social fathers. Further proscribed were such other options as attempts at cloning and research on the embryos.

The issue of ownership could not be resolved. It was decided, however, to require the parents to agree that nobody would ever be permitted to remove a frozen embryo from the premises of the Yale-New Haven Medical Center and that withdrawal of consent by either the mother or the father would be followed by destruction of the embryos. Parenthetically, the HIC acknowledged that it could not predict the outcome if these provisions were to be contested in court.

As anticipated, the most controversial issue presented by this protocol was what to do with spare embryos. First, what should be done in the event pregnancy is accomplished before all frozen embryos have been reinserted? Those who favored early destruction of spare embryos grounded their arguments in the desirability of minimizing problems associated with suspension of the time dimension (*supra*). In their view, a short period of time would limit the opportunity for parents and others to do things with the embryos that the HIC had not yet considered adequately—e.g., offer them for sale, or recruit surrogate mothers. Also, early destruction most closely approximates what occurs in nature.

Those who argued for late or no destruction of spare embryos emphasized efficiency. If the implanted embryo were lost through spontaneous abortion, one could then proceed to implant others without the need for repeat laparoscopy. Adoption of a policy of nondestruction also would facilitate having another pregnancy in the event the couple so desired and would be most respectful of the attitude held by some that membership in the human community begins at the moment of conception.

There were those who argued that the embryos should be retained long enough so that in the event a live child was born severely deformed or died shortly after birth the parents might attempt a new pregnancy with the original embryos. Opponents of this option pointed out that this could be construed as a form of insurance against an undesirable outcome. There was concern that parents might be more inclined to abandon a handicapped infant knowing that it would be relatively easy to try again. Opponents of this option argued that the institution should not do anything to encourage such an attitude or even the appearance of such an attitude.

The HIC finally decided that in the event of a successful pregnancy, spare embryos should be destroyed at 30 to 32 weeks of gestation. It is at about this time that one could be reasonably certain about the viability of the fetus.

The HIC next addressed the issue of the maximum period of time that frozen embryos could be retained while awaiting successful impregnation. Some members argued that the guidelines of the American Fertility Society that the frozen embryos should be retained as long as the mother's capability for reproduction continued should be followed. Opponents to this position again expressed concern with the problems presented by suspension of the time dimension (*supra*). The HIC decided that under no circumstances should a frozen embryo be stored for more than 2 years. In the words of the HIC's letter of approval:

> Selection of this time period reflects the following considerations. Any couple who pursues pregnancy through IVF with cryopreservation earnestly and continuously should have—within two years—come to the end of these efforts. Either they will have succeeded in bringing a pregnancy to the point at which other embryos are to be destroyed or the time will have come to concentrate on looking for reasons for the failure. Moreover, by the end of two years, such a couple probably will have used up the supply of frozen embryos. Finally, the biological consequences of very prolonged storage are unknown.

In its review of this protocol, the HIC was deliberately and self-consciously conservative. Limitation of the use of frozen embryos to reinsertion in the uterus of the donor reflected the fact that the investigators had proposed no other. The HIC explicitly precluded other uses because adequate discussion of such other uses would have prolonged greatly the review process and, thus, delayed unnecessarily the initiation of the project.

The HIC understood further that there were those who would take strong exception to its requirements regarding the disposition of spare embryos. For these reasons, in its letter of approval, it stated that it did not intend to foreclose various possibilities for all time:

> In the event the investigators wish to do any of the things that are excluded in the original protocol, they will have to request an amendment. Requests for amendment to accommodate the wishes of particular prospective parents are in order as well as requests to modify the protocol in general. For example, we are aware of the fact that some prospective parents will object for various reasons to the thawing of embryos at the specified times. If and when prospective parents mention such an objection, we trust you will mention this objection to the HIC. At this time we can consider if and how we might modify the protocol to accommodate their objections.

The Ethics Advisory Board

According to DHHS regulations, certain classes of research involving the fetus and all research involving IVF cannot be funded by DHHS without prior review and approval by the EAB. The Commission had further recommended that one class of research involving children should also be reviewed by the EAB; as noted in Chapter 10, however, the final regulations on research involving children provide for a less clearly defined system of ad hoc advisory groups.

The EAB was chartered in 1977 and first convened in 1978. Unfortunately, when its charter and funding expired on September 30, 1980, the DHHS Secretary chose not to renew it (221). Apparently, the funds used to support the EAB were necessary to fund the recently established President's Commission.

In its final report, the Presidents' Commission advanced a strong argument for reestablishing a credible national agency having a broad mandate in bioethics (576, pp. 83–86). Among the necessary functions they suggested for it are continuing review of certain categories of research involving children, fetuses, IVF, and genetic engineering. It is worth noting that in the United States from 1975 through 1983 such agencies existed virtually continously. From 1975 through 1978, there was the National Commission; from 1978 through 1980, there was the EAB; and from 1980 through March 31, 1983, there was the President's Commission. The first two agencies were charged to confine their activities to research; the third had a broader mandate to consider problems in medical practice as well as research. The absence of any such agency since 1983 has been sorely felt by the research community (221).

Fletcher and Schulman offer a convincing argument for reestablishing the EAB (221). They present in detail several unfortunate consequences of not having one. Most importantly, in their view: "Federal Regulations on fetal research are ethically sound and widely respected by investigators." However, ". . . if Federal policies on research with human subjects are understood as a moral code, it is necessary to keep their provisions under critical evaluation. Moral codes that cannot be tested and examined in the light of actual choices usually wither and die, because they lose relevance to ever new scientific questions." Several of their other concerns have been expressed earlier in this chapter. I shall close by mentioning one that has not. Because there is no clear federal mechanism for funding important advances in technology in the fields we are discussing, their development is being financed privately. After they have been developed, they are often made available only to those who can pay for them. IVF/ET is a good case in point. I agree with the conclusion of Fletcher and Schulman (221): "Families at higher genetic risk, fetuses at higher risk, pregnant women, and infertile couples are being deprived of the potential benefits of research on problems that affect their life chances. The distribution of the benefits of federally supported research has become unjust in the process and, in our view, needs correction."

Chapter 14

The Institutional Review Board

Plainly, the more rules you can invent, the less need there will be to waste time over fruitless puzzling about right and wrong. The best sort of rules are those which prohibit important, but perfectly innocent, actions, such as smoking in College courts, or walking to Madingley on Sunday without academical dress. The merit of such regulations is that, having nothing to do with right or wrong, they help to obscure these troublesome considerations in other cases, and to relieve the mind of all sense of obligation towards society.

F.M. Cornford (157, p. 10)

Evolution of the Institutional Review Board

In 1803, Thomas Percival wrote what is generally regarded as the first authoritative statement that, before proceeding with therapeutic innovation, a physician ought to consult with peers (564).

> Whenever cases occur, attended with circumstances not heretofore observed, or in which the ordinary modes of practice have been attempted without success, it is for the public good, and in a special degree advantageous for the poor (who, being the most numerous class of society, are the greatest beneficiaries of the healing art) that new remedies and new methods of chirurgical treatment should be devised. But in the accomplishment of the salutary purpose, the gentlemen of the faculty should be scrupulously and conscientiously governed by sound reason, just analogy, or well authenticated facts. And no such trials should be instituted, without a previous consultation of the physicians or surgeons, according to the nature of the case.

Not much more was said about peer review for about 150 years. The Nuremburg Code and the original Declaration of Helsinki made no mention whatever of committee or peer review; these documents placed all responsibility for the rights and welfare of subjects on the investigator. In the revision of the Declaration of Helsinki at the 29th World Medical Assembly in Tokyo in 1975, committee review is mentioned in Principle I.2:

> The design and performance of each experimental procedure involving human subjects should be clearly formulated in an experimental protocol which should be transmitted to a specially appointed independent committee for consideration, comment and guidance.

It is of interest that the authority of the committee is limited to ''consideration, comment and guidance.'' However, the Declaration of Helsinki provides for an additional general procedural protection; as discussed in Chapter 2, Principle I.8 states that the results of research conducted in violation of Helsinki's requirements should not be accepted for publication.

The first federal document requiring committee review, dated November 17, 1953, consisted of a set of guidelines entitled ''Group Consideration for Clinical Research Procedures Deviating from Accepted Medical Practice or Involving Unusual Hazard.'' These guidelines applied only to intramural research at the newly opened Clinical Center at the National Institutes of Health (NIH) (674, p. 13). Although very little is known about peer review activities in other institutions in the 1950s, it is quite clear that they were performed in at least some medical schools. In 1961, Welt reported on the results of a questionnaire he sent to each university department of medicine in the country (341, p. 889). He received responses from 66 departments; of these, 8 had procedural documents and 24 had committees to review research involving human subjects. There was a mixture of strong opinion as to whether there should be such procedural documents or committees. In 1962, a similar survey was conducted by the Law-Medicine Research Institute at Boston University (172, p. 407). Of 52 departments of medicine responding, 9 institutions had procedural documents and 5 more indicated that they were in the process of developing such documents; 22 departments had committees.

The first federal policy statement on protection of human subjects was issued by the Surgeon General of the United States Public Health Service (USPHS) on February 8, 1966. This specified that (172, p. 436 et seq.):

> No new, renewal, or continuation research or research training grant in support of clinical research and investigation involving human beings shall be awarded by the Public Health Service unless the grantee has indicated in the application the manner in which the grantee institution will provide prior review of the judgment of the principal investigator or program director by a committee of his institutional associates. This review should assure an independent determination: (1) Of the rights and welfare of the individual or individuals involved, (2) of the appropriateness of the methods used to secure informed consent, and (3) of the risks and potential medical benefits of the investigation. A description of the committee of the associates who will provide the review should be included in the application.

On July 1, 1966, a revision of this policy statement extended the requirements for prior review of research involving human beings to all USPHS grants. This revision also revoked the requirement of individual assurances of compliance and replaced it with a requirement for institution-wide assurances covering all grant proposals emanating from a single institution. Institutions that wished to use this mechanism were requested to file "general statements of assurance" of compliance with USPHS policies that were to include, among other things, a description of the review committee.

A subsequent revision of the policy statement in December 1966 notified grantee institutions that they were further responsible for assuring that investigations were ". . . in accordance with the laws of the community in which the investigations are conducted and for giving due consideration to pertinent ethical issues."

In 1968, NIH analyzed the activities of 10% of the committees operating in institutions with general statement of assurance (172, p. 443). Of 142 institutions thus examined, 73% of the committees were limited in membership to immediate peer groups; i.e., they were constituted exclusively of scientists and physicians without the addition of lawyers, clergy, or lay members. Interdisciplinary groups in which coprofessionals of the investigator were in a minority were found in only 11 institutions. A total of 20 institutions had some representatives of other professions or lay persons; of these, 18 had lawyers on the committees, 16 had lay persons, and 1 had both lawyers and clergy.

Since 1966, there have been several revisions in USPHS policy that reflect evolving federal expectations regarding the composition and duties of Institutional Review Boards (IRBs). A revision of the guidelines dated May 1, 1969, suggested for the first time that a committee composed exclusively of biomedical scientists would be inadequate to perform the functions now expected of the committee.

> . . . . [T]he committee must be composed of sufficient members of varying backgrounds to assure complete and adequate review. . . . (Further) . . . the membership should possess not only broad scientific competence to comprehend the nature of the research, but also other competencies necessary in the judgment as to the acceptability of the research in terms of institutional regulations, relevant law, standards of professional practice, and community acceptance.

The requirements for filing assurances now request specific information on each of the committee members, including "position, degrees, board certification, licensures, memberships, and other identifications of experience and competence."

On May 30, 1974, these policies and guidelines were revised and issued as Department of Health, Education, and Welfare (DHEW) regulations (45 CFR 46). The new language calling for differences in committee membership includes the following. The committee must be composed of not fewer than five persons and must include capacity to judge the proposal in terms of community attitudes. "The committee must therefore include persons whose primary concerns lie in these areas of legal, professional, and community acceptability rather than in the conduct of research, development, and service programs of the type supported by HEW." It is further required that no committee or quorum of a committee shall consist entirely of employees of the organization where the research will be conducted. Further, no committee or quorum of a committee shall be composed of a single professional or lay group.

Finally, in 1981, both the Department of Health and Human Services (DHHS) and the Food and Drug Administration (FDA) promulgated major revisions of most of their regulations that govern the protection of human research subjects. These revisions are based upon the extensive recommendations that the Commission had published 3 years earlier in its report on IRBs (527). These documents do not alter the general principles of IRB review as they had evolved over the preceeding three decades. Rather, they are concerned with some details of what the IRB is expected to accomplish and some of the procedures it must follow. The 1981 revisions dropped the distinction between general assurance and special assurance IRBs. It is now customary to refer to multiple-project assurances and single-project assurances. The latter describe IRBs designed to review only one grant or contract, often in small institutions or community hospitals; it seems unlikely that most of these will change soon to develop the more generalized competence characteristic of a multiple-project assurance IRB.

For more extensive surveys of the history of the IRB, see Curran (172) and Veatch (693).

Until 1981, a literal reading of FDA regulations supported the position that only research conducted in institutions required IRB review. The 1981 revisions required review of all FDA-regulated research regardless of where it was done. This presented a problem to many physicians who were conducting research in their private offices, usually Phase III drug trials and other similar activities. Many of these physicians had no ready access to an IRB. Thus, a new phenomenon emerged on the scene, the noninstitutional review board (NRB). For further discussion of the NRB, see Herman (290) and Meyers (492). Although there are theoretical reasons to question the validity of NRB review, I am aware of no evidence suggesting that they are not performing according to the standards expected of IRBs.

The Functions of the IRB

In the introduction to its report on IRBs, the Commission articulated the justification for and purposes of the IRB (527, pp. 1–2):

> The ethical conduct of research involving human subjects requires a balancing of society's interests in protecting the rights of subjects and in developing knowledge that can benefit the subjects or society as a whole. . . . Investigators should not have sole responsibility for determining whether research involving human subjects fulfills ethical standards. Others who are independent of the research must share this responsibility, because investigators are always in positions of potential conflict by virtue of their concern with the pursuit of knowledge as well as the welfare of the human subjects of their research.

> . . . [T]he rights of subjects should be protected by local review committees operating pursuant to Federal regulations and located in institutions where research involving human subjects is conducted. Compared to the possible alternatives of a regional or national review process, local committees have the advantage of greater familiarity with the actual conditions surrounding the conduct of research. Such committees can work closely with investigators to assure that the rights and welfare of human subjects are protected and, at the same time that the application of policies is fair to the investigators. They can contribute to the education of the research community and the public regarding the ethical conduct of research. The committees can become resource centers for information concerning ethical standards and Federal requirements and can communicate with Federal officials and other local committees about matters of common concern.

Thus, the primary purpose of the IRB is essentially as stated by the Surgeon General in 1966: to safeguard the rights and welfare of human research subjects. To this end, the IRB examines the investigators' plans to determine whether they are adequately responsive to the requirements of the six general ethical norms. The IRB frequently requires modifications in these plans that are designed to assure that the research will be conducted in accord with these ethical standards.

The attention devoted to each of the ethical norms is not uniform. Except in the very few institutions that provide compensation for research-induced injury, little attention is paid to the norm calling for compensation except to assure that prospective subjects are informed of its absence. Except when reviewing protocols designed to involve the special or other obviously vulnerable populations, the IRB spends relatively little time considering whether subjects are to be selected equitably. The responsibility for rigorous review for scientific design and investigators' competence is in the majority of cases delegated to the funding agencies (cf. Chapter 2). Most of the IRB's time is devoted to considerations of risks and hoped-for benefits and to informed consent, just as the Surgeon General directed in 1966. As I talk with IRB members from various institutions around the United States, I get the impression that the IRB focuses more of its attention on the consent form than on plans for the negotiation of informed consent; the unfortunate implications of this misdirected effort are detailed in Chapter 5. I believe that there is no more expensive or less competent redaction service available in the United States than that provided by an alarmingly large number of IRBs.

The various principles and procedures available to the IRB to accomplish its primary purpose of safeguarding the rights and welfare of human research subjects are discussed in each of the other chapters; I shall not review them here.

Most of the IRB's decisions are reached at convened meetings at which there is a quorum, as defined by regulations. Since January 1981, expedited review has been authorized by DHHS (Section 46.110); the list of approved procedures is published as a Notice following the DHHS Regulations in the Appendix.

Requirements for IRB membership functions and operations, record-keeping and so on are detailed in the regulations which are reproduced in the Appendix. Although I shall comment on some of these in passing in subsequent sections of this chapter, I shall not mention most of them. In my view, which I shall elaborate shortly, the IRB is an agent of its own institution, not a branch office of any regulatory agency. Although it must conform to federal regulations, it should develop its own set of rules and procedures that will allow it to function effectively within its own institution. Because the requirements of each institution differ, rather than attempt to develop a generally applicable guide, I offer references to descriptions of IRBs in various institutions. For descriptions of individual IRBs, see Cowan (160), Lippsett et al. (454), Brown et al. (88–90), and Cohen and Hedberg (147). For a systematic and detailed analysis of six IRBs, see Duval (199), and for an early look at the efficacy of one IRB, see Gray (271). For a large scale empirical study of many different types of IRBs, see Gray et al. (273). The activities and procedures of many IRBs are displayed regularly in the journal *IRB: A Review of Human Subjects Research*. In this journal, the case studies demonstrate how IRBs deal with problems presented by particular protocols and many articles provide procedural advice ranging from how to establish an efficient system for numbering protocols to procedures for appeals of IRB decisions.

There is one further study of IRBs that I shall mention only because it seems to have attracted the attention of persons influential in the policy arena, such as the President's Commission; this is the study reported by Goldman and Katz in 1982 (265). Substantial reasons to question the validity of their findings have been published by Brandt (81), Levine (432), and Caplan (107). Katz and Goldman have written a rebuttal to one of these criticisms (345).

Critics of IRBs often claim that the IRB is designed to protect the interests of institutions and investigators over those of subjects. Many such critics are cited in various other chapters. In general, they point to the dominance of investigators in IRB membership, the regulatory requirement for "legally effective informed consent," close ties between IRBs and hospital risk-management offices, and the like. The legal climate of the United States being what it is, it is definitely in the interests of the institution to "safeguard the rights and welfare of research subjects." In my view, the IRB should be highly attentive to the interests of institutions and investigators, as well as those of research subjects. When IRBs are functioning at their best levels, they harmonize these interests to the extent that there is no conflict.

Mashaw has written an excellent theoretical analysis of the IRB in terms of what it is expected to do, how it can go about doing it, and what limitations must be presumed (471). He concludes:

> The IRB is, after all, but one organization among many concerned with overlapping issues of propriety and competence in scientific research. If it is to do its core job well, we must live with its inevitable incompetence at other tasks. Moreover, we must also live with the rather vague regulatory standards and with the continuing inability of the Federal funding agencies to know for sure whether IRBs are functioning effectively. If we would have wise judges and paternalistic professionals, we can neither specifically direct nor objectively evaluate their behavior.

It is important to note that Mashaw uses the term "paternalistic" not in a pejorative sense, but rather in the sense of IRB members as professionals who are highly skilled in looking after the interests (rights as well as welfare) of research subjects.

For further theoretical discussion of the structure and function of IRBs, see Lackey (361) and Newton (539).

The Human Investigation Committee

I shall now describe some aspects of the structure and function of the IRB at the Yale-New Haven Medical Center; this is called the Human Investigation Committee (HIC). In common with most IRBs, we use our local "brand name" when talking with colleagues and in the community. Among other reasons, because it has been used for over 25 years, people have become familiar with it, and the generic term Institutional Review Board, coined in 1974 by Congress, conveys meaning only to the cognoscenti.

I choose the HIC as an example only because it is the IRB I know best. As noted earlier, each IRB should establish rules and procedures that facilitate its function in its own institutional setting.

The HIC is one of four IRBs at Yale University. The others are in the School of Nursing, the Veterans Administration Hospital, and the Faculty of Arts and Sciences, which includes virtually everything not covered by the first three. The HIC's jurisdiction is defined in Medical Center Policy:

> It is required by the Yale-New Haven Medical Center . . . that all research involving human subjects done by anyone on the premises of the Medical Center, or anywhere else by faculty, students, or employees of the Yale University School of Medicine or Yale-New Haven Hospital, be reviewed and approved by the HIC prior to its initiation.

Although the HIC has the authority to review all research done on the Medical Center's premises, it usually endorses approvals granted by the other three IRBs after expedited review to assure compliance with Medical Center policies.

HIC Staff

The Chairperson, the author of this book, devotes 30% of his professional effort to the HIC. The Associate Chairperson position is a full-time position. This individual shares with the Chairperson responsibility for communications with investigators and funding agencies and for identifying needs for changes in HIC Guidelines. In addition, this person performs most expedited reviews; screens student projects to determine which need HIC review; deals with complaints from investigators, subjects, and members of the informal monitoring system (*infra*); and organizing of and usually leads discussions at meetings for the continuing education of HIC members and various other groups within the institution. This is a position of great responsibility. The present incumbent, Nancy

Angoff, is known to readers of this book by virtue of frequent citations of her publications. She also holds a junior faculty position in the Department of Epidemiology and Public Health.

These two individuals are full members of the HIC. They are both authorized to sign all official documents relevant to HIC function on behalf of the University. In addition, a senior faculty member with long experience as an HIC member is designated Vice Chairperson. In the absence of the Chairperson, he assumes his duties (e.g., presiding over HIC meetings) and authority to sign official documents, grant interim approval, and so on.

The Administrative Assistant is not an HIC Member. This individual performs the functions of secretary, receptionist, and keeper of the files and log book. Because she also serves in these capacities in relation to the Chairperson's other professional activities (the 70% of his time not devoted to the HIC), it is impossible to calculate confidently the proportion of her effort devoted to the HIC.

Membership on the HIC

Individuals are appointed to membership by the Dean from a list of candidates recommended by the Medical School Council. The Medical School Council is a governance group of elected representatives of various constituencies within the institution. Each year members of the Medical School community are asked to nominate themselves for membership on various committees. From this list of volunteers, the Council selects nominees for each open committee position.

The Chairperson keeps the Council informed of the HIC's requirements for individuals with specific capabilities. If no person having the required capabilities volunteers, such persons may be recruited through the joint efforts of the Council and the HIC. No attempt is made to have all departments represented. Rather, when seeking scientific expertise, an attempt is made to have adequate representation in those research disciplines that present the greatest problems to the HIC. Thus, for example, a person having suitable expertise in oncology might be appointed from the Departments of Medicine, Pediatrics, Surgery, Pharmacology, or Therapeutic Radiology.

The term of appointment to membership is 3 years. A member may be asked to serve a second term if he or she represents an area of expertise not duplicated by others in the medical school community. Members have been asked to leave the committee before their terms are completed owing to either poor attendance at meetings or other manifestations of unwillingness or incapability to serve the committee adequately. No member has ever been asked to resign because of holding an unpopular position relevant to the purposes of the committee.

Students are selected by the Student Council, the procedures of which are less constant from year to year than those of the Medical School Council. Each year students are requested to inform the Student Council of their wish to serve on committees. In view of the time commitment that HIC membership entails, I am surprised that so many students covet this assignment.

The committee currently consists of 26 persons, of whom 15 are full-time medical school faculty; of these, 13 have M.D. degrees, 6 Ph.D. degrees, 1 a M.Div. degree, 1 LL.M., 1 M.P.H., 1 M.S., and 1 M.Ed. Many members have multiple degrees. The major responsibilities of these 15 are distributed as follows: teaching house staff, students, or both, 15; patient care, 12; patient service functions other than clinical care, 2; active research programs, 12; and major administrative responsibilities, 8. Of these 15, 11 have tenure. The other 11 committee members include a staff nurse in the Children's Clinical Research Center; a clinical pharmacist from the Hospital's Drug Information Service; a grants and contracts administrator; a consultant for minority affairs to the Medical School and Hospital; the Hospital's Medical Director of Quality Assurance; a psychological counselor who, except for IRB membership, has no affiliation with Yale; 4 medical students, 1 from each class; and 1 student from the School of Public Health. No member receives supplementary pay for service on the HIC except, of course, the one who has no other affiliation with the institution. Pay for that position is equal to that paid by NIH for service on its advisory committees (e.g., study sections). Like NIH, there is no payment for preparation time; unlike NIH, there is no reimbursement for personal expenses. Although one might question the propriety of not paying students who, of course, pay tuition to the University, one was recently quoted in a newspaper as stating "being on this committee has been my most worthwhile experience as a student."

Committee Meetings

The HIC holds its regular meetings twice each month. Each meeting lasts 2 to 3 hours. The purpose of these meetings is to review and to take action on specific protocols; these include new protocols and periodic review of and proposed amendments to approved protocols. In addition, special meetings are called as necessary to consider new policies, procedures, or regulations.

The requirements for a quorum necessary to conduct business are in accord with DHHS regulations (Section 46.108). No meeting has ever been canceled owing to lack of a quorum.

Approximately 18 years ago, the committee began to enlarge its membership so that various types of expertise and viewpoints other than what might be reasonably expected of biomedical researchers and physicians would be represented systematically. Although the committee had rarely voted on a substantive issue, it was assumed that such votes might become necessary. It was further recognized that it would never be possible to balance the various presumed vested interests without developing a committee that would be much too large to conduct meaningful business. In order to have the necessary scientific expertise available, it would always be necessary to have scientists as the majority of the committee. Thus, the following rule of procedure for voting was developed. All scientists together had one vote. All students together had one vote. The chaplain had one vote, and so on. Subsequent experience indicated that the assumptions upon which this rule was based did not obtain. Accordingly, in 1979 this rule was repealed; now each member has one vote.

We found that differences of opinion among committee members are virtually never determined by profession, discipline, or presumed constituency. The physician-investigator members of the committee are as likely to be diverse in their opinions as are the others. It is quite typical to find that on any given issue on which there are two possible positions (e.g., yes or no), those favoring the yes decision will include several scientists, some students, and several others, while the opposition is likely to be composed of a similar mixture of members. It is further typical that such differences can be resolved through discussion, explanation, or compromise. In fact, the committee rarely votes on substantive issues. Voting is usually reserved for purposes of terminating trivial discussions. The committee always attempts to achieve consensus on substantive issues and almost invariably succeeds. As I recall, the committee has voted on whether or not to approve a protocol less than twice yearly; I can recall only one occasion where more than two members voted contrary to the majority. Thus, even when it is necessary to resort to a vote, the actions of the committee represent near consensus. In order to satisfy regulatory requirements, we record votes that are not unanimous; our policy states that where votes are unrecorded, this signifies unanimity.

HIC meetings are open to the public; any person who asks to observe a meeting is permitted to do so. Although the guidelines circulated to the faculty state that regular meetings are held at a consistent time on the second and fourth Wednesday of each month, there is no public announcement of each meeting and no announcement whatever of special meetings or meetings of subcommittees. Over the years, observers have included visiting scholars; graduate students in law, sociology, divinity, and other disciplines; various members of executive and legislative branches of state government, and reporters from television and newspapers. Observers are not permitted to participate in discussions or ask questions during the meetings.

IRBs must be alert to the possibility that a protocol might be of great interest to the media. Such was the case with "Baby Fae" and with the first total artificial heart implantation in Utah. According to spokespersons for these IRBs, the presence of journalists during IRB review of these protocols was highly disruptive (640, 740). Awareness of these reports caused the HIC to close the meetings in which it discussed *in vitro* fertilization with cryopreservation of the embryo (cf. Chapter 13). For a good review of circumstances under which IRBs can or should close their meetings to the public, see Christakis and Panner (136).

In ordinary circumstances, the only people whose presence we discourage are the investigators whose protocols we are reviewing. In our early years, we invited them to present their plans or provide clarifications. This policy was abandoned as we learned of the impossibility of constraining most scientists from telling far more than one wishes to know about their interests. Investigators who are HIC members are required to leave the meeting room during review of their protocols, in accord with DHHS Regulations (Section 46.107e).

All discussions at the meeting presuppose that all members have read the protocols. Comments that betray ignorance of what is described clearly in the protocol tend to evoke various expressions of social disapproval. Requests for clarification of the technical terms and recondite ratiocinations offered by some scientists, however, are treated with respect.

The discussion of each protocol is begun by the primary reviewer, who is then followed by the second reviewer (these agents will be defined shortly). After they speak, other members may volunteer their views. Members who know they will be absent are expected to write to the Chairperson if they have any recommendations for revision; these recommendations are read at the meeting.

The review of each protocol culminates in its assignment to one of four categories; this information is transmitted to the investigators in the form of a letter from the Chairperson. The letters are written in polite tones which non-members of the institution may fail to recognize as ethical objections (345); members of the institution, by contrast, recognize that the IRB can take a firm position without resort to abrasive language.

The protocol is assigned to one of four categories, as follows.

1. *Approved*. This means that no further action is required and the investigator may proceed with the research.

2. *Approved contingent upon specific revisions*. This means that the committee has identified certain problems in either the protocol or consent form and that it has sufficient information to understand what specific revisions will make the protocol acceptable. For example, there may be a recommendation for a specified sentence in the consent form. The majority of protocols reviewed by the committee are classified in this category after initial review. Such protocols are revised by the investigator and returned to the HIC. If the specified revisions are made, the Associate Chairperson grants approval without further consulting the full committee. Letters to investigators may also include suggestions that investigators may accept of reject as they wish. Approval of a protocol is not contingent upon acceptance of suggestions.

3. *Tabled*. Any member of the committee may move to table a protocol after

its discussion has occupied more than 10 minutes. A motion to table must include a plan for subsequent action. There are three types of plans for subsequent action.

A. The investigator may be called upon to supplement the information contained in the protocol. This plan is followed if the committee decides that the cause of its inability to reach a decision is lack of information. Ordinarily, when this plan is invoked, there is agreement among the members that if this information is made available it will be possible to make a definitive decision. In this case, one member of the committee is assigned to talk with the investigator with the aim of having the necessary information made available to the committee before the meeting at which the protocol will be reconsidered.

B. The committee may agree that it has adequate information, but there are still irreconcilable differences of opinion. In this case, a subcommittee is appointed for purposes of discussing the problems further and, if necessary, negotiating some compromise. The subcommittee is appointed by the Chairperson; it ordinarily includes the most forceful exponents of the opposing points of view, as well as one or two members who are most expert in the field of research being considered. If the subcommittee finds it necessary to recommend revised procedures, it is expected to consult with the investigators to learn whether these procedures will be acceptable to them. After the subcommittee has completed its business, it reports back to a meeting of the full committee which, based upon the subcommittee's report, almost always is able to achieve consensus.

C. The committee may feel that it does not have sufficient expertise to assess some aspects of the protocol. In this case, one or more members are appointed to find a suitable consultant and to make his or her views available to the full committee at a subsequent regular meeting. Commonly, the impetus for tabling discussion of a protocol involves a combination of two or three of the above-mentioned problems.

4. *Disapproved.* This is a category that is in theory available to the HIC. However, the committee nearly never labels a protocol disapproved. The way disapproval is accomplished *de facto* is that requirements for revision are made that the investigator chooses, for any of a large variety of reasons, not to accept. At the end of each academic year, a memorandum is sent to investigators who have protocols pending and who have not taken any action that might lead to the development of an approvable protocol. The investigators are asked either to take such action or to withdraw the protocol from consideration. Thus, the ultimate disposition of these protocols is to place them in a category known as "withdrawn."

Some perspective on the frequency of use of these various categories may be provided by considering the following information (465). In the academic year

1972–1973, the HIC reviewed 145 new protocols. Of these, 24 were approved without change on first review and 16 were ultimately classified as withdrawn. Most of the remainder were initially classified in Category 2 (with a small number in Category 3), and all were ultimately approved.

We have not analyzed systematically our experience subsequently. In the academic year 1984–1985, the number of new protocols reviewed by the HIC rose to 320; owing to the availability of expedited review procedures, the number apparently approved without change on first review has increased; this reflects the fact that the Associate Chairperson often serves as both primary and expedited reviewer. However, protocols reviewed at convened meetings have fates similar to those reviewed in 1972–1973 at approximately the same rates. Chlebowski reports from UCLA School of Medicine a very similar experience in 1982–1983 (133). Of 168 new protocols, 27.9% were approved at first review "without comment," 64.4% "with comment," 7.7% "deferred," and one "disapproved." These classifications are roughly analogous to our Categories 1 through 4, respectively.

Procedures for Preparation and Review of Protocols

Each member of the Medical Center is provided with a document entitled "Guidelines for preparation of protocols for review by the Human Investigation Committee (HIC)." This contains detailed instructions for the preparation of protocols and consent forms. The investigator initiates the process by drafting a protocol and consent form. At that point he or she selects a primary reviewer:

> . . . whose field of expertise is closest to the designed research . . . if the HIC member who is asked to be primary reviewer is overburdened with requests or cannot give early consideration to the task, he or she or the chairperson of the HIC will be responsible for suggesting another member for this review. The reviewer will study the protocol and advise the investigator about revisions if necessary. The reviewer's suggestions are not binding, nor does this preliminary review insure immediate approval by the HIC, but it will promote more efficient use of the HIC to have obvious revisions made before the whole committee acts on a protocol.

After preliminary negotiations between the investigator and the primary reviewer result in a protocol suitable for the committee's consideration, the investigator presents the protocol to the chairperson of each department involved in the research for their signatures. The purpose of these signatures is to assure that each chairperson is afforded an opportunity to keep aware of what research is being conducted in the department. Some departments have their own committees which review protocols before the chairperson will sign them; this is not required by the School.

Next, a sufficient number of copies (30) are made and delivered to the HIC. Those that are received 8 days before the next scheduled regular meeting are circulated to HIC members for consideration at that meeting. Those that are received later are circulated for consideration at the next subsequent meeting.

Before they are circulated, the Associate Chairperson assigns a second reviewer. If the primary reviewer is expert in the field of study, the second reviewer is ordinarily a member having no scientific expertise or a student. In the event two types of expertise are required (e.g., oncology and cardiology), the second reviewer is selected accordingly. If the primary reviewer is (e.g.) a student, lawyer or chaplain, the second reviewer will be a scientist whose expertise is close to the field of study. Students are selected as primary reviewers for about 25% of protocols because they have a well-deserved reputation for diligent performance of this duty. When the job of primary reviewer is done well, all concerned benefit; the HIC usually has little further to add to the protocol and the investigators tend to receive prompt approval. I have published a more complete account of the primary reviewer system (421). Screening procedures having similar objectives but differing designs have been described by Cohen (144) and Blake (59).

Information Available for HIC Review The information is presented to the HIC on standard forms; the forms have four sections.

I. Face sheet. The face sheet includes the title of the project, the name(s) of investigator(s), telephone number and address of investigator responsible for correspondence, signatures of departmental chairman, and spaces for annotations of the committee actions and the dates on which they are accomplished.

II. Description of study. Information is provided under the following headings:

A. A brief statement of the purpose of the study including the hypothesis to be tested.

B. A brief synopsis of the background information and data that provide the rationale for the research. The investigator is asked to provide sufficient information so that the rationale is clear without reference to other materials; however, references to the literature may be cited to provide interested reviewers with additional information. If it is proposed to use a new drug, an old drug for a new purpose, or a regulated device, its FDA status is to be provided, along with required statements on Investigational New Drug or Device Exemptions.

C. The location of the study. Specific information is provided as to exactly which in- or out-patient facility, community, or institution will be the location of various components of the research.

D. Duration of the project. In this section there is an estimate of how long it will take to complete the entire research project. In addition, there is specified the period for which HIC approval is being requested. The HIC ordinarily will not approve projects for periods of greater than 2 years. Many agencies (e.g., DHHS and FDA) require that periods of approval not exceed 1 year.

E. Research plan. This section contains an orderly account of the proposed procedures as they directly affect the subject. To quote from the Guidelines:

> There need not be a detailed account of techniques that do not directly affect the human subject (i.e., details of *in vitro* studies). Include number and estimated length of hospitalizations, length of time for various procedures and frequency of repetition, any manipulation which may cause discomfort or inconvenience, doses and routes of administration of drugs, amount of blood to be withdrawn and plans for follow-up. If there is a point at which study procedures would be discontinued, state how they will be identified. Include measures which will be taken to overcome side effects. If letters to subjects or intermediaries, questionnaires or rating scales are to be used, append copies.

F. Economic considerations. This section provides a description of any inducements that will be offered to subjects in return for their participation, e.g., direct payment, free hospitalization, medical care, medication, food, tests, and so on. In addition, there must be a statement of costs that the subjects or their insurance carriers will be expected to assume.

G. This is a section which is to be completed only if one of the Clinical Research Centers (CRC) is to be used for the study. In this case, the following additional information is provided: an identification of which CRC is to be used, the approximate number of subjects to be admitted, an estimate of the length of stay for each subject, a statement as to why the research must be conducted in a CRC, and the names and telephone numbers of two responsible physicians.

III. Human subjects information. As a matter of convenience to investigators, this section is designed to conform to Section 2E of the United States Public Health Service Grant Application Form (PHS–398). When appropriate, this information may be typed on continuation pages provided by the USPHS or other funding agencies. The HIC requires this information whether or not the funding agency also requires it. Here are the USPHS instructions for completing this section.

> 1. Describe the characteristics of the subject population, such as their anticipated number, age ranges, sex, ethnic background, and health status. Identify the criteria for inclusion or exclusion. Explain the rationale for the use of special classes of subjects, such as fetuses, pregnant women, children, institutionalized mentally disabled, prisoners, or others who are likely to be vulnerable.
>
> 2. Identify the sources of research material obtained from individually identifiable living human subjects in the form of specimens, records, or data. Indicate whether the material or data will be obtained specifically for research purposes or whether use will be made of existing specimens, records, or data.
>
> 3. Describe plans for the recruitment of subjects and the consent procedures to be followed, including the circumstances under which consent will be sought and obtained, who will seek it, the nature of the information to be provided to prospective

subjects and the method of documenting consent. State if the institutional review board has authorized a modification or waiver of the elements of consent or the requirement for documentation of consent. The consent form, which must have institutional review board approval, should be submitted to the PHS only on request.

4. Describe any potential risks—physical, psychological, social, legal, or other—and assess their likelihood and seriousness. Where appropriate, describe alternative treatments and procedures that might be advantageous to the subjects.

5. Describe the procedures for protecting against or minimizing any potential risks, including risks to confidentiality, and assess their likely effectiveness. Where appropriate, discuss provisions for insuring necessary medical or professional intervention in the event of adverse effects to the subjects. Also, where appropriate, describe the provisions for monitoring the data collected to insure the safety of subjects.

6. Discuss why the risks to subjects are reasonable in relation to the anticipated benefits to subjects and in relation to the importance of the knowledge that may reasonably be expected to result.

IV. The consent form. The sorts of information called for by the HIC in the consent form are discussed in detail in Chapter 5.

If the investigator thinks it would be appropriate to negotiate consent orally and not document it in writing, he or she is asked to request a waiver of the requirement for documentation. In this case, he or she is asked to record on the consent form the information that will be provided to the prospective subject during the consent negotiations.

Procedures After Initial Approval of Protocol

Periodic Review. The HIC approves protocols for periods not exceeding 2 years. Many agencies including DHHS and FDA require review at 1-year intervals. When appropriate, the HIC may specify even shorter periods of approval. It is the responsibility of the investigator to initiate the process of periodic review using a form provided by the HIC. This form provides space for a brief summary of the experience of the period since approval and a statement by the investigator as to any requests to modify the protocol or consent plans based on that experience.

Four copies of this form are sent to the Associate Chairperson, who reviews the request, consulting as necessary with the Chairperson and the primary or secondary reviewer. If there are no substantive changes in the protocol, in the field of study, or in relevant HIC policy, reapproval is accomplished by expedited review unless, of course, there are applicable regulations requiring reapproval at a convened meeting. In such cases, renewal applications are presented orally at a meeting. If there are substantive changes or if any IRB member so requests, the investigator is asked to prepare 30 copies of a document that includes the necessary additions or changes for review at a HIC meeting.

Adkinson et al. recommend the use of validation stamps on consent forms (9); these stamps indicate the date on which the forms become invalid, usually 1 year after IRB approval. The HIC has accepted this suggestion in simplified form. We find it serves as a useful reminder to investigators to seek reapproval when necessary.

Amendments to Approved Protocols   The procedures for requesting amendments to approved protocols are essentially the same as those for periodic review of approved protocols. Depending upon whether the proposed amendments are substantive, the review may be either expedited or at a convened meeting (DHHS, Section 46.110b).

Reporting of Adverse Reactions   The HIC requires that certain types of adverse reactions or other untoward incidents that develop in the course of conducting research according to approved protocols be reported promptly. These include unanticipated (in that they were not mentioned in the consent form or information sheet) occurrences of physical or psychological harm and unexpected threats to privacy (e.g., lost records) or safety (e.g., contaminated intravenous solutions) of subjects. Adverse consequences of investigational therapy should be reported only if they were either unanticipated in the consent form or if the original description in the protocol underestimated substantially their probability or magnitude. Anticipated adverse consequences of research procedures must be reported if they are of sufficient magnitude to require medical therapy or to prolong hospitalization.

Hospital policy requires reporting of various additional types of adverse events to the Legal Office which shares relevant information with the IRB. For example, reports are required of all adverse events for which it could be claimed that they occurred as a consequence of negligence on the part of investigators or the institution.

Upon receipt of a report, the Chairperson assigns a HIC member, usually the primary reviewer, to look into the situation and to report to the Committee. If the reaction or incident is serious, the investigator may be requested to discontinue the study pending HIC review. Some agencies require the IRB to report some types of adverse events; e.g., DHHS requires "prompt reporting to the IRB and to the Secretary of unanticipated problems involving risks to subjects or others" (Section 46.103b4). The form provided by the HIC for reporting adverse reactions requests the name and address of individuals and agencies to whom such notification must be sent.

Institutional Endorsement   Many funding agencies require IRB approval of research involving human subjects before they will either review (DHHS) or fund (many other agencies) proposed research involving human subjects. Most agencies provide forms that must be signed by authorized persons to indicate that the necessary review and approval have been accomplished. The HIC policy is that only the Chairperson, Vice-chairperson, or Associate Chairperson may sign such forms. Further, these forms may be signed only after the investigator has signed a form indicating that there is no research involving human subjects described in the grant application that is not also described in protocols approved by the HIC. Specifically, the investigator must provide the title and number (if any) of the grant application as well as the title(s) and number(s) of protocols approved by the HIC. This policy is necessary because most grant applications are much too long to be considered in detail by the full committee.

The HIC is concerned only with those parts of grant applications that have to do with human research subjects.

HIC Records

Log Book   Each new protocol is assigned a number in order of its receipt by the HIC. This number is placed on the face sheet of the protocol; it is also placed on each bit of communication or correspondence concerning that protocol. A more sophisticated numbering system has been published by Heath (288).

Immediately upon receipt of a protocol, the following bits of information about it are entered into the log book: HIC number, title of protocol, name of the principal investigator, name of other investigator who is to receive correspondence (if other than the principal investigator), date of receipt, and names of the primary and second reviewers. Subsequently, the following information is entered in the log book: date of approval, dates of approval of amendments or of periodic reapproval, and date of withdrawal or termination.

Protocol Files   Each protocol has its own file folder. These files are arranged by HIC protocol number. There are several subdivisions in the filing system. New protocols are entered into a subdivision entitled "protocols pending review." After they are reviewed, they are entered into one of two other subdivisions; viz. "approved protocols" or "protocols pending revision." Those in the latter category are eventually entered into either of two subdivisions, "approved" or "withdrawn." In each folder there is retained a copy of the original protocol, copies of all correspondence concerning the protocol, and a copy of the final form approved by the committee and endorsed by the Chairperson.

Parenthetically, there is no administrative necessity for retaining the original protocol. However, for approximately the past 18 years, these have been retained for research purposes. Their availability permits comparison with protocols that are finally approved for purposes of identifying the effects of HIC review (465).

Minutes   After each meeting of the HIC, detailed minutes are sent to all members of the committee. These minutes contain the following information: attendance; action taken on protocols listed by categories as detailed earlier, e.g., approved, approved contingent upon specific revisions, and so on; the results of periodic review of previously approved protocols; considerations of amendments; results of expedited review procedures, and so on. These minutes merely list by category the protocols by number, title, and name of principal investigator. Detailed accounts of the actions of the HIC on each protocol are reflected in the memoranda addressed by the Chairperson to investigators, which contain the specific recommendations and suggestions of the HIC as well as understandings reached in communications between the HIC and investigators. These memoranda are considered part of the official minutes of the HIC meetings; as such they fulfill several requirements of the regulations.

Interim Approval

In some situations, it may be necessary for the investigator to secure approval to proceed before the next scheduled meeting of the HIC; the procedure for doing this is called interim approval. Interim approval is not granted except in situations that legitimately require prompt action. One such situation is presented by the unexpected availability of a patient with an unusual disease. Another is the development of an indication for an innovative therapy that has not been reviewed previously by the HIC. In each case, the main justification for interim approval is in the interests of the patient-subject. In the first case, awaiting full HIC approval may unduly prolong the patient's hospitalization. In the second case, the patient might be deprived of important therapy beyond the point at which it would be efficacious.

The investigator may request interim approval by contacting the HIC Chairperson, presenting the completed protocol along with a written request for interim approval including the reasons why it is urgent to begin the study before the next HIC meeting. The Chairperson makes the decision to grant interim approval in consultation with at least one and usually two HIC members. Interim approval usually provides authorization to proceed with one subject; the term of approval expires on the date of the next HIC meeting, when the protocol is reviewed by the entire committee. Under no circumstances can interim approval be used for purposes of meeting a grant deadline; that is, the Chairperson may never sign institutional endorsement forms based on approval by anything less than a full committee meeting except when expedited review is otherwise justified.

In true emergencies, interim approval may be granted without a written protocol (or even a consent form) or without consultation with other HIC members. Any or all procedural requirements may be waived in the interests of the patient's safety.

Appeals

At the time of this writing, Yale University has no formally established appeals mechanism. The one we had in 1976 is described in Levine (403, pp. 32–33). It ceased to exist because it was never used; there have been no requests to appeal an IRB decision in over 15 years. For a description of the diverse appeals systems established at various other institutions, see Reatig (590). The only applicable regulation is that no review group that is not itself a fully constituted IRB may approve research involving human subjects (DHHS, Section 46.112).

Credibility of the IRB

> You think (do you not?) that you have only to state a reasonable case, and people must listen to reason and act upon it at once. It is just this conviction that makes you so unpleasant. There is little hope of dissuading you; but has it occurred to you that nothing is ever done until every one is convinced that it ought to be done, and has been convinced for so long that it is now time to do something else? And are you not aware that conviction has never yet been produced by an appeal to reason, which only makes people uncomfortable?
>
> F.M. Cornford (157, p. 2)

In my opinion, the single most important factor that contributes to the successful functioning of the IRB is its credibility within the institution that it serves and within the community that its institution serves (403). Without credibility, the IRB's capacity to accomplish it purposes is sharply curtailed. It can, of course, create a record of apparent protection of the rights and welfare of human research subjects—a record that will warm the cockles of any bureaucrat's heart. It can, for example, design elegant consent forms that are not used to guide negotiations between investigators and subjects. Although it takes considerable effort, any institution can create the appearance of subject protection, but we must be aware of the distinctions between appearances and reality. Members of IRBs with credibility seem to be invited into the realities of the institution by colleagues who want their advice and assistance in fostering mutual goals (403, pp. 45–53). Inspectors without credibility, on the other hand, are shown records (appearances of reality) and then only those to which their access is authorized by regulations (190).

Let us now survey some of the more important factors relevant to the IRB's credibility.

The IRB Is an Agent of Its Own Institution

In order to function most effectively, the IRB must not only be, but also must be perceived to be, an agent of its own institution. Historically, IRBs began to function, at least in some medical centers, before this function was required by the USPHS. Further, the Commission stated in broad terms some of its reasons for preferring local to either a regional or national review process.

Federal regulations should be regarded, ideally, as minimum standards for the IRB. The IRB should do at least what they require and then whatever else is necessary to safeguard the rights and welfare of human research subjects. This concept is expressed clearly by DHHS in Section 46.103b, which calls upon the institution to assure that IRB review will be conducted in accord with:

> A statement of principles governing the institution in the discharge of its responsi-
> bilities for protecting the rights and welfare of human subjects of research conducted
> at or sponsored by the institution, regardless of source of funding. This may include
> an appropriate existing code, declaration, or statement of ethical principles, or a
> statement formulated by the institution itself. This requirement does not preempt
> provisions of these regulations applicable to Department-funded research. . . .

Let us consider some consequences of considering the IRB not as an agent of its own institution but rather, as Huff put it, as a "deputy sheriff" of a regulatory agency (321). Firstly, many commentators who perceive the IRB in this fashion are the ones who call for increasingly complex federal regulation. They simply refuse to believe that IRBs will do anything not explicitly required by law. They continually urge the development of regulations that are increasingly detailed in their prescriptions and proscriptions. In my view, the tendency to develop such regulations is most dangerous; it destroys the capability of the IRB to act with sufficient flexibility to bring about adequate resolutions to particular problems as they present themselves in peculiarly local contexts. For further elaboration of this point, see Levine (409, 415) and Huff (321). For evidence of the IRB's capacity to act thoughtfully and responsibly without explicit instruction by a regulatory algorithm, see any case published in *IRB: A Review of Human Subjects Research*.

Some IRBs, however, have needlessly undermined their own credibility at the local level through their unwillingness to assume responsibility for their actions. They have been unwilling to state to their local colleagues that the *institution* has set standards for the conduct of research involving human subjects and that the *institution* has authorized the IRB to see to it that these standards are maintained within the institution. Thus whenever a researcher complains about an IRB action, they respond, "Don't blame us; it is all the Fed's fault." In this way, IRB members may avoid the unpleasant chore of defending some of their decisions. However, this convenience is purchased at the price of the IRB's local credibility.

IRBs who have done this are confronted with several problems. They have no way of resisting federal policies that are truly inappropriate. This is because they have not established their own authority to take the initiative in determining policy. Thus, they must mindlessly go along with the letter of the law, requiring, for example, full-fledged documentation of informed consent even when it is totally pointless (314, 635). If they ever showed any flexibility, they would be found out. It would be discovered by researchers that the IRB is actually capable of independent action. Researchers might then begin to hold the IRB accountable for all its actions.

This type of IRB will be particularly threatened in situations in which the regulations call upon the IRB to make judgments. For example, FDA regulations require the IRB to determine whether a medical device is a significant risk device (cf. Chapter 3); a decision that it is not yields substantial savings and convenience to both sponsor and investigator. IRBs that have not established their authority will be subjected to heavy pressure to support sponsors' claims that their devices are not to be classified as "significant risk." If they decide otherwise, they must confront investigators and sponsors with an unpleasant reality—that they, the IRB members, have made a decision which is independent. Although they might be required by law to make a decision, they are not required to have made that particular decision.

An IRB that has undermined its credibility in this way will also find it most difficult to assert its proper authority to, for instance, impose requirements for informed consent for protocols for which DHHS regulations could be construed to permit waiver of the requirement (cf. Chapter 5) or even to review some types of research that have been exempted from coverage by DHHS regulations (Section 46.101). This difficulty, which will lead to the development of one type of double standard, is discussed further in the next section.

The IRB Should Avoid Double Standards

By double standards I mean institutional policies that impose inconsistent procedural requirements on classes of research projects that do not differ from one another in any morally relevant fashion. For example, in some institutions, only the research that is funded by agencies that require IRB review is subjected to IRB review. Other research projects which do not differ in any relevant respect are permitted to proceed without IRB review. Double standards create within the institution the often accurate impression that the procedural requirements in the view of the institution have no inherent value and, therefore, are to be evaded.

As early as 1969, Barber et al. found that the great majority (240 of 282) of IRBs they studied reviewed all clinical investigations conducted within the institution (37). Further, they found in those institutions in which all clinical investigation was reviewed, 55% of respondents said that the work of the IRB was "very well received," while in those in which the IRB reviewed less than all clinical research, only 38% of respondents reported that the committee was "very well received."

At the time Barber et al. did their studies, it was DHEW that required IRB review of all research; requirements of other funding agencies were inconsistent. By 1981, however, when nearly all philanthropies had adopted the position held earlier by DHEW, DHHS exempted research that it did not either conduct or fund, at least partially (Section 46.101a). This does not reflect an opinion that nonfunded research should not be reviewed; rather, it reflects a limitation in the statutory authority of DHHS. I, for one, applaud this limitation; those IRBs that have not undercut their authority as I discussed earlier are thus liberated to apply relevant norms and procedures to nonfunded protocols as appropriate (476, 558, 742). However, the other type of IRB will have problems.

The exemptions listed in Section 46.101b reflect an attempt to define relevant distinctions between classes of projects based upon the nature of the research. There will be disagreement as to whether the correct classes of research have been exempted. For example, many IRBs will continue to review activities covered in Section 46.101b(5), if only to ascertain that identifiers will not be linked to the documents or specimens. However, this is beside the point. The point is that, if they continue to review such activities, they must review all such activities, however they are funded.

There are those who argue that the source of funding may justify double standards of another type. For example, those who suspect the motivations of the pharmaceutical industry suggest that their protocols should be treated differently from those funded by NIH (328). For example, the IRB might require disclosure on consent forms of industrial sponsorship. I disagree (426). If we wish to consider developing local policies on disclosure of funding sources, these policies should be applied consistently to both public and private sponsors. I am aware of no good reason to single out any class of sponsors for special treatment. For further discussion of this issue, see Siris (657) and Burrell (94). The Twentieth Century Fund Task Force on the Commercialization of Scientific Research studied in detail relationships between academia and industry (689). They found that virtually all problems presented by relationships with industry were also presented by relationships with government. In their report, they provide sound recommendations for dealing with these problems.

Research conducted in cooperating institutions having dissimilar policies may create unfortunate inconsistencies in the protections afforded subjects within the same protocol. For example, protocols involving patient-subjects in both university and Veterans Administration hospitals will often have differing standards for provision of free medical therapy or compensation in the event of injury. For further discussion of this type of problem, see Lackey (361) and Newton (539). Similar problems may be encountered in collaborations with military hospitals and other institutions (735).

The IRB Must Seem To Support Defective Regulations

According to the Commission (*supra*), IRB members are expected to perform important educational functions. For example, they can explain how various IRB policies, guidelines, and regulations have their roots in important ethical principles and, therefore, should be respected. During the 1960s and early 1970s, it was a matter of satisfaction in some institutions that their IRBs were routinely following procedures that seemed important even before they were included in various guidelines and regulations. This seemed to enhance confidence within institutions that these procedures made sense and, in fact, should be followed whether or not they were required through regulation (403).

However, members of the IRB cannot create respect for guidelines and regulations that they do not themselves respect. To this end it is very important to avoid developing regulations that do not have reasonable bases. When this happens, IRB members often find it necessary to apologize for the regulations. They must acknowledge to investigators that the regulations make no sense; however, they are obliged to comply in order to secure money from the federal government to conduct their research. An example of an illegitimate policy was analyzed in detail by Holder and Levine (314). This policy required documentation of full-fledged informed consent to do research on specimens obtained at either surgery or autopsy. Fortunately, this policy was rescinded in the 1981 revision of the regulations. However, there are other requirements in these regulations that IRB members will find it difficult to explain or defend. For examples, see Chapter 5, particularly the conditions under which waivers are permitted either for disclosure of some elements of informed consent or for documentation of informed consent.

IRBs Invest Too Much Time and Energy In Doing Unimportant Things

By unimportant, I mean things that contribute little or nothing to safeguarding the rights and welfare of human research subjects. Let us consider several facets of this phenomenon.

First, the time and energy of IRB members and staff that might more fruitfully be devoted to the performance of the IRB's important functions have been diverted to trivia, largely as a consequence of excessive and, in some cases, defective regulations. In his summary of this problem, Cowan, an experienced IRB Chairperson, has expressed his concern that the entire IRB system might collapse under the "sheer weight of the bureaucracy" that is involved (160). The expedited review procedures authorized by the 1981 amendments to DHHS and FDA Regulations provide some welcome relief, but much more is needed. Consider, for example, just the record-keeping requirements for IRBs (DHHS, Section 46.115). Why must we keep "a written summary of the discussion of controverted issues?" A record of "the basis for requiring changes in . . . research?" A record of "the number of members voting for, against, and abstaining" on each issue?

It is not merely the IRB members whose time and energy are wasted. Consider the consequences of imposing on investigators the obligation to comply with some of the regulations I have called defective and some of the pointless documentation requirements detailed in Chapter 5. This creates within the institution an image of the IRB as a group that is totally preoccupied with trivia and therefore to be avoided.

A special problem has been presented to investigators who conduct studies involving multiple institutions—for examples, multi-institutional clinical trials and survey research. Apparently, each IRB interprets the regulations differently and imposes requirements for revision of protocols that may be inconsistent. A dramatic example of the problems this presented to investigators was published by Kavanagh et al. (346). Their proposal to conduct survey and interview research on genetic counseling services in 51 institutions was reviewed by IRBs in each of these institutions. Securing final authorization to proceed with the research took over a year and involved 384 distinct communications between IRBs and the investigators. Of the 51 institution, 11 decided that IRB review was not required, 28 approved the protocol as submitted, and 12 approved the protocol after requesting one or more modifications. Rothman suggests that the investigator-hours required to secure all the necessary IRB approvals to initiate a large scale epidemiologic survey may approximate 3% of an epidemiologist's active professional career (619). Section 46.114 of the DHHS Regulations is designed to reduce the bureaucratic burdens of IRB approval for research involving multiple institutions; in contrast to the 1974 Regulations it replaced, it permits "reliance upon the review of another qualified IRB, or other similar arrangements aimed at avoidance of duplication of effort." Thus, IRBs now have at their disposal the means to reduce burdens of the sort described by Rothman and by Kavanagh et al. A careful reading of Kavanagh et al., however, indicates that the IRBs with which they dealt did not use the means that were at their disposal under the 1974 Regulations.

The single most important factor engendering waste of time and energy of IRB members and staff is the requirement that all applications for DHHS funds to support research involving human subjects must be reviewed and approved by an IRB *before* DHHS will consider whether it will be funded (Section 46.103f). This means that well over one-third of the IRB's effort is devoted to negotiating with investigators the development of protocols describing research that will never be done. Actually, DHHS funds far fewer than two-thirds of the applications it receives. The estimate is corrected for the facts that some investigators, after being rejected by DHHS, have the same projects funded by other agencies and some apply to other agencies in the first place. Moreover, some projects (e.g., much student research) are supported by funds that are already available to the institution.

Would it not make more sense to call upon the IRB to review only those protocols that would actually be carried out if the IRB approves them? Certainly, the IRB has a much lower disapproval rate than does DHHS.

In recognition of the fact that current regulations waste enormous amounts of IRB members' and investigators' time and in further recognition of the distinction between appearances and reality in this field, Caplan proposes "a system of random IRB audit." IRBs could follow a policy of asking, at random, a certain percentage of . . . (investigators) . . . to comply with all of the requirements of the existing IRB regulations." (106). Although every researcher would be expected to act as required by regulations, only a small number would be required to submit to the IRB review process. Under this system, "researchers who somehow run afoul of the regulations—through failure of compliance, gross misdeeds, or subject complaints—would lose their exemption from the randomization process." All of his or her subsequent protocols would require IRB review as it is done currently. I find Caplan's proposal appealing; however, I prefer the approach I shall suggest later in this chapter.

At times, IRDs may be directed to perform illegitimate functions. For example, state governments may require IRB review of protocols and consent forms for antibody tests for the acquired immune deficiency syndrome (AIDS) virus (312). This is a standard FDA-approved diagnostic test being used for its approved purpose.

FDA regulations require IRB review and approval of "compassionate" or "humanitarian" use of investigational drugs (425, 541). In such cases, there is no intention to do any research; the sole intent is to provide the best available therapy for a patient. In almost every case, data collected in the course of such activities are useless to the sponsor in supporting applications to the FDA for marketing permits (NDAs).

It has been alleged that some institutions use their IRBs to discourage research that the institution finds distasteful. According to Murray Levine, bureaucratic tactics may be employed to "force lengthy delays in resolving issues in the hopes that the proponent of the issue will become discouraged . . . and give up pursuing it.'' (385). His allegation in the particular case he was discussing was rebutted by Lathrop, Chairperson of the IRB against which his charge was levelled (370). Institutions, of course, have the authority to decide what research will be done within the institution; however, they should not direct their IRBs to invent ethical reasons to discourage unwanted research projects (561).

The IRB Is Not a Police Force

First, let us acknowledge that the IRB cannot be certain that research is actually conducted in accord with the protocols it has approved. In fact, it has been demonstrated that in some cases investigators proceed inconsistently with the agreements they have reached with the IRB (37, 271). Consequently, several commentators have suggested that IRBs should be required by regulations to monitor the actual conduct of research in order to assure compliance with the approved protocol (271, 605).

There are several types of activities carried out by IRBs that various authors call monitoring. Health has suggested that we abandon the term and instead refer to four types of activities by more exactly descriptive names (286): 1) continuing review (periodic reapproval based upon the IRB's examination of documents prepared by investigators); 2) review of the consent process; 3) review for adherence to protocol; and 4) review to identify unapproved research. The last three classes of activities entail the dispatching of IRB members or their agents into the field at times to perform police functions. This is the object of my concern.

But first, I want it clear that I do not oppose categorically all field work by IRB members. I agree that in some cases in which there is due cause to suspect that there may be difficulty performing some essential function (e.g., negotiating informed consent), the IRB should offer assistance or, on occasion, require oversight (cf. Chapters 5 and 11). I also agree with Faden et al. that the IRB should, from time to time, evaluate itself through systematic studies of the fruits of its efforts (210). When individuals within the institution show signs of dangerously deviant behavior, I think the IRB should single them out for special attention (cf. Chapter 2).

The IRB must resist pressures to turn it into a police force; it must resist all pressures, whether they arise from within or without the IRB. The presumption that informs the current so-called monitoring activities of most IRBs is that members of the institution are to be trusted until some contrary evidence is brought forward. If the IRB is obliged to function as a police force, it can only indicate to the community of investigators that it is operating from presumptions of distrust. Presumptions of distrust cost a lot in time and energy of IRB members, most of whom have no training in police work in the first place. An even more important cost would be the loss of the basic presumption of trust itself, a point to which I shall return shortly in a more general discussion of this issue. First, I wish to point out that the IRB with credibility in its community tends to be perceived as an ally by members of the community. In such institutions one sees the development of rather extensive informal monitoring systems, that is, unsolicited reports by students, physicians, nurses, and so on (286; 403, pp. 45–47). An IRB without credibility is not likely to have this large network of volunteers assisting it. Payment of equally efficient replacements, if such could be found, would be costly indeed.

Let us now return to the general problem of presumptions of distrust. Regulations based upon presumptions of distrust are characteristically excessive in their detail and inflexibility. They tend to require various agents, e.g., IRB members, to expend enormous amounts of time and energy in developing the paper documentation of the apparent protection of the rights and welfare of human research subjects. They tend, also, to call for large numbers of advocates, auditors, and other types of persons with monitor or police functions. I am inclined to resist policies that call for the incessant harassment of the majority of investigators in the interests of finding the occasional wrongdoer. I am not merely concerned with the fact that such behavior discourages honest investigators from doing good clinical research, causing them to consider returning to laboratory bench research or some other alternative occupation. I am not merely concerned about the enormous amount of time, energy, and money wasted in such activities. I am not merely concerned about the fact that such activities don't seem to catch many wrongdoers anyhow. I am most concerned that presumptions of distrust are alien to the most fundamental presumptions of the university community. We cannot sustain our university communities if we require colleagues to police and harass each other because they think there is a high likelihood that any particular other is dishonest. Argument and dialogue, so crucial to the existence of the academic community, are impossible in an environment of distrust. I can argue with a colleague only if I can presume that he or she will only use what he or she believes to be true to support the counterargument.

Presumptions of trust are much less costly, whether the costs are expressed in terms of dollars, human resources, or the quality of our social structures. Finally, I am aware of no evidence that such presumptions are associated with a higher frequency of wrongdoing. I believe that persons generally behave as they are expected to behave; most of them live up (or down) to the community's expectations.

I recognize, of course, that not all IRBs exist in academic communities. But I speak of the type of community I know best. I believe that presumptions of trust are also essential to the existence of most other communities; for more on this point, see Ladd (362).

Some Federal Policies and Practices Preempt the Authority of the IRB

When I first identified this problem as a detriment to the IRB's credibility, I was particularly concerned with regulations proposed by DHEW in 1973–1974 that would have required review by a national Ethics Advisory Board (EAB) of all proposals to involve as research subjects persons who were then called "those having limited capacities to consent." Fortunately, the Commission recommended, and DHHS seems to be in agreement, that review by the EAB is necessary only for that small minority of research proposals in which some meaningful purpose might be served (Chapters 10 through 13). But the same problem has emerged on yet another interface between the IRB and the federal government. Specifically, Study Sections (Initial Review Groups (IRGs)) at NIH and the Alcohol, Drug Abuse and Mental Health Administration (ADAMHA), in their review of research proposals, repeatedly make decisions that are more properly in the domain of the IRB. As of April 1982, the requirement to make such decisions became official DHHS policy (60). When IRGs make such decisions, they may recommend disapproval of a research grant application on ethical grounds. This practice has at least two unfortunate consequences. First, when investigators learn that their project, which had been approved previously by the IRB, has been rejected on ethical grounds, they may question the competence of the IRB.

A second, and perhaps even more important consequence, is the impact that such decisions may have on the motivation and behavior of IRB members. IRB members do not believe that IRGs are competent to make many of these decisions. IRGs are composed exclusively of scientists who are expert in the field in which they are reviewing research. As such, they are highly qualified to make judgments about the scientific design of the proposed research. However, they do not have the additional expertise and perspectives that are necessary to make other judgments for which the IRB is responsible.

When confronted with these unfortunate reversals of their decisions, IRB members tend to lose their motivation to review protocols carefully. In addition, it is clear to IRB members that any protocol they approve might be rejected at the national level, although protocols they disapprove will never even be seen at the national level. Consequently, some IRBs might consider relaxing their standards; they may begin to operate according to the maxim, "Give each questionable protocol its day in court."

Let us now consider some anecdotal case reports.

Case 1. A study section recommended disapproval of a grant application on ethical grounds; this recommendation was upheld by the advisory council to the institute. The complete summary statement of the study section's recommendation contained no criticism whatever of the scientific method. The grounds for rejection were exclusively ethical.

The proposal was designed to develop basic physiological information on elderly persons by infusing various substances intravenously, drawing blood, collecting urine, and measuring various physiologic responses by noninvasive techniques. In a proposal prepared for the IRB, the investigators described meticulously the examinations they would use to identify subjects who were most vulnerable to harm from the intravenous infusions; those identified as vulnerable were to be excluded. They further described very careful monitoring procedures for the early detection of adverse effects; interventions were planned to forestall dangerous complications. The IRB was satisfied that adequate safety precautions had been developed.

However, the study section concluded that, although the experiments are designed to minimize risks, it cannot be guaranteed that risks are eliminated. There may be a mistake in the procedures or a failure to detect one of the diseases that provides ground for exclusion. Because the experiments are not designed to benefit the individual, a single death or serious complication would far outweigh any potential benefit to be gained in a healthy, elderly population.

Thus, it appears that this study section was applying a zero-risk standard to research on healthy, elderly subjects. I am aware of no other individual or institution that has called for a zero-risk standard in this or any other presumably vulnerable population.

Case 2. A grant application was disapproved by a study section because the investigators proposed to pay subjects slightly over the minimum wage for their time to complete questionnaires. The subjects were to include normal controls as well as those who had a particular disease. The study section said that, although it was appropriate to pay normal controls, it was not appropriate to pay patients to participate in research. It seems to me that the principle they wish to uphold is that patients must volunteer their time without reward for tasks for which normal persons should properly be paid. I seriously doubt that any competent deliberative group would choose to generalize that principle.

Although the Commission discussed this problem at several of its meetings, it took no definitive action. One of the more interesting discussions took place at the meeting of April 14, 1978 (530). At that meeting, Commissioner Donald Seldin presented a particularly disturbing case. A study section had rejected a competitive-renewal application on ethical grounds with no adverse criticism of the scientific merits of the proposal. An example of an ethical impropriety identified by the study section was that the investigator proposed to study the effects of intravenous adrenocorticotropic hormone (ACTH) infusions in normotensive children. In the grant application, this is what was proposed. However, in the protocol that had been reviewed and approved by the IRB, it was pointed out that these normotensive children were to be those referred for evaluation for adrenal steroid disorders. In no case was there any plan to infuse ACTH in a child who did not require such an infusion for diagnostic purposes. Thus, the proposal was totally in accord with the recommendations of the Commission on research involving children and particularly responsive to the requirements of Recommendation 2C that calls for the minimization of risks by using procedures performed for diagnostic or treatment purposes whenever feasible (Chapter 10).

In the course of the discussion engendered by this case, some statements were made that are relevant to the general problem of study sections rendering ethical judgments on research proposals. Seldin pointed out that the Commission had reached a conclusion that the most appropriate place for ethical review was in the institution in which the research was to be conducted. Thus, such practices were incompatible with the Commission's carefully considered judgment. He elaborated how such study section actions serve to undermine the IRB.

Seldin observed that the Commission had affirmed the long-standing tradition of having a heterogeneous IRB membership including persons who are competent to assess the legal, ethical, and social implications of any research proposal. The study section, on the other hand, is comprised exclusively of scientists because their assignment is to perform scientific review. Thus, as Seldin put it, they are ''peculiarly incompetent'' to make judgments on the ethical property of research proposals. Commissioner Albert Jonsen recalled a case in which ethical problems had been identified in a research proposal. In this case, the advisory council suggested that a special study section be formed to review the matter; the council stipulated that the special study section was to include persons from a scientific discipline not represented on the original study section, and, in addition, there was to be an ethicist. Jonsen, the ethicist appointed to the special study section, reported, ''I found no ethical problems in the protocol, no serious ethical problems. It astounded me that they thought there were. I do not know how they could have viewed things to come to that conclusion; however, the second special study section . . . found a number of scientific problems and the protocol was eventually disapproved on scientific grounds. In a sense it was a double jeopardy.''

Commissioner Joseph Brady reported that on that very morning (April 14) he had been at a meeting of his study section ''that took 2 hours longer than it should have taken.'' The study section ''was busy reviewing the ethical issues of each one of the research proposals. The reason it took longer is because I argued vociferously against that.'' Brady had protested at earlier Commission meetings that review of consent forms by study sections was an enormous waste of time. Parenthetically, it should be noted that he was a member of a study section in ADAMHA; study sections in NIH usually do not review consent forms, although they do make ethical judgments. The three cases I have presented are from NIH rather than from ADAMHA. Brady further cautioned: ''I think the whole system will just die of its own dead weight if every study section is making those judgments when the IRBs that are on the spot'' can do it better.

Brady identified an even more pernicious problem. ''The effects are even more subtle. . . . Even though (the study section) may not disapprove, a reviewer stands up and he says, 'Well, I don't like this.' He influences the priority ratings in a way that aren't as flagrant as this, but he clearly influences the priority rating and I think we have got to pass a law against that some way or other.''

In my statements on April 14, I highlighted the immediacy of interaction between the IRB and the investigator. Things can be worked out by talking with the investigator. However, when a study section takes an action of this sort, it takes a minimum of 9 months to reapply for a grant. Entire programs that have been going on for years are destroyed, often needlessly.

Seldin observed that the proposal the Commission was discussing was in its 12th year. This study section decision was, in his view, about to put a stop to a highly valuable program conducted by a very competent investigator.

(Let me digress for a moment. The story of the particular grant application discussed by the Commission had a happy ending. Shortly after this meeting, a special study section was formed. The investigator was provided with sufficient funds to continue the program until a final determination was reached. In due course, the grant application was approved and funded without any substantive modifications to resolve the apparent ethical improprieties perceived by the original study section. I wish all such stories had similarly happy endings. However, in most cases they do not.)

There should be a change in the policies of study sections. They should be directed to proceed with the assumption that the IRB has ordinarily performed competently its function of ethical review. In the event study sections detect what seems to be a serious ethical impropriety, they should telephone the IRB to learn what information was available to the IRB that justified its approval of the proposal. In particular, they should learn whether the IRB had considered the problem identified by the study section. If the IRB had identified the same problem and given it due consideration, the study section should not recommend disapproval; rather, it should make a determination of priority based on scientific merit and forward the proposal to the advisory council for final disposition. At the same time, the IRB should be put on notice that it has the obligation to show why the advisory council should not recommend disapproval on ethcial grounds.

This resolution should forestall the unfortunate delays that result from the current practice of recommending disapproval and notifying the investigator and the IRB of this action after the study section has adjourned.

Meanwhile, most investigators and IRBs accept these study section decisions without a quarrel. Most IRB members and investigators are not aware that appeals of arbitrary or capricious decisions by study sections can be successful. In addition, most investigators are intimidated. Even when the IRB wishes to appeal, the investigators plead that there be no appeal. Investigators in general do not wish to risk incurring the wrath of those who hold the power to determine whether they will receive funds to continue their research.

The late Commissioner Robert Turtle suggested that this problem might be mitigated by opening meetings of study sections to the public. However, for a variety of important reasons, the Commission had recommended that study section meetings not be opened to the public. I agree, but I think it would serve the interests of the research community if questionable ethical judgments made by study sections were made available for public scrutiny. To this end, I suggest publication of well-documented commentaries on ethical decisions made by study sections that seem to be at odds with IRB decisions. A suitable medium for such publications is *IRB: A Review of Human Subjects Research*.

For further discussion of the IRB-IRG conflict, see Van Eys (691) and Blake (60). In my view, the most frequent cause of unproductive communication from IRGs through the Office for Protection from Research Risks (OPRR) to IRBs is the fact that investigators submit their grant and contract applications before they have secured final IRB approval. Changes required by the IRB and made by investigators in order to secure IRB approval commonly are not forwarded to NIH before IRG review. Thus, we commonly receive from NIH accounts of flaws in protocols that NIH does not know have already been corrected locally.

The President's Commission provides a sensitive review of relationships between federal agencies and IRBs (574). In recognition of ongoing problems on the interface between IRBs and IRGs, they conclude (574, p. 38): "Further efforts by HHS to reduce the negative effects of study section review of ethical issues seems warranted."

Inappropriate federal intrusions into the IRB's domain are made by agents and agencies other than IRGs. Here are two recent examples.

One of the institutes at NIH designed a multi-institutional ramdomized clinical trial (RCT) to compare two forms of therapy for patients with a life-threatening condition. Following is an excerpt from the protocol: "Copies of the required consent form will be made available to each participating clinic. The Principal Investigator has the option to use, *in addition*, a consent form prepared for local use; the local form may not be substituted for the study consent form."

In my view, presentation of two different consent forms to a prospective subject can only trivialize the consent process. Because I presume that no IRB or investigator wishes to make the consent process seem trivial, I also presume that most IRBs will not approve the presentation of two consent forms to prospective subjects. In this case, the HIC successfully negotiated a single extensively revised consent form that satisfied the requirements of both the coordinating center and the HIC. Had these negotiations failed, the HIC would have been obliged either to accept the coordinating center's form or to disapprove the protocol.

In a second case, here is a statement from a consent form required by a FDA official: "If you can bear children we ask that you tell us what method of contraception you plan to use. We cannot accept you as a subject unless your plans to prevent pregnancy are acceptably secure and you have a negative pregnancy test at the screening visit. During the study we ask that you refrain from sexual activity. If you do not follow your plans for contraception exactly, please let the principal investigator know promptly."

What an embarrassment it must have been for the investigators to present this statement to prospective subjects. Fortunately, we were able to further advise prospective subjects that one of the two drugs being studied was available outside the protocol.

IRBs Must Have Adequate Membership

The Commission has recognized that it is crucial to the successful functioning of the IRB that it be composed of individuals having "sufficient maturity, experience, and competence to assure that the Board will be able to discharge its responsibilities and that its determinations will be accorded respect by investigators and the community served by the institution" (527, p. 13); this position has been affirmed in DHHS and FDA regulations. In general, this means that the IRB should have among its members senior and distinguished members of the academic faculty. In most institutions, there are very few senior professors who have as their primary scholarly interest matters germane to safeguarding the rights and welfare of human research subjects. Although many are willing to serve the academic community through membership on committees, membership on the IRB consumes a large amount of time and energy. The various factors that I have identified as detriments to the IRB's credibility each tend to create serious disincentives to service on a IRB.

Several authors have called for the addition to the IRB of various types of members who would enhance the IRB's capacity to represent and reflect the values of the community. For example, in Chapter 5, Veatch's proposal to increase the number of lay persons is discussed. In my view it will never be possible to reconstruct the IRB so that it reflects truly the values of the community; when one wishes to know the values of the community, one should consult the community directly (cf. Chapter 4) or through surrogates (cf. Chapter 9).

Barber et al. (37, pp. 196 et seq.) proposed "the invention of a new social role, that of informed outsider, a role that could more effectively represent the values and interests of all who have stakes in the proceedings of the review committees, not only the patient-subjects and future patients but the research profession itself." Several functions are specified for this "informed outsider."

> A special professional school or a training program for this kind of role could be set up, a school in which the training communicated the several kinds of knowledge that we have indicated are necessary and in which various opportunities for "case instruction" and apprenticeship experience were available. Such a school or program might be set up in close association with the law and medical schools of a given university.

Nolan has argued persuasively that the role of informed outsider can be played very effectively by medical students and other health science students. (545). "Students are more informed than most community members, and they are more like outsiders than most physicians and researchers." Further, their employment does not require the establishment of special schools as proposed by Barber et al. Over the years I have found students very effective advocates of the rights and welfare of research subjects. I believe that they are both eminently qualified and highly motivated to serve the important role of informed outsider.

Perfection Is Not the Proper Standard of Measurement of IRB Performance

In his discussion of risk and benefit assessment, Bernard Barber made an important statement on this point—a statement that I think applies equally to all other aspects of IRB function (35):

> . . . [E]ven where difficulty and strain occur in the assessment process, they (assessments) are worthwhile just because the process of making assessments has value over and above the outcome or *product* of the process. Whether routine or difficult and causing strain, the *process* . . . is important in itself. The process is in itself "consciousness-raising;" it leads to higher ethical awareness. One hopes that the *product*, now or eventually, will also be better, but that happy condition we should not expect ourselves to guarantee. We should not expect, and certainly not require perfection of risk-benefit assessment in all cases from our biomedical or behavioral review groups. Perfection in all cases is for utopias and heavenly worlds. Moreover, we should inform the general public that we do not guarantee perfection of assessment *product* but only excellence in the *process*. We should inform them that we are prepared to defend scientific research assessment, when it is conscientiously and competently carried out by professional peer review groups, against all demands for utopian perfection of *product* (emphasis in original).

In my view, demands for perfection in the outcome have the same unfortunate effects on policies and institutions as presumptions of distrust (*supra*).

Relationships with Federal Agencies

Relationships among and between investigators, IRBs, and various components of the federal government are discussed in various parts of this book. In this section I shall focus on federal activities designed to assure the IRB's compliance with federal regulations. I find it necessary to make some statements that are critical of some federal agencies. As I have pointed out in various other sections of this book, not all, perhaps not even most, time-wasting and counterproductive behavior in the field of human subject protection are federal initiatives. I would like to see all such behaviors minimized.

Both DHHS and FDA have policies designed to ensure the competent function of IRBs. DHHS policy involves accreditation based upon negotiations between each institution and the NIH (OPRR). These negotiations result in documents called "assurances" (Section 46.103). This in accord with Recommendation 2 in the Commission's Report on IRBs, which further called for DHHS to conduct:

> (ii) Compliance activities, including site visits and audits of IRB records . . . and
> (iii) educational activities to assist members of IRBs in recognizing and considering the ethical issues that are presented by research involving human subjects.

With regard to educational activities, there is currently a program of workshops sponsored by OPRR and FDA for federal officials, IRB members, institutional officials, and researchers; these workshops have been offered in various cities around the country. I have participated as faculty in several of these. In my view they are good programs necessarily designed for novices in the field. In my view, federal funds designed to educate the target audiences could be invested more prudently. "Direct costs associated with the education program (excluding staff time and travel) have been $340,000 from June 1980 through September 1982. The FDA contributed $80,000 of that amount." In that time period, these workshops had a total attendance of 1660 persons (574, pp. 14–15)).

Private agencies such as Public Responsiblity in Medicine and Research (PRIM&R) and The Hastings Center conduct workshops with similar goals at no cost to the government. Faculty for these workshops is dominated by experts who are experienced teachers from university faculties. Senior FDA and OPRR officials are included among the faculty. These programs usually have separate sessions designed to tend to the needs of novices; they are arranged in a fashion that does not exclude novices from most of the rest of the program. In my view, the most important thing that happens at these meetings takes place in the small-group discussions and during the times set aside for meals and other nonprogram activities. In these settings, the participants have opportunities to share their experiences with each other, to learn from colleagues how problems are handled in other institutions, and to establish relationships that form the bases of subsequent consultations.

FDA's approach to enhancement of IRB compliance is different from that of OPRR. There is no accreditation; rather, there are extensive grounds and procedures for disqualification, as well as lesser administrative actions (Subpart E). Rather than site visits by colleagues and experts, there are inspections by agents who wear badges and present "Notices of Inspection" that look for all the world like search warrants. It is no wonder that most academic IRBs view their relationship with NIH as being very different indeed from that with FDA. FDA's approach reflects its history as an agency having as one of its major functions the detection of miscreants (542, 565). NIH's approach, by contrast, reflects its history as an agency designed to support and nurture the conduct of biomedical research.

The President's Commission has recommended that OPRR's current practices of conducting site visits and audits of IRBs for cause be supplemented by "site visits of research institutions using experienced IRB members and staff as site visitors" (574, pp. 125–136). The President's Commission conducted a pilot study of such site visits and had good cause to be pleased with its results (574, p. 67ff). As a member of some of these pilot site visit teams, I can say that we, the site visitors, were learning as we were evaluating and educating. I have no doubt that if and when experience is gained in this field, the results will be even more pleasing than those achieved during the pilot study.

The President's Commission compared the reports from the pilot site visits with those filed by FDA inspectors who had evaluated the same IRBs: "The differences in the results of the two types of evaluation appear to reflect differences in approaches used. The assessments by the Commission's site visitors were more qualitative and oriented more toward the goals and purposes of IRB review than toward technical conformity with regulatory requirements. The FDA assessment of these IRBs appeared to depend heavily on regulatory compliance." (574, p. 113). Actually, the differences in overall evaluations were quite striking. One IRB found seriously deficient by site visitors received this report from the FDA inspector: "Our review . . . indicates that your committee was in compliance with Federal regulations at the time of this inspection" (574, pp. 111–112). For an opposite result, see page 110 (574).

The President's Commission comments (574, p. 136):

> The Commission urges all agencies, notably OPRR and FDA, to consider budgeting funds already available for education and compliance activities to carry out more site visits of randomly selected institutions. FDA should carefully consider whether the resources it devotes to IRB inspections at the institutions having assurances of compliance with OPRR could be made available for a coordinated site visit program, and whether FDA could deem those IRBs to have complied with FDA regulations.

I urge FDA to act upon this wise advice. However, as discussed in Chapter 2, I recognize the necessity for continuing routine inspections and audits of clinical investigators.

In its first two Recommendations in the IRB Report, the Commission called for the enactment of federal law or the issuance of an executive order to establish in DHHS a *single office* for the promulgation and administration of *all* federal regulations relating to the protection of human research subjects. FDA was identified explicitly as one of the agencies whose activities in this field were to be subsumed in this office. The Commission's intent was to limit both the inconsistencies in the federal regulations and the number of federal officials to which the IRB must be responsive. To this end, the Commission endorsed the Commission on Federal Paperwork's statement:

> No agency other than HEW should be permitted to paraphrase, interpret or particularize these regulations. . . . For a controversial subject of this nature there should be a mechanism for the Federal Government to speak with one voice.

Now the President's Commission, after further study and reflection, has renewed essentially the same recommendation (574, p. 133). One wonders how long the Congress and the Executive will remain unresponsive to this and some other recommendations of its most distinguished advisory bodies.

Prior Review and Approval Reconsidered

Opponents of federally mandated IRB review and approval of research involving human subjects commonly refer to this activity as "prior restraint." This rhetorical device seems to support their claim that it is unconstitutional, a violation of the First Amendment. Actually, according to Tribe (687, p. 725), the First Amendment is not an absolute bar to prior restraint, although the U.S. Supreme Court has stated repeatedly that any "system of prior restraints comes to this Court bearing a heavy presumption against its constitutional validity." Robertson (606, p. 498) opines that the federal government has the constitutional authority to require IRB review and approval of research that it conducts and supports and that this authority derives from its "conditional spending power." However, he concedes that this authority might not be so clearly established for research not funded by the federal government, particularly if the institution receives no funds whatever from the federal government to support the conduct of research involving human subjects.

I have no wish to enter this debate. I do, however, wish that all concerned would cease to call IRB review prior restraint unless they intend the proper meaning of this term.

Now let us consider whether we wish to require IRB review and approval of each and every plan to do research involving human subjects before it may be begun. Let us begin by considering the costs.

We have very little data on the costs of IRB review. Brown et al. estimated the cost of operating an IRB that reviewed only 183 protocols annually at slightly over $100,000; this IRB serves a Health Sciences Center (89). Cohen and Hedberg provide the lowest estimate I have ever seen or heard, $36,000 (147), but this was in an institution that did no biomedical research and in which over 95% of protocols were judged by the IRB to present no risk whatever to subjects. Because these two articles each omit major budgetary items included in the other, it seems fair to assume that they are both substantial underestimates. At IRB workshops, I have heard spokespersons of various institutions provide the following calculated estimates. (I am not reporting casual guesses.) One large teaching hospital (1 of 11 associated with a major medical school) spends $150,000 per year for its IRB; a separate committee that performs scientific review has an additional budget which is not included. One large state university hospital spends $250,000 annually; this also includes all protocols conducted by medical school personnel even if the hospital is not involved. One small community hospital which is a distant affiliate of a medical school in another city spends $45,000. Only the estimate provided by Brown et al. includes calculations of investigator expenses (e.g., time, photocopying, and so on). Curiously, Brown et al. think that it takes an average of 2 hours of investigator time to prepare a protocol at an average salary of $15 per hour.

Let us assume that a reasonable underestimate for operating an IRB at a "multiple project assurance" institution is $100,000 per year, and let us further assume that each has only two IRBs. (Most have more; some have more than 10.) Let us also disregard all of the special assurance IRBs, e.g., the 1200 that had to be created overnight just in response to FDA's intraocular lens regulations (741). Let us also disregard the budgets for OPRR and the FDA's much more expensive compliance programs (574, p. 61).

There were in 1978 over 650 institutions with general assurances; 650 × 2 IRBs × $100,000 = $130,000,000 per year. For this amount we could operate the Uniformed Services University of the Health Sciences for over 7 years, including even paying their students their $14,000 annual salaries (510).

All of the data and calculations presented in the preceding three paragraphs reflect conditions in 1978–1980. Subsequently, very few new data have become available (146). The reader may correct these numbers for inflation if he or she wishes. Alternatively, one can draw on information made available from the business world. For example, one noninstitutional review board (NRB) charges $750 per investigative site per protocol (709). Thus, if a drug company wishes NRB review of a study to be conducted in the offices of four physicians, the charge will be $3000. On this basis, drawing on the HIC experience, and assuming each of our protocols will be conducted in only one site ($750 per protocol), in the academic year 1984–1985 the cost of running the HIC would be calculated at $241,500; this, of course, does not include the costs of running Yale's other three IRBs or the costs for protocol preparation time (investigator and staff salaries) and so on. For present purposes, the exact number is unimportant; it suffices to understand that it is large.

What do we think we are buying with all this money? Impetus to developing the statutory requirement for IRB review was provided by a mere handful of cases that were brought to the attention of Congress in the early 1970s; most of these are cited in Chapter 4 as historical examples of violations of the fundamental ethical principles. Certainly, we should not tolerate such violations in the name of research. I suggest, however, that we can purchase adequate defenses at much lower cost by changing our approach to regulation according to the proposal that follows.

A Modest Proposal

First, we should make clear statements that there are certain sorts of offenses that our society will not tolerate. Those that are caught committing these offenses will be punished in certain specified ways. This is how we deal with murder, rape, arson, and embezzlement. Researchers, unlike arsonists, are obliged to publish an account of their activities in order to earn their rewards; thus, it should be much easier to apprehend wrongdoers. Much less detective work will be necessary. I am not suggesting that errant investigators should be dealt with by the criminal justice system, except in extreme cases. Rather, I suggest the use of procedures that have been developed to deal with other types of academic misconduct (cf. Chapter 2).

Secondly, we should establish IRBs, as advised by the Declaration of Helsinki, as independent committees that are available to provide consultation and guidance. Investigators should be advised to consult the IRB to assure themselves that they are not going to commit any punishable offense. They may also request the IRB's guidance in tending to some of the niceties of dealing with research subjects—matters more of etiquette than ethics.

Thirdly, all proposals to do research should be reviewed by some agency that is capable of judging the scientific design and the competence of the investigators. For the latter task, a licensing or certification mechanism may be the best approach.

With few exceptions, for all research involving reasonably autonomous adults, this should suffice. The exceptions are those classes of research in which there is a high probability of violating one or more fundamental ethical principles.

For these classes, prior IRB review and approval should be mandatory. These classes include research involving deceptive strategies and research in which procedures performed in the interest of research present a burden greater than mere inconvenience. Minimal risk, a standard developed for children and those institutionalized as mentally infirm, is too low a threshold for autonomous adults. The relevant burdens are risks of physical or psychological injury. Protocols presenting risks of social injury should be included only if there are unavoidable threats to confidentiality. Avoidable breaches of confidentiality should be treated as punishable offenses. On this point, Pattullo recommends going still further, excluding from the requirement for mandatory review all research other than that which presents more than minimal risk of bodily harm (560).

Prior review and approval should also be mandatory for protocols designed to introduce, test, evaluate, or compare therapeutic, diagnostic, or prophylactic maneuvers. In these cases, mandatory review and approval should be confined to assuring that there is a clear and accurate statement of alternatives; the prospective subject will be afforded ample opportunity to make a valid choice between alternatives; the prospective subject will be fully apprised of the consequences of choosing the dual role of patient-subject; and, when appropriate, there will be equitable distribution of *both* burdens and benefits. On other matters, for this class of activities the IRB's role should be advisory.

There should also be mandatory prior IRB review and approval of all plans to do research on all persons who are either incapable of consent or who, for other reasons, are not reasonably autonomous.

I further propose that all regulations that I have identified as defective should be either rescinded or, in those cases in which they are simply badly worded, rewritten so that they will provide correct guidance. All agents involved in safeguarding the rights and welfare of human research subjects should presume that all other agents are trustworthy and that they are performing their tasks competently until substantial contrary evidence is brought forward. All our policies—governmental and institutional alike—should reflect this presumption of trust.

I am aware that I am proposing a radical change in both the letter and spirit of the law. I earnestly believe that the consequences would please all but those who savor strife.

In his teachings, Jesus often called attention to distinctions between the spirit and the letter of the law. Justice was not always served best by strict adherence to the letter of the law as it was conceived by the patriarchs and interpreted by public officials. For example:

"Well then, if a child is circumcised on the Sabbath to avoid breaking the Law of Moses, why are you indignant with me for giving health on the Sabbath to the whole of a man's body? Do not judge superficially, but be just in your judgements."

John 7:23–24

References

1. Abram, M.B.; Wolf, S.M.: Public Involvement in Medical Ethics: A Model for Government Action. New Engl J Med 310 (1984) 627–632.
2. Abramowitz, S.: A Stalemate on Test-Tube Baby Research. Hastings Center Rep 14 (1) (February 1984) 5–9.
3. Abramson, N.S.; Meisel, A.; Safar, P.: Informed Consent in Resuscitation Research. JAMA 246 (1981) 2828–2830.
4. Ackerman, T.F.: Fooling Ourselves with Child Autonomy and Assent in Nontherapeutic Clinical Research. Clin Res 27 (1979) 345–348.
5. Ackerman, T.F.: Moral Duties of Investigators Toward Sick Children. IRB: Rev Human Subjects Res 3 (6) (June/July 1981) 1–5.
6. Ackerman, T.F.; Strong, C.M.: Ethics of Phase I Clinical Trials. JAMA 249 (1983) 883.
7. Adair, J.G.; Dushenko, T.W.; Lindsay, R.C.L.: Ethical Regulations and Their Impact on Research Practice. American Psychologist 40 (1985) 59–72.
8. Adams, B.R.; Shea-Stonum, M.: Toward A Theory of Control of Medical Experimentation with Human Subjects: The Role of Compensation. Case Western Reserve Law Rev 25 (Spring 1975) 604–648.
9. Adkinson, N.F.; Starklauf, B.L.; Blake, D.A.: How Can An IRB Avoid the Use of Obsolete Consent Forms? IRB: Rev Human Subjects Res 5 (1) (January/February 1983) 10.
10. Alazraki, N.: Evaluating Risk from Radiation for Research Subjects. IRB: Rev Human Subjects Res 4 (1) (January 1982) 1–3.
11. Alfidi, R.J.: Informed Consent: A Study of Patient Reaction. JAMA 216 (1971) 1325–1329.
12. American Fertility Society: Ethical Statement on In Vitro Fertilization. Fertil Steril 41 (1984) 12.
13. American Psychological Association: Ad Hoc Committee on Ethical Standards in Psychological Research. Ethical Principles in the Conduct of Research with Human Participants. American Psychological Association, Washington, 1973.
14. American Psychological Association: Ethical Principles in the Conduct of Research with Human Participants. American Psychological Association, Washington, 1982.
15. Amiel, S.A.: Pediatric Research on Diabetes: The Problem of Hospitalizing Youthful Subjects. IRB: Rev Human Subjects Res 7 (1) (January/February 1985) 4–5.
16. Anbar, M: Biological Bullets: Side Effects of Health Care Technology. In: The Machine at the Bedside: Strategies for Using Technology in Patient Care, pp. 35–45, ed. by S.J. Reiser and M. Anbar. Cambridge University Press, Cambridge, 1984.
17. Angell, M.: Patients' Preferences in Randomized Clinical Trials. New Eng J Med 310 (1984) 1385–1387.
18. Angoff, N.R.: Disclosure of the Hidden Injury. IRB: Rev Human Subjects Res 4 (9) (November 1982) 6–7.
19. Angoff, N.R.: An Inadvertent Breach of Confidentiality. IRB: Rev Human Subjects Res 6 (3) (May/June 1984) 5–6.
20. Angoff, N.R.: Against Special Protections for Medical Students. IRB: Rev Human Subjects Res 7 (5) (September/October 1985) 9–10.
21. Anisfeld, E.; Lipper, E.: Early Contact, Social Support, and Mother-Infant Bonding. Pediatrics 72 (1983) 79–83.
22. Annas, G.J.: Report on the National Commission: Good as Gold. Medicolegal News 8 (6) (December 1980) 4–7.
23. Annas, G.J.; Glantz, L.H.; Katz, B.F.: Informed Consent to Human Experimentation: The Subject's Dilemma. Ballinger Publishing Co., Cambridge, Mass., 1977.
24. Appelbaum, P.S.; Roth, L.H.: Competency to Consent to Research: A Psychiatric Overview. Paper presented at the National Institutes of Mental Health Workshop: Empirical Research on Informed Consent with Subjects of Uncertain Competence, Rockville, Md., January 12, 1981.
25. Appelbaum, P.S.; Roth, L.H.; Detre, T.:

Researchers' Access to Patients Records: An Analysis of the Ethical Problems. Clin Res 32 (1984) 399–403.

26. Arnold, J.D.: Alternatives to the Use of Prisoners in Research in the United States. In: The National Commission for the Protection of Human Subjects of Biomedical and Behavioral Research: Research Involving Prisoners: Report and Recommendations. Appendix, pp. 8.1–8.18. DHEW Publication No. (OS) 76-132, Washington, 1976.

27. Arnold, J.D.: Incidence of Injury During Clinical Pharmacology Research and Indemnification of Injured Research Subjects at the Quincy Research Center. In: President's Commission for the Study of Ethical Problems in Medicine and Biomedical and Behavioral Research: Compensating for Research Injuries: The Ethical and Legal Implications of Programs to Redress Injured Subjects. Appendix, pp. 275–302. U.S. Government Printing Office, Stock No. 040-000-00456-4, Washington, 1982.

28. Arnold, J.D.; Martin, D.C.; Boyer, S.E.: A Study of One Prison Population and Its Response to Medical Research. Ann NY Acad Sci 169 (1970) 463–470.

29. Atkins, H.; Hayward, J.L.; Klugman, D.J.; Wayte, A.B.: Treatment of Early Breast Cancer: A Report after Ten Years of a Clinical Trial. Brit Med J 2 (1972) 423–429.

30. Ayd, F.J., Jr.: Motivations and Rewards for Volunteering to be an Experimental Subject. Clin Pharmacol Therapeutics 13 (1972) 771–778.

31. Azarnoff, D.L.: Physiologic Factors in Selecting Human Volunteers for Drug Studies. Clin Pharmacol Therapeutics 13 (1972) 796–802.

32. Bailar, J.C., III: When Research Results are in Conflict. New Engl J Med 313 (1985) 1080–1081.

33. Bailey v. Lally, 481 F. Supp 203, D.C. Md, 1979.

34. Baker, M.T.; Taub, H.A.: Readability of Informed Consent Forms for Research in a Veterans Administration Medical Center. JAMA 250 (1983) 2646–2648.

35. Barber, B.: Some Perspectives on the Role of Assessment of Risk/Benefit Criteria in the Determination of the Appropriateness of Research Involving Human Subjects. In: The National Commission for the Protection of Human Subjects of Biomedical and Behavioral Research: The Belmont Report: Ethical Principles and Guidelines for the Protection of Human Subjects of Research. Appendix II, pp. 19.1–19.21, DHEW Publication (OS) 78-0014, Washington, 1978.

36. Barber, B.: Informed Consent in Medical Therapy and Research. Rutgers University Press, New Brunswick, N.J., 1980.

37. Barber, B.; Lally, J.J.; Makarushka, J.L.; Sullivan, D.: Research on Human Subjects: Problems of Social Control in Medical Experimentation. Russell Sage Foundation, New York, 1973.

38. Barclay, W.R.; McCormick, R.A.; Sidbury, J.B.; Michejda, M.; Hodgen, G.D.: The Ethics of In Utero Surgery, JAMA 246 (1981) 1550–1555.

39. Baron, C.H.: Fetal Research: The Question in the States. Hastings Center Rept 15 (2) (April 1985) 12–16.

40. Baron, R.A.: The "Costs of Deception" Revisited: An Openly Optimistic Rejoinder. IRB: Rev Human Subjects Res 3 (1) (January 1981) 8–10.

41. Bartlett, R.H.; Roloff, D.W.; Cornell, R.G.; et al.: Extracorporeal Circulation in Neonatal Respiratory Failure: A Prospective Randomized Study. Pediatrics 76 (1985) 479–487.

42. Batchelor, W.F.: AIDS: A Public Health and Psychological Emergency. Am Psychologist 39 (1984) 1279–1284.

43. Baumrind, D.: Nature and Definition of Informed Consent in Research Involving Deception. In: The National Commission for the Protection of Human Subjects of Biomedical and Behavioral Research: The Belmont Report: Ethical Principles and Guidelines for the Protection of Human Subjects of Research. Appendix II, pp. 23.1–23.71. DHEW Publication (OS) 78-0014, Washington, 1978.

44. Baumrind, D.: IRBs and Social Science Research: The Costs of Deception. IRB: Rev Human Subjects Res 1 (6) (October 1979) 1–4.

45. Baumrind, D.: Research Using Intentional Deception. Am Psychologist 40 (1985) 165–174.

46. Bayer, R.; Levine, C.; Murray, T.H.: Guidelines for Confidentiality in Research on AIDS. IRB: Rev Human Subjects Res 6 (6) (November/December 1984) 1–7.

47. Beauchamp, T.L.: Distributive Justice and Morally Relevant Differences. In: The National Commission for the Protection of Human Subjects of Biomedical and Behavioral Research: The Belmont Report: Ethical Principles and Guidelines for the Protection of Human Subjects of Research. Appendix I, pp. 6.1–6.20. DHEW Publication (OS) 78-0013, Washington, 1978.

48. Beauchamp, T.L.: The Ambiguities of "Deferred Consent." IRB: Rev Human Sub-

jects Res 2 (7) (August/September 1980) 6–8.

49. Beauchamp, T.L.; Childress, J.F.: Principles of Biomedical Ethics, Second Edition. Oxford University Press, New York, 1983.
50. Beecher, H.K.: Surgery as Placebo: A Quantitative Study of Bias. JAMA 176 (1961) 1102–1107.
51. Beecher, H.K.: Ethics and Clinical Research. New Engl J Med 274 (1966) 1354–1360.
52. Beecher, H.K.: Letter to the Editor. New Engl J Med 275 (1966) 791.
53. Beecher, H.K.: Research and the Individual: Human Studies. Little, Brown and Co., Boston, 1970.
54. Benotti, J.R.; Shannon, T.A.: How an Investigator and an IRB Cooperated in Research Design. IRB: Rev Human Subjects Research 4 (6) (June/July 1982) 6–7.
55. Berenson, C.K.; Grosser, B.I.: Total Artificial Heart Implantation. Archives Gen Psychiatry 41 (1984) 910–916.
56. Berkowitz, L.: Some Complexities and Uncertainties Regarding the Ethicality of Deception in Research with Human Subjects. In: The National Commission for the Protection of Human Subjects of Biomedical and Behavioral Research: The Belmont Report: Ethical Principles and Guidelines for the Protection of Human Subjects of Research. Appendix II, pp. 24.1–24.34. DHEW Publication (OS) 78-0014, Washington, 1978.
57. Blackburn, M.G.; Pleasure, J.R.; Sorenson, J.H.: Research in Neonatology: Improved Ratio of Risk to Benefit Through a Scientifically Based Alteration in Research Design. Clin Res 24 (1976) 317–321.
58. Blackwell, B.: For the First Time in Man. Clin Pharmacol Therapeutics 13 (1972) 812 823.
59. Blake, D.A.: An Executive Committee System for IRBs. IRB: Rev Human Subjects Res 4 (9) (November 1982) 8–9.
60. Blake, D.A.: The IRB-IRG Conflict. IRB: Rev Human Subjects Res 5 (3) (May/June 1983) 10.
61. Blaszkowski, T.P.; Insull, W., Jr. (Editors): Proceedings of the Workshop on the Development of Markers for Use as Adherence Measures in Clinical Studies. Controlled Clin Trials 5 (1984) 451–588.
62. Block, J.B.; Elashoff, R.M.: The Randomized Clinical Trial in Evaluation of New Cancer Treatment. In: Methods in Cancer Research: Cancer Drug Development, Volume XVII, Part B, pp. 259–275, ed. by V.T. DeVita, Jr., and H. Busch. Academic Press, New York 1979.

63. Blumgart, H.L.: The Medical Framework for Viewing the Problem of Human Experimentation. In: Experimentation with Human Subjects, pp. 39–65, ed. by P.A. Freund. George Braziller, New York, 1970.
64. Bok, S.: The Ethics of Giving Placebos. Scientific Am 231, (1974) 17–23.
65. Bok, S.: Fetal Research and the Value of Life. In: The National Commission for the Protection of Human Subjects of Biomedical and Behavioral Research: Research on the Fetus. Report and Recommendations. Appendix pp. 2.1–2.18, DHEW Publication No. (OS) 76-128, Washington, 1975.
66. Bok, S.: Lying: Moral Choice in Public and Private Life. Pantheon, New York, 1978.
67. Bok, S.: Secrets: On the Ethics of Concealment and Revelation. Pantheon, New York, 1982.
68. Bok, S.: The Limits of Confidentiality. Hastings Center Rept 13 (1) (February 1983) 24–31.
69. Boruch, R.F.: Should Private Agencies Maintain Federal Research Data? IRB: Rev Human Subjects Res 6 (6) (November/December 1984) 8–9.
70. Boruch, R.F.; Cecil, J.S.: Assuring the Confidentiality of Social Research Data. University of Pennsylvania Press, Philadelphia, 1979.
71. Boruch, R.F.; Ross, J.; Cecil, J.S.: Proceedings and Background Papers. Conference on Ethical and Legal Problems in Applied Social Science Research, Northwestern University, Evanston, Illinois, April 1979.
72. Boström, H.: On the Compensation for Injured Research Subjects in Sweden. In: President's Commission for the Study of Ethical Problems in Medicine and Biomedical and Behavioral Research: Compensating for Research Injuries. The Ethical and Legal Implications of Programs to Redress Injured Subjects. Appendix, pp. 309–322. U.S. Government Printing Office, Stock No. 040-000-00456-4, Washington, 1982.
73. Bowker, W.F.: "Exculpatory Language" in Consent Forms. IRB: Rev Human Subjects Res 4 (3) (March 1982) 9.
74. Brackbill, Y.; Hellegers, A.E.: Ethics and Editors. Hastings Center Rep 10 (2) (April 1980) 20–22.
75. Bracken, M.B.: The Stability of the Decision to Seek Induced Abortion. In: The National Commission for the Protection of Human Subjects of Biomedical and Behavioral Research: Research on the Fetus. Report and Recommendations. Appendix, pp. 16.1–16.23. DHEW Publication No. (OS) 76-128, Washington, 1975.

76. Bracken, M.B.; Klerman, L.V.; Bracken, M: Abortion, Adoption or Motherhood: An Empirical Study of Decision-making During Pregnancy. Am J Obstet Gynecol 130 (1978) 251–262.

77. Bracken, M.B.; Klerman L.V.; Bracken, M.: Coping with Pregnancy Resolution Among Never-Married Women. Am J Orthopsychiatry 48 (1978) 320–334.

78. Brady, J.V.: A Consent Form Does Not Informed Consent Make. IRB: Rev Human Subjects Res 1 (7) (November 1979) 6–7.

79. Brakel, S.J.; Parry, J.; Weiner, B.A.: The Mentally Disabled and the Law, third edition. American Bar Foundation Press, Chicago, 1986.

80. Brandt, A.M.: Racism and Research: The Case of the Tuskegee Syphilis Study. Hastings Center Rept 8 (6) (December 1978) 21–29.

81. Brandt, E.N., Jr.: Institutional Review Boards. JAMA 249 (1983) 2889–2890.

82. Brandt, E.N., Jr.: AIDS Research: Charting New Directions. Public Health Rept 99 (5) (1984) 433–435.

83. Branson, R.: Philosophical Perspectives on Experimentation with Prisoners. In: National Commission for the Protection of Human Subjects of Biomedical and Behavioral Research, Research Involving Prisoners: Report and Recommendations. Appendix, pp. 1.1–1.46. DHEW Publication No. (OS) 76-132, 1976.

84. Branson, R.: Prison Research: National Commission Says "No, Unless. . ." Hastings Center Rept 7 (1) (February 1977) 15–21.

85. Brennan, T.A.: Research Records, Litigation and Confidentiality: The Case of Research on Toxic Substances. IRB: Rev Human Subjects Res 5 (5) (September/October 1983) 6–9.

86. Broad, W.; Wade, N.; Betrayers of the Truth: Fraud and Deceit in the Halls of Science. Simon and Schuster, New York, 1982.

87. Brody, H.: Placebos and the Philosophy of Medicine: Clinical, Conceptual and Ethical Issues. The University of Chicago Press, Chicago, 1980.

88. Brown, J.H.U.; Schoenfeld, L.S.; Allan, P.W.: Management of an Institutional Review Board for the Protection of Human Subjects. SRA Journal, Summer (1978) 5–10.

89. Brown, J.H.U.; Schoenfeld, L.S.; Allan, P.W.: The Costs of an Institutional Review Board. J Med Ed 54 (1979) 294–299.

90. Brown, J.H.U.; Schoenfeld, L.S.; Allan, P.W.: The Philosophy of an Institutional Review Board for the Protection of Human Subjects. J Med Ed 55 (1980) 67–69.

91. Brownell, K.D.; Stunkard, A.J.: The Double-Blind in Danger: Untoward Consequences of Informed Consent. Am J Psychiatry 139 (1982) 1487–1489.

92. Buckingham, R.W.; Lack, S.A.; Mount, B.M.; MacLean, L.D.; Collins, J.T.: Living with the Dying: Use of the Technique of Participant Observation. Canadian Med Assoc J 115 (1976) 1211–1215.

93. Bulmer, M: Are Pseudo-Patient Studies Justified? J Med Ethics 8 (1982) 1097–1098.

94. Burrell, C.D.: Why Researchers Need Not Be Demoralized. IRB: Rev Human Subjects Res 5 (6) (November/December 1983) 4–5.

95. Burt, R.A.: Why We Should Keep Prisoners from the Doctors. Hastings Center Rept 5 (1) (February 1975) 25–34.

96. Burt, R.A.: The Elusive Role of "Neutral Observer" in Human Investigations. IRB: Rev Human Subjects Res 2 (1) (January 1980) 9.

97. Butler, R.N.: Why Survive? Being Old in America. Harper & Row, New York, 1975.

98. Butt, W.; Neuhauser, D.: The Machine and the Marketplace: Economic Considerations in Applying Health Care Technology. In: The Machine at the Bedside: Strategies for Using Technology in Patient Care, pp. 135–155, ed. by S.J. Reiser and M. Anbar. Cambridge University Press, Cambridge, 1984.

99. Byington, R.P.; Curb, J.D.; Mattson, M.E.: Assessment of Double-blindness at the Conclusion of the Beta-Blocker Heart Attack Trial. JAMA 253 (1985) 1733–1736.

100. Calabresi, G.: Reflections on Medical Experimentation in Humans. In: Experimentation with Human Subjects, pp. 178–196, ed. by P.A. Freund. George Braziller, New York, 1970.

101. Callahan, D.: Tradition and the Moral Life. Hastings Center Rept 12 (6) (December 1982) 23–30.

102. Canadian Medical Research Council: Ethics in Human Experimentation. Report Number 6, Ottawa, Canada, 1978.

103. Cann, C.I.; Rothman, K.J.: IRBs and Epidemiologic Research: How Inappropriate Restrictions Hamper Studies. IRB: Rev Human Subjects Res 6 (4) (July/August 1984) 5–7.

104. *Canterbury v. Spence*, 464 F 2d 72, CA DC 1972.

105. Caplan, A.L.: Organ Transplants: The Costs of Success. Hastings Center Rept 13 (6) (December 1983) 23–32.

106. Caplan, A.L.: Random Auditing—A Modest Proposal for Reforming the Regulation of Research. Clin Res 31 (1983) 142–143.

107. Caplan, A.L.: Inconsistency, Idiosyncrasy and IRBs. IRB: Rev Human Subjects Res 6 (2) (March/April 1984) 10–12.

108. Caplan, A.L.: Is There a Duty to Serve as a Subject in Biomedical Research? IRB: Rev Human Subjects Res 6 (5) (September/October 1984) 1–5.

109. Caplan, A.L.: Organ Procurement: It's Not in the Cards. Hastings Center Rept 14 (5) (October 1984) 9–12.

110. Caplan, A.L.; Lidz, C.W.; Meisel, A.; Roth, L.H.; Zimmerman, D.: Mrs. X and the Bone Marrow Transplant. Hastings Center Rept 13 (3) (June 1983) 17–19.

111. Capron, A.M.: Legal Considerations Affecting Clinical Pharmacological Studies in Children. Clin Res 21 (1973) 141–150.

112. Capron, A.M.: Legal Responsibilities Relating to the Use of Unapproved Drugs in Children. In: Clinical Pharmacology and Therapeutics: A Pediatric Perspective, pp. 305–317, ed. by B.L. Mirkin. Year Book Medical Publishers, Chicago, 1978.

113. Capron, A.M.: The Competence of Children as Self-Deciders in Biomedical Intervention: The Problems of Proxy Consent. In: Who Speaks for the Child, pp. 57–114, ed. by W. Gaylin and R. Macklin. Plenum Press, New York, 1982.

114. Capron, A.M.: A Concluding, and Possibly Final, Exchange About "Therapy" and "Research." IRB: Rev Human Subjects Res 4 (1) (January 1982) 10.

115. Capron, A.M.: Ethics of Phase I Clinical Trials. JAMA 249 (1983) 882–883.

116. Cardon, P.V.; Dommel, F.W.; Trumble, R.R.: Injuries to Research Subjects: A Survey of Investigators. New Engl J Med 295 (1976) 650–654.

117. Carson, R.A.; Frias, J.L.; Melker, R.J.: Research with Brain Dead Children. IRB: Rev Human Subjects Res 3 (1) (January 1981) 5–6.

118. Cassel, C.: Research on Senile Dementia of the Alzheimer's Type: Ethical Issues Involving Informed Consent. In: Alzheimer's Dementia: Dilemmas in Clinical Research, pp. 99–108, ed. by V.L. Melnick and N.N. Dubler. Humana Press, Clifton, N.J., 1985.

119. Cassell, E.J.: The Function of Medicine. Hastings Center Rept 7 (6) (December 1977) 16–19.

120. Cassell, E.J.: The Nature of Suffering and the Goals of Medicine. New Engl J Med 306 (1982) 639–645.

121. Cassell, E.J.: The Healer's Art. The MIT Press, Cambridge, Mass., 1985.

122. Cassell, E.J.: Siegler, M.: Changing Values in Medicine. University Publications of America, Frederick, Md., 1985.

123. Cassileth, B.R.; Lusk, E.J.; Miller, D.S.; Hurwitz, S.: Attitudes Toward Clinical Trials Among Patients and the Public. JAMA 248 (1982) 968–970.

124. Cato, A.E.; Cook, L.: Clinical Research. In: The Clinical Research Process in the Pharmaceutical Industry, pp. 217–238, ed. by G.M. Matoren. Marcel Dekker, Inc., New York, 1984.

125. Ceci, S.J.; Peters, D.; Plotkin, J.: Human Subjects Review, Personal Values, and the Regulation of Human Subjects Research. Am Psychologist 40 (1985) 994–1002.

126. Cello, J.P.; Grendell, J.H.; Crass, R.A.; et al.: Endoscopic Sclerotherapy versus Portacaval Shunt in Patients with Severe Cirrhosis and Variceal Hemorrhage. New Engl J Med 311 (1984) 1589–1594.

127. Chalmers, T.C.: The Clinical Trial. Milbank Mem Fund Quart 59 (1981) 324–339.

128. Chalmers, T.C.; Block, J.B.; Lee, S.: Controlled Studies in Clinical Cancer Research. New Engl J Med 287 (1972) 75–78.

129. Chalmers, T.C.; Celano, P.; Sacks, H.S.; Smith, H., Jr.: Bias in Treatment Assignment in Controlled Clinical Trials. New Engl J Med 309 (1983) 1358–1361.

130. Chalmers, T.C.; van Den Noort, S.; Lockshin, M.D.; Waksman, B.H.: Summary of a Workshop on the Role of Third-Party Payors in Clinical Trials of New Agents. New Engl J Med 309 (1983) 1334–1336.

131. Chervenak, F.A.; Farley, M.A.; Walters, L.; Hobbins, J.C.; Mahoney, M.J.: When is Termination of Pregnancy During the Third Trimester Morally Justifiable? New Engl J Med 310 (1984) 501–504.

132. Childress, J.F.: Compensating Injured Research Subjects: I. The Moral Argument. Hastings Center Rept 6 (6) (December 1976) 21–27.

133. Chlebowski, R.T.: How Many Protocols Are Deferred? One IRB's Experience. IRB: Rev Human Subjects Res 6 (5) (September/October 1984) 9–10.

134. Chodoff, P.: Involuntary Hospitalization of the Mentally Ill as a Moral Issue. Am J Psychiatry 141 (1984) 384–389.

135. Christakis, N.: Do Medical Student Research Subjects Need Special Protection? IRB:

Rev Human Subjects Res 7 (3) (May/June 1985) 1–4.

136. Christakis, N.; Panner, M.: Baby Fae and the Media: How the Law Allows Appropriate Access. IRB: Rev Human Subjects Res 8 (2) (March/April 1986) 5–7.

137. Christopherson, L.K.: Heart Transplants. Hastings Center Rept 12 (1) (February 1982) 18–21.

138. Clarke, A.C.: Clarke's Third Law on UFO's. Science 159 (1968) 255.

139. Clendinen, D.: The Abortion Conflict: What It Does to One Doctor. NY Times Mag (August 11, 1985) 18–29, 42.

140. Cobb, L.A.; Thomas, G.I.; Dillard, D.H.; Merendino, K.A.; Bruce, R.A.: An Evaluation of Internal Mammary Artery Ligation by a Double Blind Technique. New Engl J Med 260 (1959) 1115–1118.

141. *Cobbs v. Grant*, 502 P 2d 1, Cal 1972.

142. Cohen, C.: Medical Experimentation on Prisoners. Perspectives Biol Med 21 (1978) 357–372.

143. Cohen, D.: Bathroom Behaviors: A Watershed in Ethical Debate. Hastings Center Rept 8 (2) (April 1978) 13.

144. Cohen, J.M.: The Benefits of Professional Staff for IRBs. IRB: Human Subjects Res 3 (6) (June/July 1981) 8–9.

145. Cohen, J.M.: Extra Credit for Research Subjects. IRB: Rev Human Subjects Res 4 (8) (October 1982) 10–11.

146. Cohen, J.M.: The Costs of IRB Review. In: Human Subjects Research: A Handbook for Institutional Review Boards, pp. 39–47, ed. by R.A. Greenwald, M.K. Ryan, and J.E. Mulvihill. Plenum Press, New York, 1982.

147. Cohen, J.M.; Hedberg, W.B.: The Annual Activity of a University IRB. IRB: Rev Human Subjects Res 2 (5) (May 1980) 5–6.

148. Cohen, S.N.: Development of Drug Therapy for Children. Federation Proceedings 36 (1977) 2356–2358.

149. Cole, T.R.: The "Enlightened" View of Aging: Victorian Morality in a New Key. Hastings Center Rept 13 (3) (June 1983) 34–40.

150. Commission on the Federal Drug Approval Process: Final Report. U.S. Government Printing Office No. 98-901-0, Washington, October 1982.

151. Committee to Consider the Social, Ethical and Legal Issues Arising From In Vitro Fertilization. Report of the Disposition of Embryos Produced by In Vitro Fertilization. Melbourne, Australia, 1984.

152. The Committee of Inquiry into Human Fertilization and Embryology: Report. Her Majesty's Stationery Office, London, 1984.

153. Comptroller General of The United States: Services for Patients Involved in National Institutes of Health-Supported Research: How Should They Be Classified and Who Should Pay for Them? U.S. General Accounting Office Publication No. HRD-78-21, Washington, December 22, 1977.

154. Comroe, J.H., Jr.; Dripps, R.D.: Scientific Basis for the Support of Biomedical Science. Science 192 (1976) 105–111.

155. Conference on the Role of the Individual and the Community in the Research, Development and Use of Biologicals with Criteria for Guidelines: A Memorandum. Bull World Health Organization 54 (1976) 645–655.

156. Connell, A.: The Nature of Responsibility. In: Ethical Responsibility in Medicine, pp. 1–22, ed. by V. Edmunds and C.G. Scorer. E. and S. Livingstone, Edinburgh, 1967.

157. Cornford, F.M.: Microcosmographia Academica: Being a Guide for the Young Academic Politician, (eighth edition, 1970.) Bowes and Bowes, Ltd., London, 1908.

158. Coulehan, J.L.; Eberhard, S.; Kapner, L.; Taylor, F.; Rogers, K.; Garry, P.: Vitamin C and Acute Illness in Navajo School Children. New Engl J Med 295 (1976) 973–977.

159. Council for Science and Society: Human Protection: Ethical Aspects of the New Techniques: Report of a Working Party. Oxford University Press, Oxford, 1984.

160. Cowan, D.H.: Human Experimentation: The Review Process in Practice. Case-Western Reserve Law Rev 25 (1975) 533–564.

161. Cowan, D.H.: Scientific Design, Ethics and Monitoring: IRB Review of Randomized Clinical Trials. IRB: Rev Human Subjects Res 2 (9) (November 1980) 1–4.

162. Cowan, D.H.: The Ethics of Clinical Trials of Ineffective Therapy. IRB: Rev Human Subjects Res 3 (5) (May 1981) 10–11.

163. Cowan, D.H.: Should Information Be Collected for a Tumor Registry, and Should It Be Available for Research? IRB: Rev Human Subjects Res 4 (3) (March 1982) 8.

164. Cowan, D.H.; Adams, B.R.: Ethical and Legal Considerations for IRBs: Research with Medical Records. IRB: Rev Human Subjects Res 1 (8) (December 1979) 1–4 and 8.

165. Crawford-Brown, D.J.: Truth and Meaning in the Determination of Radiogenic Risk. IRB: Rev Human Subjects Res 5 (5) (September/October 1983) 1–5 and 12.

166. Culliton, B.J.: Mo Cell Case Has its First Court Hearing. Science 226 (1984) 813–814.

167. Culliton, B.J.: Court Rules in Patient

Privacy Case. Science 229 (1985) 360–361.

168. Culver, C.M.: Should We Research Doctor-Patient Sex? IRB: Rev Human Subjects Res 3 (5) (May 1981) 7–8.

169. Culver, C.M.; Gert, B.: Philosophy in Medicine: Conceptual and Ethical Issues in Medicine and Psychiatry. Oxford University Press, New York, 1982.

170. Cupples, B.; Gochnauer, M.: The Investigator's Duty Not to Deceive. IRB: Rev Human Subjects Res 7 (5) (September/October 1985) 1–6.

171. Curran, J.W.: AIDS—Two Years Later. New Engl J Med 309 (1983) 609–611.

172. Curran, W.J.: Government Regulation of the Use of Human Subjects in Medical Research: The Approaches of Two Federal Agencies. In: Experimentation with Human Subjects, pp. 402–454, ed. by P.A. Freund. George Braziller, New York, 1970.

173. Curran, W.J.: Ethical Issues in Short Term and Long Term Psychiatric Research. In: Medical, Moral, and Legal Issues in Mental Health Care, ed. by F.J. Ayd. Williams & Wilkins, Baltimore, 1974.

174. Curran, W.J.: Reasonableness and Randomization in Clinical Trials: Fundamental Law and Governmental Regulation. New Engl J Med 300 (1979) 1273–1275.

175. Curran, W.J.: AIDS Research and "The Window of Opportunity." New Engl J Med 312 (1985) 903–904.

176. Davis, D.S.: "Dear Mrs. X. . ." IRB: Rev Human Subjects Res 5 (6) (November-December 1983) 6–9.

177. de Sola Pool, I.: Protecting Human Subjects of Research: An Analysis of Proposed Amendments to HEW Policy. Political Sci 12 (1979) 452–455.

178. DeBakey, L.: Ethically Questionable Data: Publish or Reject? Clin Res 22 (1974) 113–121.

179. DeBakey, L.: The Scientific Journal: Editorial Policies and Practices. C.V. Mosby Co., St. Louis, 1976.

180. DeBakey, L.: Literacy: Mirror of Society. Technical Writing Commun 8 (1978) 279–319.

181. Department of Health, Education, and Welfare: Protection of Human Subjects: Policies and Procedures. Fed Register 38 (221) (November 16, 1973) 31738–31749.

182. Department of Health, Education, and Welfare: The NIH Normal Volunteer Program. DHEW Publication No. (NIH) 5-74, Washington, 1975.

183. Department of Health, Education, and Welfare: Secretary's Task Force on the Compensation of Injured Research Subjects: Report. DHEW Publication No. (OS) 77-003, Appendix A, DHEW Publication No. (OS) 77-004, Appendix B, DHEW Publication No. (OS) 7-005, Washington, 1977.

184. Department of Health, Education, and Welfare: Proposed Regulations on Research Involving Prisoners. Fed Register 43 (3) (January 5, 1978) 1050–1053.

185. Department of Health, Education, and Welfare: Protection of Human Subjects: Research Involving Children. Fed Register 43 (141) (July 21, 1978) 31786–31794.

186. Department of Health, Education, and Welfare: Informed Consent: Definition Amended to Include Advice on Compensation: Interim Final Regulation. Fed Register 43 (214) (November 3, 1978) 51559.

187. Department of Health, Education, and Welfare: Protection of Human Subjects: Additional Protections Pertaining to Biomedical and Behavioral Involving Prisoners as Subjects. Fed Register 43 (222) (November 16, 1978) 53652–53656.

188. Department of Health, Education, and Welfare: Protection of Human Subjects: Proposed Regulations on Research Involving Those Institutionalized as Mentally Infirm. Fed Register 43 (223) (November 17, 1978) 53950–53956.

189. Department of Health, Education, and Welfare: Proposed Regulations Amending Basic HEW Policy for Protection of Human Subjects. Fed Register 44 (158) (August 14, 1979) 47688–47698.

190. Department of Health and Human Services: IRB Compliance Activity Workshop. November 7, 1980, transcript, 149 pp.

191. Department of Health and Human Services: Additional Protections for Children Involved as Subjects in Research. Fed Register 48 (46) (March 8, 1983) 9814–9820.

192. Department of Health and Human Services: Recombinant DNA Research: Request for Public Comment on "Points to Consider in the Design and Submission of Human Somatic-Cell Gene Therapy Protocols." Fed Register 50 (14) (January 22, 1985) 2939–2945.

193. Diamond, G.A.; Forrester, J.S.: Clinical Trials and Statistical Verdicts: Probable Grounds for Appeal. Ann Intern Med 98 (1983) 385–394.

194. Dieck, G.S.: Posture Pictures, Permission and Privacy Protection. IRB: Rev Human Subjects Res 3 (10) (December 1981) 6–7.

195. Dillon, W.P.; Lee, R.V.; Tronolone, M.J.; Buckwald, S.; Foote, R.J.: Life Support and Maternal Brain Death During Pregnancy. JAMA 248 (1982) 1089–1091.

196. Drane, J.F.: The Many Faces of Com-

petency. Hastings Center Report 15 (No. 2) (April 1985) 17–21.

197. Dresser, R.: Bound to Treatment: The Ulysses Contract. Hastings Center Rept 14 (3) (June 1984) 13–16.

198. Dubler, N.N.: The Burdens of Research in Prisons. IRB: Rev Human Subjects Res 4 (9) (November 1982) 9–10.

199. DuVal, B.S., Jr.: The Human Subjects Protection Committee: An Experiment in Decentralized Federal Regulation. American Bar Foundation Res J Summer (3) (1979) 571–688.

200. Dyer, A.R.: Informed Consent and the Nonautonomous Person. IRB: Rev Human Subjects Res 4 (7) (August/September 1982) 1–4.

201. Editorial: Research Regulations: Watching the Watchful Watchers Watch. Hastings Center Rept 6 (1) (February 1976) 2–3.

202. Editorial: Vitamins, Neural-Tube Defects and Ethics Committees. Lancet 1 (1980) 1061–1062.

203. Eichwald, E.J.; Woolley, F.R.; Cole, B.; Beamer, V.: Insertion of the Total Artificial Heart. IRB: Rev Human Subjects Res 3 (7) (August/September 1981) 4–5.

204. Ellenberg, S.S.: Randomization Designs in Comparative Clinical Trials. New Engl J Med 310 (1984) 1404–1408.

205. Engelhardt, H.T., Jr.: A Study of the Federal Government's Ethical Obligations to Provide Compensation for Persons Injured in the Course of Their Participation in Research Supported by Funds Administered by the Secretary of Health, Education and Welfare. In: HEW Secretary's Task Force on the Compensation of Injured Research Subjects: Report, Appendix A, pp. 45–63. DHEW Publication No. (OS) 77-004, Washington 1977.

206. Epstein, L.C.; Lasagna, L.: Obtaining Informed Consent: Form or Substance? Arch Intern Med 123 (1969) 682–688.

207. Ethics Advisory Board, Department of Health and Human Services: Report. Fed Register 44 (118) (June 18, 1979) 35033–35058.

208. Ethics Advisory Board, Department of Health and Human Services: The Request of the National Institutes of Health for a Limited Exemption from the Freedom of Information Act. Report Submitted to the Secretary, DHHS, May 21, 1980.

209. Faden, R.R.; Lewis, C.; Becker, C.; Faden, A.I.; Freeman, J.: Disclosure Standards and Informed Consent. J Health Politics, Policy Law 6 (1981) 255–284.

210. Faden, R.R.; Lewis, C.; Rimer, B.: Monitoring Informed Consent Procedures:

An Exploratory Record Review. IRB: Rev Human Subjects Res 2 (8) (October 1980) 9–10.

211. *Fante and the Upjohn Company v. the Department of Health and Human Services et al.*, Civil Action No. 80-72778, U.S. District Court, E.D. Mich, 1980.

212. Farley, M.A.: Obligating Features of Personhood. Unpublished Manuscript, Yale University, 1982.

213. Feinstein, A.R.: Clinical Judgment. Williams & Wilkins, Baltimore, 1967.

214. Feinstein, A.R.: Clinical Biostatistics: XXVI. Medical Ethics and the Architecture of Clinical Research. Clin Pharmacol Therapeutics 15 (1974) 316–334.

215. Feinstein, A.R.: Clinical Biostatistics. C.V. Mosby, St. Louis, 1977.

216. Feinstein, A.R.: An Additional Basic Science for Clinical Medicine: II. The Limitations of Randomized Trials. Ann Intern Med 99 (1983) 544–550.

217. Feinstein, A.R.: An Additional Basic Science for Clinical Medicine: IV. The Development of Clinimetrics. Ann Intern Med 99 (1983) 843–848.

218. Finkel, M.J.: Proposed Regulations for Study of New Drugs in Children. In: Clinical Pharmacology and Therapeutics: A Pediatric Perspective, pp. 299–304, ed. by B.L. Mirkin. Year Book Medical Publishers, Chicago, 1978.

219. Fletcher, J.: Fetal Research: An Ethical Appraisal. In: The National Commission for the Protection of Human Subjects of Biomedical and Behavioral Research: Research on the Fetus: Report and Recommendations. Appendix, pp. 3.1–3.14. DHEW Publication No. (OS) 76-128, Washington, 1975.

220. Fletcher, J.C.; Dommel, F.W. Jr.; Cowell, D.D.: Consent to Research with Impaired Human Subjects. IRB: Rev Human Subjects Res 7 (6) (November/December 1985) 1–6.

221. Fletcher, J.C.; Schulman, J.D.: Fetal Research: The State of the Question. Hastings Center Rept 15 (2) (April 1985) 6–12.

222. Food and Drug Administration: General Considerations for the Clinical Evaluation of Drugs. DHEW Publication No. (FDA) 77-3040, Washington, 1977.

223. Food and Drug Administration: Protection of Human Subjects: Proposed Establishment of Regulations. Fed Register 43 (88) (May 5, 1978) 19417–19422.

224. Food and Drug Administration: Protection of Human Subjects: Proposed Establishment of Regulations. Fed Register 44 (80) (April 24, 1979) 24106–24111.

225. Food and Drug Administration: Protection of Human Subjects: Standards for Institutional Review Boards for Clinical Investigations. Fed Register 44 (158) (August 14, 1979) 47699–47712.

226. Food and Drug Administration: Protection of Human Subjects: Informed Consent. Fed Register 44 (158) (August 14, 1979) 47713–47729.

227. Food and Drug Administration: Medical Devices: Procedures for Investigational Device Exemptions. Fed Register 45 (13) (January 18, 1980) 3732–3759.

228. Food and Drug Administration: Protection of Human Subjects: Prisoners Used as Subjects in Research. Fed Register 45 (106) (May 30, 1980) 36386–36392.

229. Food and Drug Administration: Protection of Human Subjects: Prisoners Used as Research Subjects: Reproposal of Regulations. Fed Register 46 (243) (December 18, 1981) 61666–61671.

230. Food and Drug Administration: FDA: Regulatory Requirements for Medical Devices: A Workshop Manual. DHHS Publication FDA-83-4165, Rockville, Md., 1983.

231. Food and Drug Administration: Import/Export of Medical Devices: A Workshop Manual. DHHS Publication FDA–83–4167, Rockville, Md., 1983.

232. Fost, N.: A Surrogate System for Informed Consent. JAMA 233 (1975) 800–803.

233. Fost, N.: Consent as a Barrier to Research. New Engl J Med 300 (1979) 1272–1273.

234. Fost, N.: Research on the Brain Dead. J Pediatrics 96 (1980) 54–56.

235. Fost, N.; Chudwin, D.; Wikler, D.: The Limited Moral Significance of "Fetal Viability." Hastings Center Rept 10 (6) (December 1980) 10–13.

236. Fost, N.; Cohen, S.: Ethical Issues Regarding Case Reports: To Publish or Perish the Thought. Clin Res 24 (1976) 269–273.

237. Fost, N.; Robertson, J.A.: Deferring Consent with Incompetent Patients in an Intensive Care Unit. IRB: Rev Human Subjects Res 2 (7) (August/September 1980) 5–6.

238. Fox, R.C.: Experiment Perilous. The Free Press, Glencoe, Illinois, 1959.

239. Fox, R.C.: Some Social and Cultural Factors in American Society Conducive to Medical Research on Human Subjects. Clin Pharmacol Therapeutics 1 (1960) 423–443.

240. Fox, R.C.; Swazey, J.P.: The Courage to Fail: A Social View of Organ Transplants and Dialysis, second edition. University of Chicago Press, Chicago, 1978.

241. Fox, R.C.; Swazey, J.P.: Medical Morality Is Not Bioethics—Medical Ethics in China and the United States. Perspectives Biol Med 27 (1984) 336–360.

242. Fox, R.C.; Willis, D.P.: Personhood, Medicine and American Society. Milbank Mem Fund Quart 61 (1) (Winter 1983) 127–147.

243. Frank, S.; Agich, G.J.: Nontherapeutic Research on Subjects Unable to Grant Consent. Clin Res 33 (1985) 459–464.

244. Frankena, W.K.: Ethics, second edition. Prentice-Hall, Inc., Englewood Cliffs, N.J., 1973.

245. Fredrickson, D.S.: Welcoming Remarks. Clin Pharmacol Therapeutics 25 (1979) 630–631.

246. Freedman, B.: A Moral Theory of Informed Consent. Hastings Center Rept 5 (4) (August 1975) 32–39.

247. Freedman, B.: The Validity of Ignorant Consent to Medical Research. IRB: Rev Human Subjects Res 4 (2) (February 1982) 1–5.

248. Freedman, B.: Man Bites Dog: A Bioethicist's Deception. IRB: Rev Human Subjects Res 5 (5) (September/October 1983) 8–9.

249. Freedman, B: Withdrawing Data as a Substitute for Consent. IRB: Rev Human Subjects Res 5 (6) (November/December 1983) 10.

250. Freireich, E.J.; Gehan, E.A.: The Limitations of the Randomized Clinical Trial. In: Methods in Cancer Research: Cancer Drug Development, Volume XVII, Part B, pp. 277–310, ed. by V.T. DeVita, Jr., and H. Busch. Academic Press, New York, 1979.

251. Freund, P.A.: Ethical Problems in Human Experimentation. New Engl J Med 273 (1965) 687–692.

252. Freund, P.A.: Experimentation with Human Subjects. George Braziller, New York, 1970.

253. Fried, C.: Medical Experimentation: Personal Integrity and Social Policy. American Elsevier Company, New York, 1974.

254. Friedman, L.; Demets, D.: The Data Monitoring Committee: How It Operates and Why. IRB: Rev Human Subjects Res 3 (4) (April 1981) 6–8.

255. Gamble, H.F.: Students, Grades and Informed Consent. IRB: Rev Human Subjects Res 4 (5) (May 1982) 7–10.

256. Gaylin, W.: Harvesting the Dead: The Potential for Recycling Human Bodies. Harper's Magazine 249 (September 1974) 23–30.

257. Gaylin, W.: Competence: No Longer All or None. In: Who Speaks for the Child, pp. 27–54, ed. by W. Gaylin and R. Macklin.

Plenum Press, New York, 1982.

258. Gaylin, W.; Macklin, R., Editors: Who Speaks for the Child: The Problems of Proxy Consent. Plenum Press, New York, 1982.

259. Geller, D.M.: Alternatives to Deception: Why, What and How? In: The Ethics of Social Research: Surveys and Experiments, pp. 39–55, ed. by J.E. Sieber. Springer-Verlag, New York, 1982.

260. Giammona, M.; Glantz, S.A.: Poor Statistical Design in Research on Humans: The Role of Committees on Human Research. Clin Res 31 (1983) 572–578.

261. Gilligan, C.: In a Different Voice: Psychological Theory and Women's Development. Harvard University Press, Cambridge, Mass., 1982.

262. Glantz, L.H.: Commentary: Property Rights and Excised Tissue. IRB: Rev Human Subjects Res 1 (6) (October 1979) 5–6.

263. Goffman, E.: Asylums. Doubleday & Co., Garden City, N.Y., 1961.

264. Goldiamond, I.: On the Usefulness of Intent for Distinguishing Between Research and Practice, and Its Replacement by Social Contingency: Implications for Standard and Innovative Procedures, Coercion and Contractual Relations. In: The National Commission for the Protection of Human Subjects of Biomedical and Behavioral Research: The Belmont Report: Ethical Principles and Guidelines for the Protection of Human Subjects of Research. Appendix II, pp. 14.1–14.73. DHEW Publication (OS) 78-0014, Washington, 1978.

265. Goldman, J.; Katz, M.D.: Inconsistency and Institutional Review Boards. JAMA 248 (1982) 197–202.

266. Goldstein, J.: On the Right of the "Institutionalized Mentally Infirm" to Consent to or Refuse to Participate as Subjects in Biomedical and Behavioral Research. The National Commission for the Protection of Human Subjects of Biomedical and Behavioral Research: Research Involving Those Institutionalized as Mentally Infirm: Report and Recommendations. Appendix, pp. 2.1–2.39. DHEW Publication No. (OS) 78-0007, Washington, 1978.

267. Gordis, L.: Conceptual and Methodologic Problems in Measuring Patient Compliance. In: Compliance in Health Care, ed. by R.B. Haynes, D.W. Taylor, and D.L. Sackett. Johns Hopkins University Press, Baltimore, 1979.

268. Gordis, L.: Gold, E.: Privacy, Confidentiality and the Use of Medical Records in Research. Science 207 (1980) 153–156.

269. Gordon, R.S., Jr.: The Design and Conduct of Randomized Clinical Trials. IRB: Rev Human Subjects Res 7 (1) (January/February 1985) 1–3 and 12.

270. Gordon, R.S., Jr.; Fletcher, J.C.: Can Strict Randomization Be Ethically Acceptable? Clin Res 31 (1983) 23–25.

271. Gray, B.H.: Human Subjects in Medical Experimentation. Wiley-Interscience, New York, 1975.

272. Gray, B.H.: Changing Federal Regulations of IRBs, Part III: Social Research and the Proposed DHEW Regulations. IRB: Rev Human Subjects Res 2 (1) (January 1980) 1–5 and 12.

273. Gray, B.H.; Cook, R.A.; Tannenbaum, A.S.: Research Involving Human Subjects. Science 201 (1978) 1094–1101.

274. Grobstein, C.: The Moral Use of "Spare" Embryos. Hastings Center Rept 12 (3) (June 1982) 5–6.

275. Grobstein, C.; Flower, M.: Gene Therapy: Proceed with Caution. Hastings Center Rept 14 (2) (April 1984) 13–17.

276. Grobstein, C.; Flower, M.; Mendeloff, J.: Frozen Embryos: Policy Issues. New Engl J Med 313 (1985) 1584–1588.

277. Grundner, T.M.: On the Readability of Surgical Consent Forms. New Engl J Med 302 (1980) 900–902.

278. Grundner, T.M.: How to Make Consent Forms More Readable. IRB: Rev Human Subjects Res 3 (7) (August/September 1981) 9–10.

279. Gutheil, T.G.; Applebaum, P.S.: Substituted Judgment: Best Interests in Disguise. Hastings Center Rept 13 (3) (June 1983) 8–11.

280. Hallisey, P.L.: The Fetal Patient and the Unwilling Mother: A Standard for Judicial Intervention. Pacific Law J 14 (1983) 1065–1094.

281. Harrison, M.R.; Filly, R.A.; Golbus, M.S.; et al.: Fetal Treatment 1982. New Engl J Med 307 (1982) 1651–1652.

282. Harvey, M.; Levine, R.J.: Risk of Injury Associated with Twenty Invasive Procedures Used in Human Experimentation and Assessment of Reliability of Risk Estimates. In: President's Commission for the Study of Ethical Problems in Medicine and Biomedical and Behavioral Research: Compensating for Research Injuries: The Ethical and Legal Implications of Programs to Redress Injured Subjects, Appendix, pp. 73–171. U.S. Government Printing Office, Stock No. 040-000-00456-4, Washington, 1982.

283. Harvey, M.; Levine, R.J.: The Risk of

Research Procedures: Methodologic Problems and Proposed Standards. Clin Res 31 (1983) 126–139.

284. The Hastings Center: A Fifteenth Anniversary Symposium: Autonomy—Paternalism—Community. Hastings Center Rept 14 (5) (October 1984) 5–49.

285. Havighurst, C.C.: Mechanisms for Compensating Persons Injured in Human Experimentation. In: HEW Secretary's Task Force on the Compensation of Injured Research Subjects: Report, Appendix A, DHEW Publication No. (OS) 77-004, pp. 81–132, Washington, 1977.

286. Heath, E.J.: The IRB's Monitoring Function: Four Concepts of Monitoring. IRB: Rev Human Subjects Res 1 (5) (August/September 1979) 1–3 and 12.

287. Heath, E.J.: In A "No Risk" Protocol, Does the Purpose Count? IRB: Rev Human Subjects Res 1 (6) (October 1979) 5.

288. Heath, E.J.: What is a Workable Protocol Numbering System? IRB: Rev Human Subjects Res 2 (9) (November 1980) 8.

289. Herceg-Baron, R.: Parental Consent and Family Planning Research Involving Minors. IRB: Rev Human Subjects Res 3 (9) (November 1981) 5–8.

290. Herrman, S.S.: The Noninstitutional Review Board: A Case History. IRB: Rev Human Subjects Res 6 (1) (January/February 1984) 1–3 and 12.

291. Hershey, N.: Law and Privacy Protection. IRB: Rev Human Subjects Res 3 (6) (June/July 1981) 10.

292. Hershey, N.: Using Patient Records for Research: The Response from Federal Agencies and the State of Pennsylvania. IRB: Rev Human Subjects Res 3 (8) (October 1981) 7–8.

293. Hershey, N.: IRB Jurisdiction and Limits on IRB Actions. IRB: Rev Human Subjects Res 7 (2) (March/April 1985) 7–9.

294. Hill, A.B.: Medical Ethics and Controlled Trials. Brit Med J 1 (1963) 1043–1049.

295. Hodges, R.E.; Bean, W.B.: The Use of Prisoners for Medical Research. JAMA 202 (1967) 513–515.

296. Holden, C.: Ethics in Social Science Research. Science 206 (1979) 537–540.

297. Holder, A.R.: Medical Malpractice Law, second edition. John Wiley and Sons, New York, 1978.

298. Holder, A.R.: Consent to the Use of An Investigational Cardiac Assist Device. IRB: Rev Human Subjects Res 1 (1) (March 1979) 6–7.

299. Holder, A.R.: What Commitment is Made by a Witness to a Consent Form? IRB: Rev Human Subjects Res 1 (7) (November 1979) 7–8.

300. Holder, A.R.: Job Applicants as Research Subjects: The Case of a Rubella Vaccine Trial. IRB: Rev Human Subjects Res 2 (2) (February 1980) 5–7.

301. Holder, A.R.: FDA's Final Regulations: IRBs and Medical Devices. IRB: Rev Human Subjects Res 2 (6) (June/July 1980) 1–4.

302. Holder, A.R.: Can Teenagers Participate in Research Without Parental Consent? IRB: Rev Human Subjects Res 3 (2) (February 1981) 5–7.

303. Holder, A.R.: Videotaping on a Psychiatric Unit. IRB: Rev Human Subjects Res 3 (3) (March 1981) 4–5.

304. Holder, A.R.: Can Amniocentesis Be Performed Solely for Research? IRB: Rev Human Subjects Res 3 (6) (June/July 1981) 6–7.

305. Holder, A.R.: Do Researchers and Subjects Have a Fiduciary Relationship? IRB: Rev Human Subjects Res 4 (1) (January 1982) 6–7.

306. Holder, A.R.: Contraceptive Research: Do Sex Partners Have Rights? IRB: Rev Human Subjects Res 4 (2) (February 1982) 6–7.

307. Holder, A.R.: Involuntary Commitment, Incompetency and Consent. IRB: Rev Human Subjects Res 5 (2) (March/April 1983) 6–8.

308. Holder, A.R.: Teenagers and Questionnaire Research. IRB: Rev Human Subjects Res 5 (3) (May/June 1983) 4–6.

309. Holder, A.R.: Research on Unemployment: When Statutes Create Vulnerability. IRB: Rev Human Subjects Res 6 (2) (March/April 1984) 6.

310. Holder, A.R.: The Unlicensed Physician in the Research Institution. IRB: Rev Human Subjects Res 7 (3) (May/June 1985) 5–6.

311. Holder, A.R.: When Researchers are Served Subpoenas. IRB: Rev Human Subjects Res 7 (4) (July/August 1985) 5–7.

312. Holder, A.R.: Is This a Job for the IRB? The Case of the ELISA Assay. IRB: Rev Human Subjects Res 7 (6) (November/December 1985) 7–8.

313. Holder, A.R.: Legal Issues in Pediatrics and Adolescent Medicine, second edition. Yale University Press, New Haven, 1985.

314. Holder, A.R.; Levine, R.J.: Informed Consent for Research on Specimens Ob-

tained at Autopsy or Surgery: A Case Study in the Overprotection of Human Subjects. Clin Res 24 (2) (1976) 68–77.

315. Hollenberg, N.K.; Dzau, V.J.; Williams, G.H.: Letter to the Editor: Are Uncontrolled Clinical Studies Ever Justified? New Engl J Med 303 (1980) 1067.

316. Hollister, L.E.: Prediction of Therapeutic Uses of Psychotherapeutic Drugs from Experiences with Normal Volunteers. Clin Pharmacol Therapeutics 13 (1972) 803–808.

317. Horwitz, R.I.; Feinstein, A.R.: Improved Observational Method for Studying Therapeutic Efficacy: Suggestive Evidence that Lidocaine Prophylaxis Prevents Death in Acute Myocardial Infarction. JAMA 246 (1981) 2455–2459.

318. Houston, M.C.: Evaluating Risk When Antihypertensive Medications are Abruptly Discontinued in Research Subjects. IRB: Rev Human Subjects Res 6 (1) (January/February 1984) 9–11.

319. Houston, J.; Whittemore, A.S.; Hoover, J.J.; Panos, M.; et al.: How Blind Was the Patient in AMIS? Clin Pharmacol Therapeutics 32 (1982) 543–553.

320. Howe, E.G., III; Kark, J.A.; Wright, D.G.: Studying Sickle Cell Trait in Healthy Army Recruits: Should the Research Be Done? Clin Res 31 (1983) 119–125.

321. Huff, T.A.: The IRB as Deputy Sheriff: Proposed FDA Regulation of the Institutional Review Board. Clin Res 27 (1979) 103–108.

322. Humphreys, L: Tearoom Trade: Impersonal Sex in Public Places. Aldine Publishing Co., Chicago, 1970.

323. Ingelfinger, F.J.: Informed (but Uneducated) Consent. New Engl J Med 287 (1972) 466–470.

324. Ingelfinger, F.J.: Ethics of Experimentation on Children. New Engl J Med 288 (1973) 791–792.

325. Ingelfinger, F.J.: The Unethical in Medical Ethics. Ann Intern Med 83 (1975) 264–269.

326. Ivy, A.C.: The History and Ethics of the Use of Human Subjects in Medical Experiments. Science 108 (1948) 1–5.

327. Janis, I.L.: Stress Inoculation in Health Care: Theory and Research. In: Stress Reduction and Prevention, pp. 67–99, ed. by D. Meichenbaum and M.E. Jaremko. Plenum Press, New York, 1983.

328. Jellinek, M.S.: IRBs and Pharmaceutical Company Funding of Research. IRB: Rev Human Subjects Res 4 (8) (October 1982) 9.

329. Jonas, H.: Philosophical Reflections on Experimenting with Human Subjects. In: Experimentation with Human Subjects, pp. 1–31, ed. by P.A. Freund. George Braziller, New York, 1970.

330. Jones, H.W. Jr.: Ethics of *In Vitro* Fertilization: 1984. Ann New York Acad Sci 442 (1985) 577–582.

331. Jonsen, A.R.: A Map of Informed Consent. Clin Res 23 (1975) 277–279.

332. Jonsen, A.R.: Do No Harm. Ann Intern Med 88 (1978) 827–832.

333. Jonsen, A.R.: Ethical Issues in Compliance. In: Compliance in Health Care, ed. by R.B. Haynes, D.W. Taylor, and D.L. Sackett. Johns Hopkins University Press, Baltimore, 1979.

334. Jonsen, A.R.: A Concord in Medical Ethics. Ann Intern Med 99 (1983) 261–264.

335. Jonsen, A.R.: Watching the Doctor. New Engl J Med 308 (1983) 1531–1535.

336. Jonsen, A.R.; Eichelman, B.: Ethical Issues in Psychopharmacologic Treatment. In: Legal and Ethical Issues in Human Research and Treatment, pp. 143–158, ed. by D.M. Gallant and R. Force. Spectrum Publications, New York, 1978.

337. Jonsen, A.R.; Yesley, M.: Rhetoric and Research Ethics: An Answer to Annas. Medicolegal News 8 (6) (December 1980) 8–13.

338. Kahn, N.; Siris E.S.: Information, Not Unnecessary Alarm. IRB: Rev Human Subjects Res 7 (3) (May/June 1985) 9.

339. Kapp, M.B.: Children's Assent for Participation in Pediatric Research Protocols. Clin Pediatrics 22 (1983) 275–278.

340. Katz, J.: Letter to the Editor. New Engl J Med 275 (1966) 790.

341. Katz, J.: Experimentation with Human Beings. Russell Sage Foundation, New York, 1972.

342. Katz, J.: The Regulation of Human Research—Reflections and Proposals. Clin Res 21 (1973) 785–791.

343. Katz, J.: The Silent World of Doctor and Patient. The Free Press, New York, 1984.

344. Katz, J.; Capron, A.M.: Catastrophic Disease—Who Decides What? Russell Sage Foundation, New York, 1975.

345. Katz, M.D.; Goldman, J.: Reply: Compelling Evidence for New Policies. IRB: Rev Human Subjects Res 6 (1) (January/February 1984) 6–8.

346. Kavanagh, C.; Matthews, D.; Sorenson, J.R.; Swazey, J.P.: We Shall Overcome: Multi-institutional Review of a Genetic Counseling Study. IRS: Rev Human Subjects Res 1 (2) (April 1979) 1–3 and 12.

347. Kay, E.M.: Legislative History of Title

II—Protection of Human Subjects of Biomedical and Behavioral Research—of the National Research Act: PL-93-348. Unpublished manuscript prepared for the Commission, 1975.

348. Kelman, H.C.: The Human Use of Human Subjects: The Problem of Deception in Social Psychological Experiments. Psycholog Bull 67 (1967) 1–11.

349. Kelman, H.C.: Privacy and Research with Human Beings. J Soc Issues 33 (1977) 169–195.

350. Kelsey, J.R.: Privacy and Confidentiality in Epidemiological Research Involving Patients. IRB: Rev Human Subjects Res 3 (2) (February 1981) 1–4.

351. Kerenyi, T.D.; Chitkara, U.: Selective Birth in Twin Pregnancy with Discordancy for Down's Syndrome. New Engl J Med 304 (1981) 1525–1527.

352. King, L.S.: Medical Thinking: An Historical Preface. Princeton University Press, Princeton, New Jersey, 1982.

353. Kinney, E.L.; Trautmann, J.; Gold, J.A.; Vesell, E.S.; Zelis, R.: Underrepresentation of Women in New Drug Trials. Ann Intern Med 95 (1981) 495–499.

354. Kjelsberg, M.: Memorandum to MRFIT Clinical Center Principal Investigators. Minneapolis, February 21, 1979.

355. Knight, J.A.: Exploring the Compromise of Ethical Principles in Science. Perspect Biol Med 27 (1984) 432–442.

356. Kohlberg, L.: The Philosophy of Moral Development: Moral Stages and the Idea of Justice. Harper and Row, San Francisco, 1981.

357. Koocher, G.P.: Bathroom Behaviors and Human Dignity. J Personality Soc Psychology 35 (1977) 120–121.

358. Kopelman, L.: Ethical Controversies in Medical Research: The Case of XYY Screening. Perspect Biol Med 21 (1978) 196–204.

359. Kopelman, L.: Estimating Risk in Human Research. Clin Res 29 (1981) 1–8.

360. Kopelman, L.: Randomized Clinical Trials, Consent and the Therapeutic Relationship. Clin Res 31 (1983) 1–11.

361. Lackey, D.P.: Which Subjects Should an IRB Protect? Two Moral Models. IRB: Rev Human Subjects Res 4 (7) (August/September 1982) 5–6.

362. Ladd, J.: The Idea of Community. New Engl J, New England Chapter of the American Institute of Planners 1 (1) (August 1972) 6–43.

363. Ladd, J.: Legalism and Medical Ethics. J Med Philosophy 4 (1979) 70–80.

364. Ladd, J.: Philosophy and the Moral Professions. In: Social Controls and the Medical Profession, pp. 11–30, ed. by J.P. Swazey and S.R. Scher. Oelgeschlager, Gunn and Hain, Boston, 1985.

365. Landesman, S.H.; Ginzburg, H.M.; Weiss, S.H.: The AIDS Epidemic. New Engl J Med 312 (1985) 521–525.

366. Lappé, M.: Ethics at the Center of Life: Protecting Vulnerable Subjects. Hastings Center Rept 8 (5) (October 1978) 11–13.

367. Lasagna, L.: Drug Evaluation Problems in Academic and Other Contexts. Ann New York Acad Sci 169 (1970) 503–508.

368. Lasagna, L.: Prisoner Subjects and Drug Testing. Fed Proc 36 (1977) 2349–2351.

369. Lasagna, L.; von Felsinger, J.M.: The Volunteer Subject in Research. Science 120 (1954) 359–361.

370. Lathrop, V.G.: Careful Review, Not Bureaucratic Delay. IRB: Rev Human Subjects Res 5 (4) (July/August 1983) 9–10.

371. Lebacqz, K.: The National Commission and Research in Pharmacology: An Overview. Fed Proc 36 (1977) 2344–2348.

372. Lebacqz, K.: Reflections on the Report and Recommendations of the National Commission: Research on the Fetus. Villanova Law Rev 22 (1977) 357–366.

373. Lebacqz, K.: Ethical Issues in Psychopharmacological Research. In: Legal and Ethical Issues in Human Research and Treatment: Psychopharmacologic Considerations, pp. 113–138, ed. by D.M. Gallant and R. Force. Spectrum Publications, New York, 1978.

374. Lebacqz, K.: Beyond Respect for Persons and Beneficence: Justice in Research. IRB: Rev Human Subjects Res 2 (7) (August/September 1980) 1–4.

375. Lebacqz, K.: Justice, Choice and the Language of Research. Clin Res 31 (1983) 26–27.

376. Lebacqz, K.: Professional Ethics: Power and Paradox. Abingdon Press, Nashville, 1985.

377. Lebacqz, K.; Levine, R.J.: Respect for Persons and Informed Consent to Participate in Research. Clin Res 25 (1977) 101–107.

378. Lebacqz, K.; Levine, R.J.: Informed Consent in Human Research: Ethical and Legal Aspects. In: Encyclopedia of Bioethics, pp. 754–762, ed. by W.T. Reich. The Free Press, New York, 1978.

379. Leikin, S.L.: An Ethical Issue in Biomedical Research: The Involvement of Minors in Informed and Third Party Consent. Clin Res 31 (1983) 34–40.

380. Levine, C.: Ethics Advisory Board Ap-

proves Waivers for Fetoscopy Research. IRB: Rev Human Subjects Res 1 (2) (April 1979) 8.

381. Levine, C.: Commentary: Teenager, Research and Family Involvement. IRB: Rev Human Subjects Res 3 (9) (November 1981) 8.

382. Levine, C.: Premenstrual Syndrome: Do Raging Hormones Lead to Crime? Hastings Center Rept 12 (4) (August 1982) 2.

383. Levine, C.: NIH Modifies Gene Therapy Research Guidelines. Hastings Center Rept 15 (3) (June 1985) 2–3.

384. Levine, C.; Bermel, J. (editors): AIDS: The Emerging Ethical Dilemmas. Hastings Center Rept 15 (4), Supplement, (August 1985) 1–31.

385. Levine, M.: IRB Review as a "Cooling Out" Device, IRB: Rev Human Subjects Res 5 (4) (July/August 1983) 8–9.

386. Levine, R.J.: Variation in Gastric Histamine Levels and Effects of Histidine Decarboxylase Inhibition in Rats from Different Sources. Biochem Pharmacol 15 (1966) 403–405.

387. Levine, R.J.: Ethical Considerations in the Publication of the Results of Research Involving Human Subjects. Clin Res 21 (1973) 763–767.

388. Levine, R.J.: Guidelines for Negotiating Informed Consent with Prospective Human Subjects of Experimentation. Clin Res 22 (1974) 42–46.

389. Levine, R.J.: Letter to M.S. Yesley. Reproduced in Agenda Book 5 of the Commission, March 23, 1975.

390. Levine, R.J.: Symposium on Definitions of Fetal Life. Clin Res 23 (1975) 103–105.

391. Levine, R.J.: Viability and Death of the Human Fetus: Biologic Definitions. Clin Res 23 (1975) 211–216.

392. Levine, R.J.: Boundaries Between Research Involving Human Subjects and Accepted and Routine Professional Practices. In: Human Experimentation, pp. 3–20, ed. by R.L. Bogomolny. Southern Methodist University Press, Dallas, 1976.

393. Levine, R.J.: The Role of Assessment of Risk-Benefit Criteria in the Determination of the Appropriateness of Research Involving Human Subjects. Bioethics Dig 1 (3) (July 1976) 1–17.

394. Levine, R.J.: The Impact on Fetal Research of the Report of the National Commission for the Protection of Human Subjects of Biomedical and Behavioral Research. Villanova Law Rev 22 (1977) 367–383.

395. Levine, R.J.: Informed Consent to Participate in Research (in 2 parts). Bioethics Digest 1 (11) (March 1977) 1–13; 1 (12) (April 1977) 1–16.

396. Levine, R.J.: Introduction to the Symposium: Recommendations of the National Commission for the Protection of Human Subjects of Biomedical and Behavioral Research: Impact on Research in Pharmacology. Fed Proc 36 (1977) 2341–2343.

397. Levine, R.J.: Non-developmental Research on Human Subjects: The Impact of the Recommendations of the National Commission for the Protection of Human Subjects of Biomedical and Behavioral Research. Fed Proc 36 (1977) 2359–2364.

398. Levine, R.J.: Appropriate Guidelines for the Selection of Human Subjects for Participation in Biomedical and Behavioral Research. In Appendix I to The National Commission for the Protection of Human Subjects of Biomedical and Behavioral Research: The Belmont Report: Ethical Principles and Guidelines for the Protection of Human Subjects of Research, pp. 4.1–4.103. DHEW Publication No. (OS) 78-0013, Washington, 1978.

399. Levine, R.J.: Biomedical Research. In: Encyclopedia of Bioethics, pp. 1481–1492, ed. by W.T. Reich. The Free Press, New York, 1978.

400. Levine, R.J.: The Boundaries Between Biomedical or Behavioral Research and the Accepted and Routine Practice of Medicine. In: The National Commission for the Protection of Human Subjects of Biomedical and Behavioral Research: The Belmont Report: Ethical Principles and Guidelines for the Protection of Human Subjects of Research, Appendix I, pp. 1.1–1.44. DHEW Publication (OS) 78-0013, Washington, 1978.

401. Levine, R.J.: Commentary on Research and the Law: A Decade of Distrust: Terminological Inexactitude. In: Legal and Ethical Issues in Human Research and Treatment: Psychopharmacologic Considerations, pp. 85–98, ed. by D.M. Gallant and R. Force. Spectrum Publications, New York, 1978.

402. Levine, R.J.: Drug Evaluation in Children: Ethical, Legal and Regulatory Considerations. In: Clinical Pharmacology and Therapeutics: A Pediatric Perspective, pp. 275–278, ed. by B.L. Mirkin. Year Book Medical Publishers, Chicago, 1978.

403. Levine, R.J.: The Institutional Review Board. In: The National Commission for the Protection of Human Subjects of Biomedical and Behavioral Research, Institutional Review Boards: Report and Recommendations, Appendix, pp. 4.1–4.73. DHEW Publication No. (OS) 78-0009, Washington, 1978.

404. Levine, R.J.: The Nature and Definition

of Informed Consent in Various Research Settings. In: The National Commission for the Protection of Human Subjects of Biomedical and Behavioral Research: The Belmont Report: Ethical Principles and Guidelines for the Protection of Human Subjects of Research, Appendix I, pp. 3.1–3.91. DHEW Publication (OS) 78-0013, Washington, 1978.

405. Levine, R.J.: On the Relevance of Ethical Principles and Guidelines Developed for Research to Health Services Conducted or Supported by the Secretary. In: The National Commission for the the Protection of Human Subjects of Biomedical and Behavioral Research, Report and Recommendations: Ethical Guidelines for the Delivery of Health Services by DHEW. Appendix, pp. 2.1–2.36. DHEW Publication No. (OS) 78-0011, Washington, 1978.

406. Levine, R.J.: Regulation of Drug Abuse Research Involving Human Subjects. Proceedings of the Fortieth Annual Scientific Meeting, Committee on Problems of Drug Dependence, pp. 45–55, 1978.

407. Levine, R.J.: Research Involving Children: The National Commission's Report. Clin Res 26 (1978) 61–66.

408. Levine, R.J.: The Role of Assessment of Risk-Benefit Criteria in the Determination of the Appropriateness of Research Involving Human Subjects. In: The National Commission for the Protection of Human Subjects of Biomedical and Behavioral Research: The Belmont Report: Ethical Principles and Guidelines for the Protection of Human Subjects of Research, Appendix I, pp. 2.1–2.59. DHEW Publication (OS) 78-0013, Washington, 1978.

409. Levine, R.J.: Changing Federal Regulations of IRBs: The Commission's Recommendations and the FDA's Proposals. IRB: Rev Human Subjects Res 1 (1) (March 1979) 1–3 and 12.

410. Levine, R.J.: Advice on Compensation: One IRB's Response to DHEW's "Interim Final Regulation." IRB: Rev Human Subjects Res 1 (1) (March 1979) 5.

411. Levine, R.J.: Advice on Compensation: More Responses to DHEW's "Interim Final Regulation." IRB: Rev Human Subjects Res 1 (2) (April 1979) 5–7.

412. Levine, R.J.: Clarifying the Concepts of Research Ethics. Hastings Center Rept 9 (3) (June 1979) 21–26.

413. Levine, R.J.: Deceiving Dentists: Health Care Providers as "Subjects at Risk." IRB: Rev Human Subjects Res 1 (5) (August/September 1979) 7–8.

414. Levine, R.J.: What Should Consent

Forms say about Cash Payments? IRB: Rev Human Subjects Res 1 (6) (October1979) 7–8.

415. Levine, R.J.: Changing Federal Regulations of IRBs, Part II: DHEW's and FDA's Proposed Regulations. IRB: Rev Human Subjects Res 1 (7) (November 1979) 1–5 and 12.

416. Levine, R.J.: Regulation of the Use of Human Tissues and Body Fluids as Research Materials: Current Modifications. Biochem Pharmacol 28 (1979) 1893–1895.

417. Levine, R.J.: Medical Students as Social Scientists: Are There Role Conflicts? IRB: Rev Human Subjects Res 2 (1) (January 1980) 6–8.

418. Levine, R.J.: The Senate's Proposed Statutory Definition of "Voluntary and Informed Consent." IRB: Rev Human Subjects Res 2 (4) (April 1980) 8–10.

419. Levine, R.J.: Ethical Considerations in the Development and Application of Compliance Strategies for the Treatment of Hypertension. In: Patient Compliance to Prescribed Antihypertensive Medication Regimens: A Report to the National Heart, Lung and Blood Institute, pp. 229–248, ed. by R.B. Haynes, (NIH Publication No. 81–2102), Bethesda, 1980.

420. Levine, R.J.: The Impact of Institutional Review Boards on Clinical Research. Perspec Biol Med 23 (1980) S98–S114.

421. Levine, R.J.: A Primary Reviewer System. IRB: Rev Human Subjects Res 3 (6) (June/July 1981) 9–10.

422. Levine, R.J.: Can or Should the IRB Assume the FDA's Functions at Early Stages of the IND Process? IRB: Rev Human Subjects Res 3 (10) (December 1981) 4–5.

423. Levine, R.J.: Response to Capron. IRB: Rev Human Subjects Res 4 (1) (January 1982) 10.

424. Levine, R.J.: Research on Prisoners: Why Not? IRB: Rev Human Subjects Res 4 (5) (May 1982) 6.

425. Levine, R.J.: What Are the IRB's Obligations to Review the Use of Drugs for Purposes That FDA Has Not Approved? IRB: Rev Human Subjects Res 4 (4) (April 1982) 8–9.

426. Levine, R.J.: Commentary on Jellinek. IRB: Rev Human Subjects Res 4 (8) (October 1982) 9–10.

427. Levine, R.J.: Reply to Dubler on the Burdens of Research in Prisons. IRB: Rev Human Subjects Res 4 (9) (November 1982) 10–11.

428. Levine, R.J.: Consent to Incomplete Disclosure As An Alternative to Deception. IRB: Rev Human Subjects Res 4 (10) (De-

cember 1982) 9–11.

429. Levine, R.J.: Research Involving Children: An Interpretation of the New Regulations. IRB: Rev Human Subjects Res 5 (4) (July/August 1983) 1–5.

430. Levine, R.J.: Informed Consent in Research and Practice: Similarities and Differences. Arch Intern Med 143 (1983) 1229–1231.

431. Levine, R.J.: No Code! Commentary. Bioethics Reporter (1983) 10–11.

432. Levine, R.J.: Inconsistency and IRBs: Flaws in the Goldman-Katz Study. IRB: Rev Human Subjects Res 6 (1) (January/February 1984) 4–6.

433. Levine, R.J.: What Kind of Subjects Can Understand This Protocol? IRB: Rev Human Subjects Res 6 (5) (September/October 1984) 6–8.

434. Levine, R.J.: Total Artificial Heart Implantation—Eligibility Criteria. JAMA 252 (1984) 1458–1459.

435. Levine, R.J.: The IRB and the Virtuous Investigator. IRB: Rev Human Subjects Res 7 (1) (January/February 1985) 8.

436. Levine, R.J.: The Use of Placebos in Randomized Clinical Trials. IRB: Rev Human Subjects Res 7 (2) (March/April 1985) 1–4.

437. Levine, R.J.: Referral of Patients with Cancer for Participation in Randomized Clinical Trials: Ethical Considerations. Ca—Cancer J Clinicians, 36 (1986) 95–99.

438. Levine, R.J.: Research That Could Yield Marketable Products from Human Materials: The Problem of Informed Consent. IRB: Rev Human Subjects Res 8 (1) (January-Febuary 1986) 6–7.

439. Levine, R.J.: Legal Constraints on Do Not Resuscitate Decisions In: Proceedings of The Conference on Science and Morality (at Nazareth College of Rochester, N.Y., November 5, 1982), ed. by D.T. Zallen (to be published).

440. Levine, R.J.; Cohen, E.D.: The Hawthorne Effect. Clin Res 22 (1974) 111–112.

441. Levine, R.J.; Lebacqz, K.: Some Ethical Considerations in Clinical Trials. Clin Pharmacol Therapeutics 25 (1979) 728–741.

442. Levine, R.J.; Nolan, K.A.: Do Not Resuscitate Decisions: A Policy. Connecticut Med 47 (1983) 511–512.

443. Levinson, S.: Under Cover: The Hidden Costs of Infiltration. Hastings Center Rept 12 (4) (August 1982) 29–37.

444. Lewis, J.H.: An IRB's Consent Form Survey. IRB: Rev Human Subjects Res 6 (4) (July/August 1984) 10.

445. Lewis, M.: The Ethics of Inducing Para-

noia in an Experimental Setting. IRB: Rev Human Subjects Res 3 (10) (December 1981) 9.

446. Lidz, C.W.; Meisel, A.; Osterweis, M.; et al.: Barriers to Informed Consent. Ann Intern Med 99 (1983) 539–543.

447. Lidz. C.W.; Meisel, A.; Zerubavel, E.; Carter, M.; Sestak, R.M.; Roth, L.H.: Informed Consent: A Study of Decisionmaking in Psychiatry. The Guilford Press, New York, 1984.

448. Lind, S.E.: Can Patients Be Asked to Pay for Experimental Treatment? Clin Res 32 (1984) 393–398.

449. The Lipid Research Clinics Program: The Coronary Primary Prevention Trial: Design and Implementation. J Chron Dis 32 (1979) 609–631.

450. Lipsett, M.B.: Therapeutic and Nontherapeutic Research. Hastings Center Rept 9 (6) (December 1979) 47–48.

451. Lipsett, M.B.: On the Nature and Ethics of Phase I Clinical Trials of Cancer Chemotherapies. JAMA 248 (1982) 941–942.

452. Lipsett, M.B.: Ethics of Phase I Clinical Trials. JAMA 249 (1983) 883.

453. Lipsett, M.B.; Fletcher, J.C.: Do Vitamins Prevent Neural Tube Defects (and Can We Find Out Ethically?) Hastings Center Rept 13 (4) (August 1983) 5–8.

454. Lipsett, M.B.; Fletcher, J.C.; Secundy, M.: Research Review at NIH. Hastings Center Rept 9 (1) (February 1979) 18–21.

455. Louis, T.A.; Lavori, P.W.; Bailar, J.C., III; Polansky, M.: Crossover and Self-Controlled Designs in Clinical Research. New Engl J Med 310 (1984) 24–31.

456. Luy, M.L.L.: Package Insert Roulette: The Catch-22 of Prescribing. Mod Med 44 (5) (1976) 23.

457. MacIntyre, A.: After Virtue: A Study in Moral Theory. University of Notre Dame Press, Notre Dame, Indiana, 1981.

458. MacKay, C.R.: Ethical Issues in Research Design and Conduct: Developing a Test to Detect Carriers of Huntington's Disease. IRB: Rev Human Subjects Res 6 (4) (July/August 1984) 1–5.

459. Macklin, R.: "Due" and "Undue" Inducements: On Paying Money to Research Subjects. IRB: Rev Human Subjects Res 3 (5) (May 1981) 1–6.

460. Macklin, R.: Response: Beyond Paternalism. IRB: Rev Human Subjects Res 4 (3) (March 1982) 6–7.

461. Macklin, R.: Personhood in the Bioethics Literature. Milbank Mem Fund Quart 61 (1) (Winter 1983) 35–57.

462. Mahler, D.M.: Issues of Timing of In-

formed Consent in Behavioral Research: A Study of Mother-Infant Bonding. IRB: Rev Human Subjects Res 8 (3) (May/June 1986) 7–11.

463. Mahoney, M.J.: Implications for Restrictions on Fetal Research for Biomedical Advance. Clin Res 23 (1975) 229–232.

464. Makarushka, J.L.: The Requirement for Informed Consent in Research on Human Subjects: The Problem of Uncontrolled Consequences of Health-Related Research. Clin Res 24 (1976) 64–67.

465. Marcy, S.E.: A Systems Study of a University Committee for Protection of Human Subjects of Experimentation. M.P.H. Dissertation, 77 pp., Yale University, May 1974.

466. Marini, J.L: Methodology and Ethics: Research on Human Aggression. IRB: Rev Human Subjects Res 2 (5) (May 1980) 1–4.

467. Marini, J.L.; Sheard, M.H.; Bridges, C.I.: An Evaluation of "Informed Consent" with Volunteer Prisoner Subjects. Yale J Biol Med 49 (1976) 427–437.

468. Marquis, D.: Leaving Therapy to Chance. Hastings Center Rept 13 (4) (August 1983) 40–47.

469. Marshall, C.L.; Marshall, C.P.: Poverty and Health in the United States. In: Encyclopedia of Bioethics, pp. 1316–1321, ed. by W.T. Reich. The Free Press, New York, 1978.

470. Martin, D.C.; Arnold, J.D.; Zimmerman, T.F.; Richart, R.H.: Human Subjects in Clinical Research—A Report of Three Studies. New Engl J Med 279 (1968) 1426–1431.

471. Mashaw, J.L.: Thinking About Institutional Review Boards. In: The President's Commission for the Study of Ethical Problems in Medicine and Biomedical and Behavioral Research: Whistleblowing in Biomedical Research: Policies and Procedures for Responding to Reports of Misconduct, pp. 3–22. U.S. Government Printing Office, Stock No. 040-000-00458-1, Washington, 1982.

472. Massé, F.X.; Miller, T.: Exposure to Radiation and Informed Consent. IRB: Rev Human Subjects Res 7 (4) (July/August 1985) 1–4.

473. Mattson, M.E.; Curb, J.D.; McArdle, R.; et al.: Participation in a Clinical Trial: The Patient's Point of View. Controlled Clin Trials 6 (1985) 156–167.

474. Maugh, T.H., II: Blood Substitute Passes Its First Test. Science 206 (1979) 205.

475. McCann, D.J.; Pettit, J.R.: A Report on Adverse Effects Insurance for Human Subjects. In: President's Commission for the Study of Ethical Problems in Medicine and Biomedical and Behavioral Research: Compensating for Research Injuries: The Ethical and Legal Implications of Programs to Redress Injured Subjects, Appendix, pp. 241–274. U.S. Government Printing Office, Stock No. 040-000-00456-4, Washington, 1982.

476. McCarthy, C.R.: Response to de Sola Pool. IRB: Rev Human Subjects Res 3 (10) (December 1981) 8–9.

477. McCarthy, C.R.: Regulatory Aspects of the Distinction Between Research and Practice. IRB: Rev Human Subjects Res 6 (3) (May/June 1984) 7–8.

478. McCartney, J.J.: Encephalitis and Ara-A: An Ethical Case Study. Hastings Center Rept 8 (6) (December 1978) 5–7.

479. McCormick, R.A.: Proxy Consent in the Experimentation Situation. Perspect Biol Med 18 (2) (1974) 2–20.

480. McCrae, R.R.: Are Stress Questionnaires Stressful? IRB: Rev Human Subjects Res 4 (4) (April 1982) 1–2.

481. McNeil, B.J.; Weichselbaum, R.; Pauker, S.G.: Speech and Survival: Tradeoffs Between Quality and Quantity of Life in Laryngeal Cancer. New Engl J Med 305 (1981) 982–987.

482. McNeil, B.J.; Pauker, S.G.; Sox, H.C., Jr.; Tversky, A.: On the Elicitation of Preferences for Alternative Therapies. New Engl J Med 306 (1982) 1259–1262.

483. McNeil, B.J.; Weichselbaum, R.; Pauker, S.G.: Fallacy of the Five-Year Survival in Lung Cancer. New Engl J Med 299 (1978) 1397–1401.

484. Mead, M.: Research with Human Beings: A Model Derived from Anthropological Field Practice. In: Experimentation with Human Subjects, pp. 152–177, ed. by P.A. Freund. George Braziller, New York, 1970.

485. Meinert, C.L.: Funding for Clinical Trials. Controlled Clin Trials 3 (1982) 165–171.

486. Meisel, A.: What Would It Mean to be Competent Enough to Consent To or Refuse Participation in Research: A Legal Overview. Paper presented at the National Institutes of Mental Health Workshop, "Empirical Research on Informed Consent with Subjects of Uncertain Competence," Rockville, Maryland, January 12, 1981.

487. Meisel, A.: Commentary on Grundner. IRB: Rev Human Subjects Res 4 (1) (January 1982) 9.

488. Meisel, A.; Roth, L.H.: What We Do and Do Not Know About Informed Consent. JAMA 246 (1981) 2473–2477.

489. Melnick, V.L.; Dubler, N.N.: Alz-

heimer's Dementia: Dilemmas in Clinical Research. Humana Press, Clifton, New Jersey, 1985.

490. Melton, G.B.; Koocher, G.P.; Saks, M.J. (editors): Children's Competence to Consent. Plenum Press, New York, 1983.

491. Meyers, K.: Drug Company Employees as Research Subjects: Programs, Problems and Ethics. IRB: Rev Human Subjects Res 1 (8) (December 1979) 5–6.

492. Meyers, K.: Is Local Review Being Circumvented? IRB: Rev Human Subjects Res 4 (7) (August/September 1982) 6.

493. Middlemist, R.D.; Knowles, E.S.; Matter, C.F.: Personal Space Invasion in the Lavatory: Suggestive Evidence for Arousal. J Personality Soc Psychology 33 (1976) 541–546.

494. Middlemist, R.D.; Knowles, E.S.; Matter, C.F.: What To Do and What to Report: A Reply to Koocher. J Personality Soc Psychology 35 (1977) 122–124.

495. Miers, M.L.: Current NIH Perspectives on Misconduct in Science. Am Psychologist 40 (1985) 831–835.

496. Milgrim, S.: Issues in the Study of Obedience: A Reply to Baumrind. Am Psychologist 19 (1964) 848–852.

497. Milgrim, S.: Obedience to Authority: An Experimental View. Harper and Row, New York, 1974.

498. Milgrim, S.: Subject Reaction: The Neglected Factor in the Ethics of Experimentation. Hastings Center Rept 7 (5) (October 1977) 19–23.

499. Miller, B.L.: Autonomy and the Refusal of Lifesaving Treatment. Hastings Center Rept 11 (4) (August 1981) 22–28.

500. Miller, B.L.: Autonomy and Proxy Consent. IRB: Rev Human Subjects Res 4 (10) (December 1982) 1–8.

501. Miller, B.L.: Autonomy and Proxy Consent. In: Alzheimer's Dementia: Dilemmas in Clinical Research, pp. 239–263, ed. by V.L. Melnick and N.N. Dubler. Humana Press, Clifton, New Jersey, 1985.

502. Miller, R.; Willner, H.S.: The Two-Part Consent Form: A Suggestion for Promoting Free and Informed Consent. New Engl J Med 290 (1974) 964–966.

503. Miller, R.A.; Schaffner, K.F.; Meisel, A.: Ethical and Legal Issues Related to the Use of Computer Programs in Clinical Medicine. Ann Intern Med 102 (1985) 529–536.

504. Mirkin, B.L.: Drug Therapy and the Developing Human: Who Cares? Clin Res 23 (1975) 106–113.

505. Mirkin, B.L.: Impact of Public Policy on the Development of Drugs for Pregnant Women and Children. Clin Res 23 (1975) 233–237.

506. Mitford, J.: Experiments Behind Bars: Doctors, Drug Companies and Prisoners. Atlantic Monthly 231 (1) (January 1973) 64–73.

507. Mitford, J.: Kind and Usual Punishment: The Prison Business. Vintage Press, New York, 1974.

508. Moe, K.: Should the Nazi Research Data Be Cited? Hastings Center Rept 14 (6) (December 1984) 5–7.

509. Moore, F.D.: Therapeutic Innovation: Ethical Boundaries in the Initial Clinical Trials of New Drugs and Surgical Procedures. In: Experimentation with Human Subjects, pp. 358–378, ed. by P.A. Freund. George Braziller, New York, 1970.

510. Moore, F.D.: First Class. New Engl J Med 302 (1980) 1202–1203.

511. Moore, J.D.: The Deer Lodge Research Unit. Clin Pharmacol Therapeutics 13 (1972) 833–834.

512. Moore, R.S.; Hofmann, A.D. (editors): American Academy of Pediatrics Conference on Consent and Confidentiality in Adolescent Health Care. American Academy of Pediatrics, Evanston, Illinois, 1982.

513. Morgan, P.P.: Randomized Clinical Trials Need to be More Clinical. JAMA 253 (1985) 1782–1783.

514. Morris, R.C.: Legal Problems of Emergency and Outpatient Care. Connecticut Med 38 (1974) 543–547.

515. Murphy, E.A.: The Analysis and Interpretation of Experiments: Some Philosophical Issues. J Med Philosophy 7 (1982) 307–325.

516. Murphy, W.D.; Thomasma, D.C.: The Ethics of Research on Court-Ordered Evaluation and Therapy for Exhibitionism. IRB: Rev Human Subjects Res 3 (9) (November 1981) 1–4.

517. Murray, J.C.; Pagon, R.A.: Informed Consent for Research Publication of Patient-Related Data. Clin Res 32 (1984) 404–408.

518. Murray, T.H.: Learning to Deceive. Hastings Center Rept 10 (2) (April 1980) 11–14.

519. Murray, T.H.: Was This Deception Necessary? IRB: Rev Human Subjects Res 2 (10) (December 1980) 7–8.

520. Murray, T.H.: Who Owns the Body? On the Ethics of Using Human Tissue for Commercial Purposes. IRB: Rev Human Subjects Res 8 (1) (January/February 1986) 1–5.

521. Murray, W.B.; Buckingham, R.W.: Implications of Participant Observation in

Medical Studies. Can Med Assoc J 115 (1976) 1187–1190.

522. The National Cancer Institute: What Are Clinical Trials All About? A Booklet for Patients with Cancer. NIH Publication No. 85-2706, Bethesda, Maryland, 1984.

523. The National Commission for the Protection of Human Subjects of Biomedical and Behavioral Research: Research on the Fetus: Report and Recommendations. DHEW Publication No. (OS) 76-127, Appendix, DHEW Publication No. (OS) 76-128, Washington, 1975.

524. The National Commission for the Protection of Human Subjects of Biomedical and Behavioral Research: Research Involving Prisoners: Report and Recommendations. DHEW Publication No. (OS) 76-131, Appendix, DHEW Publication No. (OS) 76-132, Washington, 1976.

525. The National Commission for the Protection of Human Subjects of Biomedical and Behavioral Research: Research Involving Children: Report and Recommendations. DHEW Publication No. (OS) 77-0004, Appendix, DHEW Publication No. (OS) 77-0005, Washington, 1977.

526. The National Commission for the Protection of Human Subjects of Biomedical and Behavioral Research: The Belmont Report: Ethical Principles and Guidelines for the Protection of Human Subjects of Research. DHEW Publication No. (OS) 78-0012, Appendix I, DHEW Publication No. (OS) 78-0013, Appendix II, DHEW Publication (OS) 78-0014, Washington, 1978.

527. The National Commission for the Protection of Human Subjects of Biomedical and Behavioral Research: Institutional Review Boards: Report and Recommendations. DHEW Publication No. (OS) 78-0008, Appendix, DHEW Publication No. (OS) 78-0009, Washington, 1978.

528. The National Commission for the Protection of Human Subjects of Biomedical and Behavioral Research: Report and Recommendations: Ethical Guidelines for the Delivery of Health Services by DHEW Publication No. (OS) 78-0010, Appendix, DHEW Publication No. (OS) 78-0011, Washington, 1978.

529. The National Commission for the Protection of Human Subjects of Biomedical and Behavioral Research: Research Involving Those Institutionalized as Mentally Infirm: Report and Recommendations. DHEW Publication No. (OS) 78-0006, Appendix, DHEW Publication No. (OS) 78-0007, Washington, 1978.

530. The National Commission for the Protection of Human Subjects of Biomedical and Behavioral Research: Transcript of the 41st Meeting, April 14–15, 1978. Report No. NCPHS-78/04. Available from the National Technical Information Service, U.S. Department of Commerce, Springfield, Virginia.

531. The National Institute on Aging: Protection of Elderly Research Subjects. DHEW Publication No. (NIH) 79-1801, Washington, 1979.

532. The National Institutes of Health: Guidelines and Information for the General Clinical Research Centers Program of the Division of Research Resources. Bethesda, Maryland, June 1982.

533. Nejelski, P.M.: Social Research in Conflict with Law and Ethics. Ballinger, Cambridge, Massachusetts, 1976.

534. Nelkin, D.: The Politics of Personhood. Milbank Mem Fund Quart 61 (1) (Winter 1983) 101–112.

535. Nelson, L.J.; Mills, J.: Ethics and Research Involving Celibate Religious Groups. Clin Res 26 (1978) 322–329.

536. Nelson, R.B.: Are Clinical Trials Pseudoscience? Forum Med 2 (9) (September 1979) 594–600.

537. Newton, L.H.: Inducement, Due and Otherwise. IRB: Rev Human Subjects Res 4 (3) (March 1982) 4–6.

538. Newton, L.H.: Dentists and Pseudo-Patients: Further Meditations on Deception in Research. IRB: Rev Human Subjects Res 4 (8) (October 1982) 6–8.

539. Newton, L.H.: For Whose Protection? Another Moral Model. IRB: Rev Human Subjects Res 5 (3) (May/June 1983) 9–10.

540. Newton, L.H.: Agreement to Participate in Research: Is That a Promise? IRB: Rev Human Subjects Res 6 (2) (March/April 1984) 7–9.

541. Nightingale, S.L.: Drug Regulation and Policy Formulation. Milbank Mem Fund Quart 59 (1981) 412–444.

542. Nightingale, S.L.: The Food and Drug Administration's Role in the Protection of Human Subjects. IRB: Rev Human Subjects Res 5 (1) (January/February 1983) 6–9.

543. Noble, G.R. (Program Chairman): International Conference on Acquired Immunodeficiency Syndrome. Ann Intern Med 103 (1985) 653–781.

544. Nolan, K.A.: "Protecting" Medical Students from the Risks of Research. IRB: Rev Human Subjects Res 1 (5) (August/September 1979) 9.

545. Nolan, K.A.: Student Members: "In-

formed Outsiders'' on IRBs. IRB: Rev Human Subjects Res 2 (8) (October 1980) 1–4.

546. Nolan, K.A.: In Death's Shadow: Foundations and Guidelines for Decisions Not to Resuscitate. M.D. Dissertation, 153 pp., Yale University, 1982.

547. Novick, A.: At Risk for AIDS: Confidentiality in Research and Surveillance. IRB: Rev Human Subjects Res 6 (6) (November/December 1984) 10–11.

548. Oates, J.A.: A Scientific Rationale for Choosing Patients Rather than Normal Subjects for Phase I Studies. Clin Pharmacol Therapeutics 13 (1972) 809–811.

549. Office for Protection from Research Risks: Guidance for IRBs for AIDS Studies. OPRR Reports, December 26, 1984.

550. O'Neil, R.: Determining Proxy Consent. J Med Philosophy 8 (1983) 389–403.

551. Ostfeld, A.M.: Older Research Subjects: Not Homogeneous, Not Especially Vulnerable. IRB: Rev Human Subjects Res 2 (8) (October 1980) 7–8.

552. Outka, G.: Social Justice and Equal Access to Health Care. Perspect Biol Med 18 (1975) 185–203.

553. Ozar, D.T.: The Case Against Thawing Unused Frozen Embryos. Hastings Center Rept 15 (4) (August 1985) 7–12.

554. Parsons, T.: The Social System. The Free Press, New York, 1951.

555. Parsons, T.: Definitions of Health and Illness in the Light of American Values and Social Structure. In: Patients, Physicians and Illness: A Sourcebook in Behavioral Science and Health, pp. 107–127, ed. by E.G. Jaco. The Free Press, New York, 1972.

556. Pattullo, E.L.: Who Risks What in Social Research? IRB: Rev Human Subjects Res 2 (3) (March 1980) 1–3 and 12.

557. Pattullo, E.L.: Transforming a Personal Inquiry Into a Research Project. IRB: Rev Human Subjects Res 3 (4) (April 1981) 5–6.

558. Pattullo, E.L.: How General an Assurance? IRB: Rev Human Subjects Res 3 (5) (May 1981) 8–9.

559. Pattullo, E.L.: The Limits of the ''Right'' of Privacy. IRB: Rev Human Subjects Res 4 (4) (April 1982) 3–5.

560. Pattullo, E.L.: Institutional Review Boards and the Freedom to Take Risks. New Engl J Med 307 (1982) 1156–1159.

561. Pattullo, E.L.: No More Echelons of Review. IRB: Rev Human Subjects Res 5 (4) (July/August 1983) 11.

562. Pattullo, E.L.: The Wages of Sin. IRB: Rev Human Subjects Res 7 (5) (September/October 1985) 7–8.

563. Pence, G.E.: Children's Dissent to Research—A Minor Matter? IRB: Rev Human Subjects Res 2 (10) (December 1980) 1–4.

564. Percival, T.: Medical Ethics. Russell, London, 1803 (A more generally available edition was edited by C.D. Leake, The Williams & Wilkins Co., Baltimore, 1927).

565. Petricciani, J.C.: An Overview of FDA, IRBs and Regulations. IRB: Rev Human Subjects Res 3 (10) (December 1981) 1–3 and 11.

566. Phillips, W.R.: Patients, Pills and Professionals: The Ethics of Placebo Therapy. Pharos 44 (1) (Winter 1981) 21–25.

567. Pollin, W.; Perlin, S.: Psychiatric Evaluation of ''Normal Control'' Volunteers. Am J Psychiatry 115 (1958) 129–133.

568. Powledge, T.M.: A Report from the Del Zio Trial. Hastings Center Rept 8 (5) (October 1978) 15–17.

569. The President's Commission for the Study of Ethical Problems in Medicine and Biomedical and Behavioral Research: Protecting Human Subjects: The Adequacy and Uniformity of Federal Rules and Their Implementation. U.S. Government Printing Office, Stock No. 040-000-00452-1, Washington, 1981.

570. The President's Commission for the Study of Ethical Problems in Medicine and Biomedical and Behavioral Research: Compensating for Research Injuries: The Ethical and Legal Implications of Programs to Redress Injured Subjects. U.S. Government Printing Office, Stock No. 040-000-00455-6, Washington, 1982; the Appendix to this Report is Stock No. 040-000-00456-4.

571. The President's Commission for the Study of Ethical Problems in Medicine and Biomedical Research: Making Health Care Decisions: The Ethical and Legal Implications of Informed Consent in the Patient-Practitioner Relationship. U.S. Government Printing Office, Stock No. 040-000-00459-9, Washington, 1982. Two appendices, Volumes Two and Three, numbered respectively 040-000-00468-8 and 040-000-00469-6.

572. The President's Commission for the Study of Ethical Problems in Medicine and Biomedical and Behavioral Research: Splicing Life: The Social and Ethical Issues of Genetic Engineering with Human Beings. U.S. Government Printing Office, Stock No. 040-000-00464-5, Washington, 1982.

573. The President's Commission for the Study of Ethical Problems in Medicine and

Biomedical and Behavioral Research: Whistle-blowing in Biomedical Research: Policies and Procedures for Responding to Reports of Misconduct. U.S. Government Printing Office, Stock No. 040-000-00458-1, Washington, 1982.

574. The President's Commission for the Study of Ethical Problems in Medicine and Biomedical and Behavioral Research: Implementing Human Research Regulations: The Adequacy and Uniformity of Federal Rules and of Their Implementation. U.S. Government Printing Office, Stock No. 040-000-00471-8, Washington, 1983.

575. The President's Commission for the Study of Ethical Problems in Medicine and Biomedical and Behavioral Research: Screening and Counseling for Genetic Conditions: The Ethical, Social and Legal Implications of Genetic Screening, Counseling and Education Programs. U.S. Government Printing Office, Stock No. 040-000-00461-1, Washington, 1983.

576. The President's Commission for the Study of Ethical Problems in Medicine and Biomedical and Behavioral Research: Summing Up. U.S. Government Printing Office, Stock No. 040-000-00475-1, Washington, 1973.

577. Preus, A.: Respect for the Dead and Dying. J Med Philosophy 9 (1984) 409–415.

578. Privacy Protection Study Commission: Personal Privacy in an Information Society. U.S. Government Printing Office No. 052-003-00395-3, Washington, 1977.

579. Quigley, M.M.; Andrews, L.B.: Human In Vitro Fertilization and the Law. Fertil Steril 42 (1984) 348–355.

580. Rabkin, M.T.: Will the DRG Decimate Clinical Research? Clin Res 32 (1984) 345–347.

581. Rachels, J.: Barney Clark's Key. Hastings Center Rept 13 (2) (April 1983) 17–19.

582. Ramsey, P.: The Patient as Person. Yale University Press, New Haven, 1970.

583. Ramsey, P.: The Enforcement of Morals: Nontherapeutic Research on Children. Hastings Center Rept 6 (4) (August 1976) 21–30.

584. Ramsey, P.: Manufacturing Our Offspring: Weighing the Risks. Hastings Center Rept 8 (5) (October 1978) 7–9.

585. Ramsey, P.: "Unconsented Touching" and the Autonomy Absolute. IRB: Rev Human Subjects Res 2 (10) (December 1980) 9–10.

586. Ratzan, R.M.: "Being Old Makes You Different:" The Ethics of Research With Elderly Subjects. Hastings Center Rept 10 (5) (October 1980) 32–42.

587. Ratzan, R.M.: Cautiousness, Risk and Informed Consent in Clinical Geriatrics. Clin Res 30 (1982) 345–353.

588. Rawls, J.: A Theory of Justice. Harvard University Press, Cambridge, 1971.

589. Reatig, N.: Confidentiality Certificates: A Measure of Privacy Protection. IRB: Rev Human Subjects Res 1 (3) (May 1979) 1–4 and 12.

590. Reatig, N.: Can Investigators Appeal Adverse IRB Decisions? IRB: Rev Human Subjects Res 2 (3) (March 1980) 8–9.

591. Reatig, N.: Government Regulations Affecting Psychopharmacology Research in the United States: Implications for the Future. In: Human Psychopharmacology: Research and Clinical Practice, Vol. II, ed. by G.D. Burrows and J.S. Werry. JAI Press, Inc., Greenwich, Connecticut, 1981.

592. Reilly, P.: When Should an Investigator Share Raw Data with the Subjects? IRB: Rev Human Subjects Res 2 (9) (November 1980) 4–5 and 12.

593. Reiser, S.J.; Dyck, A.J.; Curran, W.J.: Ethics in Medicine: Historical Perspectives and Contemporary Concerns. MIT Press, Cambridge, 1977.

594. Reiss, A., Jr.: Selected Issues in Informed Consent and Confidentiality with Special Reference to Behavioral-Social Science Research-Inquiry. In: The National Commission for the Protection of Human Subjects of Biomedical and Behavioral Research: The Belmont Report: Ethical Principles and Guidelines for the Protection of Human Subjects of Research, Appendix I, pp. 25.1–25.165, DHEW Publication (OS) 78-0013, Washington, 1978.

595. Relman, A.S.: Lessons from the Darsee Affair. New Engl J Med 308 (1983) 1415–1417.

596. Renaud, M.; Beauchemin, J.; Lalonde, C.; Poirer, H.; Berthiaume, S.: Practice Settings and Prescribing Profiles: The Simulation of Tension Headaches to General Practitioners Working in Different Practice Settings in the Montreal Area. Amer J Pub Health 70 (1980) 1068–1073.

597. Renaud, M.: The Ethics of Consumer Protection Research. Amer J Public Health 70 (1980) 1098–1099.

598. *Reyes v. Wyeth Laboratories*, 498 F 2d 1264, CCA 5, 1974.

599. Richman, J.; Weissman, M.M.; Klerman, G.L.; Neu, C.; Prusoff, B.A.: Ethical

Issues in Clinical Trials: Psychotherapy Research in Acute Depression. IRB: Rev Human Subjects Res 2 (2) (February 1980) 1–4.

600. Riskin, L.L.: IRB Review of Psychotherapist-Patient Sexual Relations. IRB: Rev Human Subjects Res 3 (10) (December 1981) 7–8.

601. Rivlin, A.M.; Timpane, P.M.: Ethical and Legal Issues of Social Experimentation. The Brookings Institution, Washington, 1975.

602. Robb, S.S.: All the Adverse Effects of Drawing Blood. IRB: Rev Human Subjects Res 7 (3) (May/June 1985) 7–9.

603. Robertson, J.A.: Compensating Injured Research Subjects: II. The Law. Hastings Center Rept 6 (6) (December 1976) 29–31.

604. Robertson, J.A.: *In Vitro* Conception and Harm to the Unborn. Hastings Center Rept 8 (5) (October 1978) 13–14.

605. Robertson, J.A.: Ten Ways to Improve IRBs. Hastings Center Rept 9 (1) (February 1979) 29–33.

606. Robertson, J.A.: The Law of Institutional Review Boards. UCLA Law Rev 26 (1979) 484–549.

607. Robertson, J.A.: Research on the Brain Dead. IRB: Rev Human Subjects Res 2 (4) (April 1980) 4–6.

608. Robertson, J.A.: Taking Consent Seriously: IRB Intervention in the Consent Process. IRB: Rev Human Subjects Res 4 (5) (May 1982) 1–5.

609. Robertson, J.A.: The Right to Procreate and In Utero Fetal Therapy. J Legal Med 3 (1982) 333–366.

610. Roginsky, M.S.; Handley, A.: Ethical Implications of Withdrawal of Experimental Drugs at the Conclusion of Phase III Trials. Clin Res 26 (1978) 384–388.

611. Rolleston, F.: Was This IRB Too Casual? IRB: Rev Human Subjects Res 3 (8) (October 1981) 11–12.

612. Rosenfeld, A.: At Risk for Huntington's Disease: Who Should Know What and When? Hastings Center Rept 14 (3) (June 1984) 5–8.

613. Rosenhan, D.L.: On Being Sane in Insane Places. Science 179 (1973) 250–258.

614. Ross, L.; Lepper, M.R.; Hubbard, M.: Perserverance and Self-Perception and Social Perception: Biased Attributional Processes in the Debriefing Paradigm. J Personality Soc Psychology 32 (1975) 880–892.

615. Roth, H.P.; Gordon, R.S., Jr.: Proceedings of the National Conference on Clinical Trials Methodology. Clin Pharmacol Therapeutics 25 (1979) 629–766.

616. Roth, L.H.; Lidz, C.W.; Meisel, A.; Soloff, P.H.; et al.: Competency to Decide about Treatment or Research: An Overview of Some Empiricial Data. Internat J Law Psychiatry 5 (1982) 29–50.

617. Rothman, D.J.: Behavior Modification in Total Institutions. Hastings Center Rept 5 (1) (February 1975) 17–24.

618. Rothman, D.J.: Were Tuskegee and Willowbrook "Studies in Nature?" Hastings Center Rept 12 (2) (April 1982) 5–7.

619. Rothman, K.J.: The Rise and Fall of Epidemiology, 1950–2000 A.D. New Engl J Med 304 (1981) 600–602.

620. Royston, I.: Cell Lines from Human Patients: Who Owns Them? A Case Report. Clin Res 33 (1985) 442–443.

621. Rutstein, D.D.: The Ethical Design of Human Experiments. In: Experimentation with Human Subjects, pp. 383–401, ed. by P.A. Freund. George Braziller, New York, 1970.

622. Sackett, D.L.: The Competing Objectives of Randomized Trials. New Engl J Med 303 (1980) 1059–1060.

623. Sacks, H.; Kupfer, S.; Chalmers, T.C.: Letter to the Editor: Are Uncontrolled Clinical Studies Ever Justified? New Engl J Med 303 (1980) 1067.

624. Sadler, A.M.; Sadler, B.L.; Bliss, A.A.: The Physician's Assistant—Today and Tomorrow. Ballinger, Cambridge, Massachusetts, 1975.

625. Sass, H.-M.: Reichsrundschreiben 1931: Pre-Nuremberg German Regulations Concerning New Therapy and Human Experimentation. J Med Philosophy 8 (1983) 99–111.

626. Schafer, A.: The Ethics of the Randomized Clinical Trial. New Engl J Med 307 (1982) 719–724.

627. Schafer, A.: The Randomized Clinical Trial: For Whose Benefit? IRB: Rev Human Subjects Res 7 (2) (March/April 1985) 4–6.

628. Schmutte, G.T.: Using Students as Subjects Without Their Knowledge. IRB: Rev Human Subjects Res 2 (10) (December 1980) 5–6.

629. Schoeman, F.: Children's Competence and Children's Rights. IRB: Rev Human Subjects Res 4 (6) (June/July 1982) 1–6.

630. Schoeman, F.: Parental Discretion and Children's Rights: Background and Implications for Medical Decision-Making. J Med Philosophy 10 (1985) 45–61.

631. Segers, M.C.: Can Congress Settle the Abortion Issue? Hastings Center Rept 12 (3) (June 1982) 20–28.

632. Seppälä, M.: The World Collaborative

Report on *In Vitro* Fertilization and Embryo Replacement: Current State of the Art in January, 1984. Ann NY Acad Sci 442 (1985) 558–563.

633. Seppälä, M.; Edwards, R.G., editors: *In Vitro* Fertilization and Embryo Transfer. Ann NY Acad Sci 442 (1985).

634. Shannon, T.A.: Should Medical Students Be Research Subjects? IRB: Rev Human Subjects Res 1 (2) (April 1979) 4.

635. Shannon, T.A.: Consent in a Neonatal Screening Program. IRB: Rev Human Subjects Res 1 (3) (May 1979) 5–6.

636. Shannon, T.A.; Ockene, I.S.: Approving High Risk, Rejecting Low Risk: The Case of Two Cases. IRB: Rev Human Subjects Res 7 (1) (January/February 1985) 6–8.

637. Shapiro, A.K.: A Contribution to the History of the Placebo Effect. Behav Sci 5 (1960) 109–135.

638. Shapiro, M.F.; Charrow, R.P.: Scientific Misconduct in Investigational Drug Trials. New Engl J Med 312 (1985) 731–736.

639. Shaw, L.W.; Chalmers, T.C.: Ethics in Cooperative Trials. Ann NY Acad Sci 169 (1970) 487–495.

640. Sheldon, R.: The IRB's Responsibility to Itself. Hastings Center Rept 15 (1) (February 1985) 11–12.

641. Shirkey, H.C.: Therapeutic Orphans. J Pediatrics 72 (1968) 119–120.

642. Short, E.M.: L'Affair Darsee. Clin Res 31 (1983) 448–449.

643. Shubin, S.: Research Behind Bars: Prisoners as Experimental Subjects. The Sciences 21 (January 1981) 10–13 and 29.

644. Sieber, J.E.: How Humanism and Determinism Differ: Understanding Risk in Psychological Research. IRB: Rev Human Subjects Res 4 (3) (March 1982) 1–3 and 12.

645. Sieber, J.E.: Deception in Social Research I: Kinds of Deception and the Wrongs They May Involve. IRB: Rev Human Subjects Res 4 (9) (November 1982) 1–5.

646. Sieber, J.E.: The Ethics of Social Research: Surveys and Experiments. Springer-Verlag, New York, 1982.

647. Sieber, J.E.: Deception in Social Research II: Evaluating the Potential for Harm or Wrong. IRB: Rev Human Subjects Res 5 (1) (January/February 1983) 1–6.

648. Sieber, J.E.: Deception in Social Research III: The Nature and Limits of Debriefing. IRB: Rev Human Subjects Res 5 (3) (May/June 1983) 1–4.

649. Sieber, J.E.: Withdrawing Data as a Substitute for Consent. IRB: Rev Human Subjects Res 5 (6) (November/December 1983) 10–11.

650. Siegler, M.: Confidentiality in Medicine—A Decrepit Concept. New Engl J Med 307 (1982) 1518–1521.

651. Siegler, M.; Wikler, D.: Brain Death and Live Birth. JAMA 248 (1982) 1089–1091.

652. Silverman, W.A.: Human Experimentation: A Guided Step into the Unknown. Oxford University Press, New York, 1985.

653. Singer, E.: More on the Limits of Consent Forms. IRB: Rev Human Subjects Res 2 (3) (March 1980) 7.

654. Singer, E.; Frankel, M.R: Informed Consent Procedures in Telephone Interviews. Amer Soc Rev 47 (1982) 416–426.

655. Singer, P.: Making Laws on Making Babies. Hastings Center Rept 14 (4) (August 1985) 5–6.

656. Singer, P.; Wells, D.: *In Vitro* Fertilisation: The Major Issue. J Medical Ethics 9 (1983) 192–195.

657. Siris, E.S.: In Search of Funding: The Clinical Investigator and the Drug Company. IRB: Rev Human Subjects Res 5 (6) (November/December 1983) 1–4.

658. Small, R.D.; Schor, S.S.: Bayesian and Non-Bayesian Methods of Inference. Ann Intern Med 99 (1983) 857–859.

659. Smith, C.P.: How (Un)acceptable is Research Involving Deception? IRB: Rev Human Subjects Res 3 (8) (October 1981) 1–4.

660. Smith, D.H.: On Being Queasy. IRB: Rev Human Subjects Res 2 (4) (April 1980) 6–7.

661. Soble, A.: Deception in Social Science Research: Is Informed Consent Possible? Hastings Center Rept 8 (5) (October 1978) 40–46.

662. Somerville, M.A.: Selective Birth in Twin Pregnancy. New Engl J Med 305 (1981) 1218–1219.

663. Sordillo, P.P.; Schaffner, K.F.: The Last Patient in a Drug Trial. Hastings Center Rept 11 (6) (December 1981) 21–23.

664. Soskis, D.A.: Schizophrenic and Medical Inpatients as Informed Drug Consumers. Arch Gen Psychiatry 35 (1978) 645–647.

665. Spiro, H.M.: Constraint and Consent: On Being a Patient and a Subject. New Engl J Med 293 (1975) 1134–1135.

666. Sprague, R.L.: Obtaining Consent in a Clinical Setting. IRB: Rev Human Subjects Res 7 (2) (March/April 1985) 10.

667. Stanley, B.H.: Informed Consent and Competence: A Review of Empirical Research. Paper presented at the National In-

stitutes of Mental Health Workshop, "Empirical Research on Informed Consent with Subjects of Uncertain Competence." Rockville, Maryland, January 12, 1981.

668. Stanley, B.H.; Guido, J.; Stanley, M.; Shortell, D.: The Elderly Patient and Informed Consent. JAMA 252 (1984) 1302–1306.

669. Stanley, B.H.; Stanley, M.: Testing Competency in Psychiatric Patients. IRB: Rev Human Subjects Res 4 (8) (October 1982) 1–6.

670. Stanley, B.H.; Stanley, M.; Schwartz, N.; Lautin, A.; Kane, J.: The Ability of the Mentally Ill to Evaluate Research Risks. IRCS Medical Sci 8 (1980) 657–658.

671. Steinfels, M.O.: *In Vitro* Fertilization: "Ethically Acceptable" Research. Hastings Center Rept 9 (3) (June 1979) 5–8.

672. Strain, L.A.; Chappell, N.L.: Problems and Strategies: Ethical Concerns in Survey Research With the Elderly. Gerontologist 22 (1982) 526–531.

673. Student Council of New York University School of Medicine: Proceedings of the Symposium on Ethical Issues in Human Experimentation: The Case of Willowbrook State Hospital Research. Urban Affairs Health Program, The New York University Medical Center, New York, 1973.

674. Subcommittee on Health of the Committee on Labor and Public Welfare, United States Senate: Federal Regulation of Human Experimentation, 1975. U.S. Government Printing Office No. 45-273-0, 1975.

675. Swazey, J.P.: Role Conflicts of Social Scientists. IRB: Rev Human Subjects Res 2 (1) (January 1980) 8.

676. Swazey, J.P. (Chairperson): The Problem of Personhood: Biomedical, Social, Legal and Policy Views. Milbank Mem Fund Quart 61 (1) (Winter 1983).

677. Swazey, J.P.; Scher, S.R., editors: Social Controls and the Medical Profession. Oelgeschlager, Gunn and Hain, Boston, 1985.

678. Symposium: Research on the Fetus. Villanova Law Rev 22 (2) (1977) 297–417.

679. Taub, H.A.; Baker, M.T.: A Reevaluation of Informed Consent in the Elderly: A Method of Improving Comprehension Through Direct Testing. Clin Res 32 (1984) 17–21.

680. Taylor, K.M.; Margolese, R.G.; Soskolne, C.L.: Physicians' Reasons for Not Entering Eligible Patients in a Randomized Clinical Trial of Surgery for Breast Cancer. New Engl J Med 310 (1984) 1363–1367.

681. Thomasma, D.C.: Beyond Medical Paternalism and Patient Autonomy: A Model of Physician Conscience for the Physician-Patient Relationship. Ann Intern Med 98 (1983) 243–248.

682. Thomasma, D.C.; Mauer, A.M.: Ethical Complications of Clinical Therapeutic Research on Children. Soc Sci Med 16 (1982) 913–919.

683. Tooley, M.: Abortion and Infanticide. Oxford University Press, New York, 1984.

684. Toulmin, S.: Fetal Experimentation: Moral Issues and Institutional Controls. In: The National Commission for the Protection of Human Subjects of Biomedical and Behavioral Research: Research on the Fetus. Report and Recommendations, Appendix, pp. 10.1–10.26, DHEW Publication No. (OS) 76-128, Washington, 1975.

685. Toulmin, S.: *In Vitro* Fertilization: Answering the Ethical Objections. Hastings Center Rept 8 (5) (October 1978) 9–11.

686. Toulmin, S.: The Tyranny of Principles. Hastings Center Rept 11 (6) (December 1981) 31–39.

687. Tribe, L.H.: American Constitutional Law. The Foundation Press, Mineola, New York, 1978.

688. Tuskegee Syphilis Study Ad Hoc Advisory Panel: Final Report. United States Public Health Service, Washington, D.C., April 1973.

689. Twentieth Century Fund Task Force on the Commercialization of Scientific Research: The Science Business. Priority Press, New York, 1984.

690. VanDeVeer, D.: Experimentation on Children and Proxy Consent. J Med Philosophy 6 (1981) 281–293.

691. Van Eys, J.: The Concept of the IRB and Bureaucratic Reality: An Exchange of Letters. IRB: Rev Human Subjects Res 6 (4) (July/August 1984) 8–10.

692. Veatch, R.M.: Experimental Pregnancy: The Ethical Complexities of Experimentation with Oral Contraceptives. Hastings Center Rept 1 (1) (June 1971) 2–3.

693. Veatch, R.M.: Human Experimentation Committees: Professional or Representative? Hastings Center Rept 5 (5) (October 1975) 31–40.

694. Veatch, R.M.: Ethical Principles in Medical Experimentation. In: Ethical and Legal Issues of Social Experimentation, pp. 21–59, ed. by A.M. Rivlin and P.M. Timpane. The Brookings Institution, Washington, 1975.

695. Veatch, R.M.: Case Studies in Medical Ethics. Harvard University Press, Cambridge, 1977.

696. Veatch, R.M.: Three Theories of In-

formed Consent: Philosophical Foundations and Policy Implications. In: The National Commission for the Protection of Human Subjects of Biomedical and Behavioral Research: The Belmont Report: Ethical Principles of Guidelines for the Protection of Human Subjects of Research, Appendix II, pp. 26.1–26.66, DHEW Publication (OS) 78-0014, Washington, 1978.

697. Veatch, R.M.: Longitudinal Studies, Sequential Design and Grant Renewals: What to do with Preliminary Data. IRB: Rev Human Subjects Res 1 (4) (June/July 1979) 1–3.

698. Veatch, R.M.: Commentary: Beyond Consent to Treatment. IRB: Rev Human Subjects Res 3 (2) (February 1981) 7–8.

699. Veatch, R.M.: The Ethics of Research Involving Radiation. IRB: Rev Human Subjects Res 4 (1) (January 1982) 3–5.

700. Veatch, R.M.: Limits to the Right of Privacy: Reason, Not Rhetoric. IRB: Rev Human Subjects Res 4 (4) (April 1982) 5–7.

701. Veatch, R.M.: Maternal Brain Death: An Ethicist's Thoughts. JAMA 248 (1982) 1102–1103.

702. Veatch, R.M.: Justice and Research Design: The Case for a Semi-Randomization Clinical Trial. Clin Res 31 (1983) 12–22.

703. Wagner, M.S.: The Measure of Investigators' Reputations. IRB: Rev Human Subjects Res 7 (4) (July/August 1985) 8.

704. Walters, L.: Ethical and Public Policy Issues in Fetal Research. In: The National Commission for the Protection of Human Subjects of Biomedical and Behavioral Research: Research on the Fetus. Report and Recommendations, Appendix, pp. 8.1–8.18, DHEW Publication No. (OS) 76-128, Washington, 1975.

705. Walters, L.: Human *In Vitro* Fertilization: A Review of the Ethical Issues. Hastings Center Rept 9 (4) (August 1979) 23–43.

706. Walters, L.: Biomedical Ethics. JAMA 247 (1982) 2942–2944.

707. Walters, L.; et al.: Final Report of the Ad Hoc Committee: The Use of Medical Records for Research at Georgetown University. IRB: Rev Human Subjects Res 3 (3) (March 1981) 1–3.

708. Walters, W.A.W.: Personhood and the Human Embryo. Paper presented to The Second Hannah Conference in the History of Medicine, "Moral Priorities in Medical Research," the University of Western Ontario. London, Ontario, Canada, November, 1984.

709. Ward, C.M.: Letter. Philadelphia As-

sociation for Clinical Trials, Philadelphia, December 30, 1985.

710. Ward, R.; Krugman, S.; Giles, J.P.; Jacobs, A.M.; Bodansky, O.: Infectious Hepatitis: Studies of Its Natural History and Prevention. New Engl J Med 258 (1958) 407–416.

711. Wartofsky, M.W.: On Doing It for Money. In: The National Commission for the Protection of Human Subjects of Biomedical and Behavioral Research: Research Involving Prisoners: Report and Recommendations, Appendix, pp. 3.1–3.24, DHEW Publication No. (OS) 76-132, Washington, 1976.

712. Warwick, D.P.: Tearoom Trade: Means and Ends in Social Research. Hastings Center Studies 1 (1) (1973) 24–38.

713. Warwick, D.P.: Social Scientists Ought to Stop Lying. Psychology Today 8 (February 1975) 38ff.

714. Warwick, D.P.: Contraceptives in the Third World. Hastings Center Rept 5 (4) (August 1975) 9–12.

715. Wasserstrom, R.: Ethical Issues Involved in Experimentation on the Nonviable Human Fetus. In: The National Commission for the Protection of Human Subjects of Biomedical and Behavioral Research: Research on the Fetus: Report and Recommendations. Appendix, pp. 9.1–9.10, DHEW Publication No. (OS) 76-128, Washington, 1975.

716. Watkins, J.C.: Selected Readings: AIDs—the Ethical, Legal and Social Considerations. Public Responsibility in Medicine and Research, Boston, 1985.

717. Way, W.L.: Placebo Controls. New Engl J Med 311 (1984) 413–414.

718. Wei, L.J.: Durham, S.: The Randomized Play-the-Winner Rule in Medical Trials. J Amer Stat Assoc 73 (1978) 840–843.

719. Weintraub, M.: Improving the Ethics of Clinical Trials: The Case of an "Add-On" Study. IRB: Rev Human Subjects Res 5 (3) (May/June 1983) 7–8.

720. Weiss, G.B.: Who Pays for Clinical Research? Clin Res 27 (1979) 297–299.

721. Weiss, R.J.: The Use and Abuse of Deception. Amer J Public Health 70 (1980) 1097–1098.

722. Weithorn, L.A.: Children's Capacities to Decide About Participation in Research. IRB: Rev Human Subjects Res 5 (2) (March/April 1983) 1–5.

723. Weithorn, L.A.: Involving Children in Decisions Involving Their Own Welfare. In: Children's Competence to Consent, pp. 235–260, ed. by G.B. Melton, G.P. Koocher,

and M.J. Saks. Plenum Press, New York, 1983.

724. Wells, S.H.; Kennedy, P.M.; Kenny, J.; Reznikoff, M.; Sheard, M.H.: Pharmacological Testing in a Correctional Institution. Charles C Thomas, Springfield, Illinois, 1975.

725. West, C.: Philosophical Perspective on the Participation of Prisoners in Experimental Research. In: The National Commission for the Protection of Human Subjects of Biomedical and Behavioral Research: Research Involving Prisoners: Report and Recommendations, Appendix, pp. 2.1–2.22, DHEW Publication No. (OS) 76-132, Washington, 1976.

726. Whitbeck, C.: A Different Reality: Feminist Ontology. In: Beyond Domination: New Perspectives on Women and Philosophy, pp. 64–88, ed. by C.C. Gould. Rowman and Allanheld, Totowa, New Jersey, 1983.

727. Whitley, R.J.; Alford, C.A.: Encephalitis and Adenine Arabinoside: An Indictment Without Fact. Hastings Center Rept 9 (4) (August 1979) 4 and 44–46.

728. Wieranga, E.: Proxy Consent and Conterfactual Wishes. J Med Philosophy 8 (1983) 405–416.

729. Wigodsky, H.S.: Fraud and Misrepresentation in Research—Whose Responsibility? IRB: Rev Human Subjects Res 6 (2) (March/April 1984) 1–5.

730. Wikler, D.I.: Ought We Try to Save Aborted Fetuses? Ethics 90 (October 1975) 58–65.

731. Wikler, D.I.: The Central Ethical Problem in Human Experimentation and Three Solutions. Clin Res 26 (1978) 380–388.

732. Wilkins, L.T.: Putting "Treatment" on Trial. Hastings Center Rept 5 (1) (February 1975) 35–48.

733. Williams, P.C.: Success in Spite of Failure: Why IRBs Falter in Reviewing Risks and Benefits. IRB: Rev Human Subjects Res 6 (3) (May/June 1984) 1–4.

734. Winston, M.E.; Winston, S.M.; Appelbaum, P.S.; Rhoden, N.K.: Can a Subject Consent to a "Ulysses Contract?" Hastings Center Rept 12 (4) (August 1982) 26–28.

735. Winter, P.E.: Human Subject Research Review in the Department of Defense. IRB: Rev Human Subjects Res 6 (3) (May/June 1984) 9–10.

736. Wolfe, J.E.; Bone, R.C.: Informed Consent in Critical Care Medicine. Clin Res 25 (1977) 53–56.

737. Woodford, F.P.: Ethical Experimentation and the Editor. New Engl J Med 286 (1972) 892.

738. Woodward, W.E.: Informed Consent of Volunteers: A Direct Measurement of Comprehension and Retention of Information. Clin Res 27 (1979) 248–252.

739. Woolley, F.R.: Ethical Issues in the Implantation of the Total Artificial Heart. New Engl J Med 310 (1984) 292–296.

740. The Working Group on Mechancial Circulatory Support of the National Heart, Lung and Blood Institute: Artificial Heart and Assist Devices: Directions, Needs, Costs, Societal and Ethical Issues. National Heart, Lung and Blood Institute, Bethesda, Maryland, 1985.

741. Worthen, D.M.: Reflections on the FDA's Intraocular Lens Regulations. IRB: Rev Human Subjects Res 2 (4) (April 1980) 1–3.

742. Yale University: Assurance of Compliance with DHHS Regulations for Protection of Human Research Subject. Weekly Bulletin and Calendar, Yale University, New Haven, April 29–May 6, 1985.

743. Yankauer, A.: Commentary (on Brackbill and Hellegers, 1980). Hastings Center Rept 10 (2) (April 1980) 23–24.

744. Yankauer, A.: On Deception, Dentists and Editors. IRB: Rev Human Subjects Res 5 (2) (March/April 1983) 9.

745. Young, M.J.; Williams, S.V.; Eisenberg, J.M.: The Technological Strategist: Employing Techniques of Clinical Decision Making. In: The Machine at the Bedside: Strategies for Using Technology in Patient Care, pp. 153–176, ed. by S.J. Reiser and M. Anbar. Cambridge University Press, Cambridge, 1984.

746. Yoongner, S.J.; Allen, M.; Bartlett, E.T.; et al.: Psychosocial and Ethical Implications of Organ Retrieval. New Engl J Med 313 (1985) 321–324.

747. Zaner, R.M.: A Criticism of Moral Conservatism's View of *In Vitro* Fertilization and Embryo Transfer. Perspect Biol Med 27 (1984) 200–212.

748. Zarafonetis, C.J.D.; Riley, P.A., Jr.; Willis, P.W., III; Power, L.H.; Werbelow, J.; Farhat, L.; Beckwith, W.; Marks, B.H.: Clinically Significant Adverse Effects in a Phase I Testing Program. Clin Pharmacol Therapeutics 24 (1978) 127–132.

749. Zeisel, H.: Reducing the Hazards of Human Experiments Through Modifications in Research Design. Ann NY Acad Sci 169 (1970) 475–486.

750. Zelen, M.: Play-the-Winner Rule and

the Controlled Clinical Trial. J Amer Stat Assoc 64 (1969) 131–146.

751. Zelen, M.: A New Design for Randomized Clinical Trials. New Engl J Med 300 (1979) 1242–1245.

752. Zelen, M.: Alternatives to Classic Randomized Trials. Surgical Clin North America 61 (1981) 1425–1432.

753. Zimbardo, P.G.: The Ethics of Introducing Paranoia in an Experimental Setting. IRB: Rev Human Subjects Res 3 (10) (December 1981) 10–11.

754. Zimmer, A.W.; Calkins, E.; Hadley, E.; Ostfeld, A.M.; Kaye, J.M.; Kaye, D.: Conducting Clinical Research in Geriatric Populations. Ann Intern Med 103 (1985) 276–283.

Appendix 1

Department of Health and Human Services Rules and Regulations[1] 45 CFR 46[2]

Subpart A—Basic HHS Policy for Protection of Human Research Subjects

Sec.

46.101 To what do these regulations apply?

46.102 Definitions.

46.103 Assurances.

46.104 Section reserved.

46.105 Section reserved.

46.106 Section reserved.

46.107 IRB membership.

46.108 IRB functions and operations.

46.109 IRB review of research.

46.110 Expedited review procedures for certain kinds of research involving no more than minimal risk, and for minor changes in approved research.

46.111 Criteria for IRB approval of research.

46.112 Review by institution.

46.113 Suspension or termination of IRB approval of research.

46.114 Cooperative research.

46.115 IRB records.

46.116 General requirements for informed consent.

46.117 Documentation of informed consent.

46.118 Applications and proposals lacking definite plans for involvement of human subjects.

46.119 Research undertaken without the intention of involving human subjects.

46.120 Evaluation and disposition of applications and proposals.

46.121 Investigational new drug or device 30-day delay requirement.

46.122 Use of federal funds.

46.123 Early termination of research funding; evaluation of subsequent applications and proposals.

46.124 Conditions.

Subpart B—Additional Protections Pertaining to Research, Development, and Related Activities Involving

---

[1] There are other Federal regulations applicable to research involving human subjects, many of which are cited throughout this book.

[2] Title 45; Code of Federal Regulations; Part 46: Revised as of March 8, 1983.

Subpart A—Basic HHS Policy for Protection of Human Research Subjects

Source: 46 FR 8386, January 26, 1981, 48 FR 9269, March 4, 1983.

§ 46.101 To what do these regulations apply?

(a) Except as provided in paragraph (b) of this section, this subpart applies to all research involving human subjects conducted by the Department of Health and Human Services or funded in whole or in part by a Department grant, contract, cooperative agreement or fellowship.

(1) This includes research conducted by Department employees, except each Principal Operating Component head may adopt such nonsubstantive, procedural modifications as may be appropriate from an administrative standpoint.

(2) It also includes research conducted or funded by the Department of Health and Human Services outside the United States, but in appropriate circumstances, the Secretary may, under paragraph (e) of this section waive the applicability of some or all of the requirements of these regulations for research of this type.

(b) Research activities in which the only involvement of human subjects will be in one or more of the following categories are exempt from these regulations unless the research is covered by other subparts of this part:

(1) Research conducted in established or commonly accepted educational settings, involving normal educational practices, such as (i) research on regular and special education instructional strategies, or (ii) research on the effectiveness of or the comparison among instructional techniques, curricula, or classroom management methods.

(2) Research involving the use of educational tests (cognitive, diagnostic, aptitude, achievement), if information taken from these sources is recorded in such a manner that subjects cannot be identified, directly or through identifiers linked to the subjects.

(3) Research involving survey or interview procedures, except where all of the following conditions exist: (i) responses are recorded in such a manner that the human subjects can be identified, directly or through identifiers linked to the subjects, (ii) the subject's responses, if they became known outside the research, could reasonably place the subject at risk of criminal or civil liability or be damaging to the subject's financial standing or employability, and (iii) the research deals with sensitive aspects of the subject's own behavior, such as illegal conduct, drug use, sexual behavior, or use of alcohol. All research involving survey or interview procedures is exempt, without exception, when the respondents are elected or appointed public officials or candidates for public office.

(4) Research involving the observation (including observation by participants) of public behavior, except where all of the following conditions exist: (i) observations are recorded in such a manner that the human subjects can be identified, directly or through identifiers linked to the subjects, (ii) the observations recorded about the individual, if they became known outside the research, could reasonably place the subject at risk of criminal or civil liability or be damaging to the subject's financial standing or employability, and (iii) the research deals with sensitive aspects of the subject's own behavior such as illegal conduct, drug use, sexual behavior, or use of alcohol.

(5) Research involving the collection or study of existing data, documents, records, pathological specimens, or diagnostic specimens, if these sources are publicly available or if the information is recorded by the investigator in such a manner that subjects cannot be identified, directly or through identifiers linked to the subjects.

(6) Unless specifically required by statute (and except to the extent specified in paragraph (i)), research and demonstration projects which are conducted by or subject to the approval of the Department of Health and Human Services, and which are designed to study, evaluate, or otherwise examine: (i) programs under the Social Security Act, or other public benefit or service programs; (ii) procedures for obtaining benefits or services under those programs; (iii) possible changes in or alternatives to those programs or procedures; or (iv) possible changes in methods or levels of payment for benefits or services under those programs.

(c) The Secretary has final authority to determine whether a particular activity is covered by these regulations.

(d) The Secretary may require that specific research activities or classes of research activities conducted or funded by the Department, but not otherwise covered by these regulations, comply with some or all of these regulations.

(e) The Secretary may also waive applicability of these regulations to specific research activities or classes of research activities, otherwise covered by these regulations. Notices of these actions will be published in the *Federal Register* as they occur.

(f) No individual may receive Department funding for research covered by these regulations unless the individual is affiliated with or sponsored by an institution which assumes responsibility for the research under an assurance satisfying the requirements of this part, or the individual makes other arrangements with the Department.

(g) Compliance with these regulations will in no way render inapplicable pertinent federal, state, or local laws or regulations.

(h) Each subpart of these regulations contains a separate section describing to what the subpart applies. Research which is covered by more than one subpart shall comply with all applicable subparts.

(i) If, following review of proposed research activities that are exempt from these regulations under paragraph (b)(6), the Secretary determines that a research or demonstration project presents a danger to the physical, mental, or emotional well-being of a participant or subject of the research or demonstration project, then federal funds may not be expended for such a project without the written, informed consent of each participant or subject.

§ 46.102 Definitions.

(a) "Secretary" means the Secretary of Health and Human Services and any other officer or employee of the Department of Health and Human Services to whom authority has been delegated.

(b) "Department" or "HHS" means the Department of Health and Human Services.

(c) "Institution" means any public or private entity or agency (including federal, state, and other agencies).

(d) "Legally authorized representative" means an individual or judicial or other body authorized under applicable law to consent on behalf of a prospective subject to the subject's participation in the procedure(s) involved in the research.

(e) "Research" means a systematic investigation designed to develop or contribute to generalizable knowledge. Activities which meet this definition constitute "research" for purposes of these regulations, whether or not they are supported or funded

under a program which is considered research for other purposes. For example, some "demonstration" and "service" programs may include research activities.

(f) "Human subject" means a living individual about whom an investigator (whether professional or student) conducting research obtains (1) data through intervention or interaction with the individual, or (2) identifiable private information. "Intervention" includes both physical procedures by which data are gathered (for example, venipuncture) and manipulations of the subject or the subject's environment that are performed for research purposes. "Interaction" includes communication or interpersonal contact between investigator and subject. "Private information" includes information about behavior that occurs in a context in which an individual can reasonably expect that no observation or recording is taking place, and information which has been provided for specific purposes by an individual and which the individual can reasonably expect will not be made public (for example, a medical record). Private information must be individually identifiable (i.e., the identity of the subject is or may readily be ascertained by the investigator or associated with the information) in order for obtaining the information to constitute research involving human subjects.

(g) "Minimal risk" means that the risks of harm anticipated in the proposed research are not greater, considering probability and magnitude, than those ordinarily encountered in daily life or during the performance of routine physical or psychological examinations or tests.

(h) "Certification" means the official notification by the institution to the Department in accordance with the requirements of this part that a research project or activity involving human subjects has been reviewed and approved by the Institutional Review Board (IRB) in accordance with the approved assurance on file at HHS. (Certification is required when the research is funded by the Department and not otherwise exempt in accordance with § 46.101(b)).

§ 46.103 Assurances.

(a) Each institution engaged in research covered by these regulations shall provide written assurance satisfactory to the Secretary that it will comply with the requirements set forth in these regulations.

(b) The Department will conduct or fund research covered by these regulations only if the institution has an assurance approved as provided in this section, and only if the institution has certified to the Secretary that the research has been reviewed and approved by an IRB provided for in the assurance, and will be subject to continuing review by the IRB. This assurance shall at a minimum include:

(1) A statement of principles governing the institution in the discharge of its responsibilities for protecting the rights and welfare of human subjects of research conducted at or sponsored by the institution, regardless of source of funding. This may include an appropriate existing code, declaration, or statement of ethical principles, or a statement formulated by the institution itself. This requirement does not preempt provisions of these regulations applicable to Department-funded research and is not applicable to any research in an exempt category listed in § 46.101.

(2) Designation of one or more IRBs established in accordance with the requirements of this subpart, and for which provisions are made for meeting space and sufficient staff to support the IRB's review and recordkeeping duties.

(3) A list of the IRB members identified by name; earned degrees; representative capacity; indications of experience such as board certifications, licenses, etc., sufficient to describe each member's chief anticipated contributions to IRB deliberations; and any employment or other relationship between each member and the institution;

for example: full-time employee, part-time employee, member of governing panel or board, stockholder, paid or unpaid consultant. Changes in IRB membership shall be reported to the Secretary.[1]

(4) Written procedures which the IRB will follow (i) for conducting its initial and continuing review of research and for reporting its findings and actions to the investigator and the institution; (ii) for determining which projects require review more often than annually and which projects need verification from sources other than the investigators that no material changes have occurred since previous IRB review; (iii) for insuring prompt reporting to the IRB of proposed changes in a research activity, and for insuring that changes in approved research, during the period for which IRB approval has already been given, may not be initiated without IRB review and approval except where necessary to eliminate apparent immediate hazards to the subject; and (iv) for insuring prompt reporting to the IRB and to the Secretary[1] of unanticipated problems involving risks to subjects or others.

(c) The assurance shall be executed by an individual authorized to act for the institution and to assume on behalf of the institution the obligations imposed by these regulations, and shall be filed in such form and manner as the Secretary may prescribe.

(d) The Secretary will evaluate all assurances submitted in accordance with these regulations through such officers and employees of the Department and such experts or consultants engaged for this purpose as the Secretary determines to be appropriate. The Secretary's evaluation will take into consideration the adequacy of the proposed IRB in light of the anticipated scope of the institution's research activities and the types of subject populations likely to be involved, the appropriateness of the proposed initial and continuing review procedures in light of the probable risks, and the size and complexity of the institution.

---

[1]Reports should be filed with the Office for Protection from Research Risks, National Institutes of Health, Department of Health and Human Services, Bethesda, Maryland 20205.

(e) On the basis of this evaluation, the Secretary may approve or disapprove the assurance, or enter into negotiations to develop an approvable one. The Secretary may limit the period during which any particular approved assurance or class of approved assurances shall remain effective or otherwise condition or restrict approval.

(f) Within 60 days after the date of submission to HHS of an application or proposal, an institution with an approved assurance covering the proposed research shall certify that the application or proposal has been reviewed and approved by the IRB. Other institutions shall certify that the application or proposal has been approved by the IRB within 30 days after receipt of a request for such a certification from the Department. If the certification is not submitted within these time limits, the application or proposal may be returned to the institution.

§ 46.104 [Reserved]

§ 46.105 [Reserved]

§ 46.106 [Reserved]

§ 46.107 IRB membership.

(a) Each IRB shall have at least five members, with varying backgrounds to promote complete and adequate review of research activities commonly conducted by the institution. The IRB shall be sufficiently qualified through the experience and expertise of its members, and the diversity of the members' backgrounds including consideration of the racial and cultural backgrounds of members and sensitivity to such issues as community attitudes, to promote respect for its advice and counsel in safeguarding the rights and welfare of human subjects. In addition to possessing the professional competence necessary to review specific research activities, the IRB shall be able to ascertain the acceptability of proposed research in terms of institutional commitments and regulations, applicable law, and standards of professional conduct and practice. The IRB shall there-

fore include persons knowledgeable in these areas. If an IRB regularly reviews research that involves a vulnerable category of subjects, including but not limited to subjects covered by other subparts of this part, the IRB shall include one or more individuals who are primarily concerned with the welfare of these subjects.

(b) No IRB may consist entirely of men or entirely of women, or entirely of members of one profession.

(c) Each IRB shall include at least one member whose primary concerns are in nonscientific areas; for example: lawyers, ethicists, members of the clergy.

(d) Each IRB shall include at least one member who is not otherwise affiliated with the institution and who is not part of the immediate family of a person who is affiliated with the institution.

(e) No IRB may have a member participating in the IRB's initial or continuing review of any project in which the member has a conflicting interest, except to provide information requested by the IRB.

(f) An IRB may, in its discretion, invite individuals with competence in special areas to assist in the review of complex issues which require expertise beyond or in addition to that available on the IRB. These individuals may not vote with the IRB.

§ 46.108 IRB functions and operations.

In order to fulfill the requirements of these regulations each IRB shall:

(a) Follow written procedures as provided in § 46.103(b)(4).

(b) Except when an expedited review procedure is used (see § 46.110), review proposed research at convened meetings at which a majority of the members of the IRB are present, including at least one member whose primary concerns are in nonscientific areas. In order for the research to be approved, it shall receive the approval of a majority of those members present at the meeting.

(c) Be responsible for reporting to the appropriate institutional officials and the

Secretary[1] any serious or continuing noncompliance by investigators with the requirements and determinations of the IRB.

§ 46.109 IRB review of research.

(a) An IRB shall review and have authority to approve, require modifications in (to secure approval), or disapprove all research activities covered by these regulations.

(b) An IRB shall require that information given to subjects as part of informed consent is in accordance with § 46.116. The IRB may require that information, in addition to that specifically mentioned in § 46.116, be given to the subjects when in the IRB's judgment the information would meaningfully add to the protection of the rights and welfare of subjects.

(c) An IRB shall require documentation of informed consent or may waive documentation in accordance with § 46.117.

(d) An IRB shall notify investigators and the institution in writing of its decision to approve or disapprove the proposed research activity, or of modifications required to secure IRB approval of the research activity. If the IRB decides to disapprove a research activity, it shall include in its written notification a statement of the reasons for its decision and give the investigator an opportunity to respond in person or in writing.

(e) An IRB shall conduct continuing review of research covered by these regulations at intervals appropriate to the degree of risk, but not less than once per year, and shall have authority to observe or have a third party observe the consent process and the research.

§ 46.110 Expedited review procedures for certain kinds of research involving no more than minimal risk, and for minor changes in approved research.

(a) The Secretary has established, and published in the *Federal Register*, a list of categories of research that may be reviewed by the IRB through an expedited review procedure. The list will be amended, as appropriate, through periodic republication in the *Federal Register*.

(b) An IRB may review some or all of the research appearing on the list through an expedited review procedure, if the research involves no more than minimal risk. The IRB may also use the expedited review procedure to review minor changes in previously approved research during the period for which approval is authorized. Under an expedited review procedure, the review may be carried out by the IRB chairperson or by one or more experienced reviewers designated by the chairperson from among members of the IRB. In reviewing the research, the reviewers may exercise all of the authorities of the IRB except that the reviewers may not disapprove the research. A research activity may be disapproved only after review in accordance with the nonexpedited procedure set forth in § 46.108(b).

(c) Each IRB which uses an expedited review procedure shall adopt a method for keeping all members advised of research proposals which have been approved under the procedure.

(d) The Secretary may restrict, suspend, or terminate an institution's or IRB's use of the expedited review procedure when necessary to protect the rights or welfare of subjects.

§ 46.111 Criteria for IRB approval of research.

(a) In order to approve research covered by these regulations the IRB shall determine that all of the following requirements are satisfied:

(1) Risks to subjects are minimized: (i) By using procedures which are consistent with sound research design and which do not unnecessarily expose subjects to risk, and (ii) whenever appropriate, by using procedures already being performed on the

---

[1]Reports should be filed with the Office for Protection from Research Risks, National Institutes of Health, Department of Health and Human Services, Bethesda, Maryland 20205.

subjects for diagnostic or treatment purposes.

(2) Risks to subjects are reasonable in relation to anticipated benefits, if any, to subjects, and the importance of the knowledge that may reasonably be expected to result. In evaluating risks and benefits, the IRB should consider only those risks and benefits that may result from the research (as distinguished from risks and benefits of therapies subjects would receive even if not participating in the research). The IRB should not consider possible long-range effects of applying knowledge gained in the research (for example, the possible effects of the research on public policy) as among those research risks that fall within the purview of its responsibility.

(3) Selection of subjects is equitable. In making this assessment the IRB should take into account the purposes of the research and the setting in which the research will be conducted.

(4) Informed consent will be sought from each prospective subject or the subject's legally authorized representative, in accordance with, and to the extent required by § 46.116.

(5) Informed consent will be appropriately documented, in accordance with, and to the extent required by § 46.117.

(6) Where appropriate, the research plan makes adequate provision for monitoring the data collected to insure the safety of subjects.

(7) Where appropriate, there are adequate provisions to protect the privacy of subjects and to maintain the confidentiality of data.

(b) Where some or all of the subjects are likely to be vulnerable to coercion or undue influence, such as persons with acute or severe physical or mental illness, or persons who are economically or educationally disadvantaged, appropriate additional safeguards have been included in the study to protect the rights and welfare of these subjects.

§ 46.112 Review by institution.

Research covered by these regulations that has been approved by an IRB may be subject to further appropriate review and approval or disapproval by officials of the institution. However, those officials may not approve the research if it has not been approved by an IRB.

§ 46.113 Suspension or termination of IRB approval of research.

An IRB shall have authority to suspend or terminate approval of research that is not being conducted in accordance with the IRB's requirements or that has been associated with unexpected serious harm to subjects. Any suspension or termination of approval shall include a statement of the reasons for the IRB's action and shall be reported promptly to the investigator, appropriate institutional officials, and the Secretary.[1]

§ 46.114 Cooperative research.

Cooperative research projects are those projects, normally supported through grants, contracts, or similar arrangements, which involve institutions in addition to the grantee or prime contractor (such as a contractor with the grantee, or a subcontractor with the prime contractor). In such instances, the grantee or prime contractor remains responsible to the Department for safeguarding the rights and welfare of human subjects. Also, when cooperating institutions conduct some or all of the research involving some or all of these subjects, each cooperating institution shall comply with these regulations as though it received funds for its participation in the project directly from the Department, except that in complying with these regulations institutions may use joint review, reliance upon the review of another qualified IRB, or similar arrange-

[1]Reports should be filed with the Office for Protection from Research Risks, National Institutes of Health, Department of Health and Human Services, Bethesda, Maryland 20205.

ments aimed at avoidance of duplication of effort.

§ 46.115 IRB records.

(a) An institution, or where appropriate an IRB, shall prepare and maintain adequate documentation of IRB activities, including the following:

(1) Copies of all research proposals reviewed, scientific evaluations, if any, that accompany the proposals, approved sample consent documents, progress reports submitted by investigators, and reports of injuries to subjects.

(2) Minutes of IRB meetings which shall be in sufficient detail to show attendance at the meetings; actions taken by the IRB; the vote on these actions including the number of members voting for, against, and abstaining; the basis for requiring changes in or disapproving research; and a written summary of the discussion of controverted issues and their resolution.

(3) Records of continuing review activities.

(4) Copies of all correspondence between the IRB and the investigators.

(5) A list of IRB members as required by § 46.103(b)(3).

(6) Written procedures for the IRB as required by § 46.103(b)(4).

(7) Statements of significant new findings provided to subjects, as required by § 46.116(b)(5).

(b) The records required by this regulation shall be retained for at least 3 years after completion of the research, and the records shall be accessible for inspection and copying by authorized representatives of the Department at reasonable times and in a reasonable manner.

§ 46.116 General requirements for informed consent.

Except as provided elsewhere in this or other subparts, no investigator may involve a human being as a subject in research covered by these regulations unless the investigator has obtained the legally effective informed consent of the subject or the sub-

ject's legally authorized representative. An investigator shall seek such consent only under circumstances that provide the prospective subject or the representative sufficient opportunity to consider whether or not to participate and that minimize the possibility of coercion or undue influence. The information that is given to the subject or the representative shall be in language understandable to the subject or the representative. No informed consent, whether oral or written, may include any exculpatory language through which the subject or the representative is made to waive or appear to waive any of the subject's legal rights, or releases or appears to release the investigator, the sponsor, the institution or its agents from liability for negligence.

(a) Basic elements of informed consent. Except as provided in paragraph (c) or (d) of this section, in seeking informed consent the following information shall be provided to each subject:

(1) A statement that the study involves research, an explanation of the purposes of the research and the expected duration of the subject's participation, a description of the procedures to be followed, and identification of any procedures which are experimental;

(2) A description of any reasonably foreseeable risks or discomforts to the subject;

(3) A description of any benefits to the subject or to others which may reasonably be expected from the research;

(4) A disclosure of appropriate alternative procedures or courses of treatment, if any, that might be advantageous to the subject;

(5) A statement describing the extent, if any, to which confidentiality of records identifying the subject will be maintained;

(6) For research involving more than minimal risk, an explanation as to whether any compensation and an explanation as to whether any medical treatments are available if injury occurs and, if so, what they consist of, or where further information may be obtained;

(7) An explanation of whom to contact for answers to pertinent questions about the research and research subjects' rights, and whom to contact in the event of a research-related injury to the subject; and

(8) A statement that participation is voluntary, refusal to participate will involve no penalty or loss of benefits to which the subject is otherwise entitled, and the subject may discontinue participation at any time without penalty or loss of benefits to which the subject is otherwise entitled.

(b) Additional elements of informed consent. When appropriate, one or more of the following elements of information shall also be provided to each subject:

(1) A statement that the particular treatment or procedure may involve risks to the subject (or to the embryo or fetus, if the subject is or may become pregnant) which are currently unforeseeable;

(2) Anticipated circumstances under which the subject's participation may be terminated by the investigator without regard to the subject's consent;

(3) Any additional costs to the subject that may result from participation in the research;

(4) The consequences of a subject's decision to withdraw from the research and procedures for orderly termination of participation by the subject;

(5) A statement that significant new findings developed during the course of the research which may relate to the subject's willingness to continue participation will be provided to the subject; and

(6) The approximate number of subjects involved in the study.

(c) An IRB may approve a consent procedure which does not include, or which alters, some or all of the elements of informed consent set forth above, or waive the requirement to obtain informed consent provided the IRB finds and documents that:

(1) The research or demonstration project is to be conducted by or subject to the approval of state or local government of-

ficials and is designed to study, evaluate, or otherwise examine: (i) programs under the Social Security Act, or other public benefit or service programs; (ii) procedures for obtaining benefits or services under those programs; (iii) possible changes in or alternatives to those programs or procedures; or (iv) possible changes in methods or levels of payment for benefits or services under those programs; and

(2) The research could not practicably be carried out without the waiver or alteration.

(d) an IRB may approve a consent procedure which does not include, or which alters, some or all of the elements of informed consent set forth above, or waive the requirements to obtain informed consent provided the IRB finds and documents that:

(1) The research involves no more than minimal risk to the subjects;

(2) The waiver or alteration will not adversely affect the rights and welfare of the subjects;

(3) The research could not practicably be carried out without the waiver or alteration; and

(4) whenever appropriate, the subjects will be provided with additional pertinent information after participation.

(e) The informed consent requirements in these regulations are not intended to preempt any applicable federal, state, or local laws which require additional information to be disclosed in order for informed consent to be legally effective.

(f) Nothing in these regulations is intended to limit the authority of a physician to provide emergency medical care, to the extent the physician is permitted to do so under applicable federal, state, or local law.

§ 46.117 Documentation of informed consent.

(a) Except as provided in paragraph (c) of this section, informed consent shall be documented by the use of a written consent form approved by the IRB and signed by the subject or the subject's legally authorized representative. A copy shall be given to the person signing the form.

(b) Except as provided in paragraph (c) of this section, the consent form may be either of the following:

(1) A written consent document that embodies the elements of informed consent required by § 46.116. This form may be read to the subject or the subject's legally authorized representative, but in any event, the investigator shall give either the subject or the representative adequate opportunity to read it before it is signed; or

(2) A "short form" written consent document stating that the elements of informed consent required by § 46.116 have been presented orally to the subject or the subject's legally authorized representative. when this method is used, there shall be a witness to the oral presentation. Also, the IRB shall approve a written summary of what is to be said to the subject or the representative. Only the short form itself is to be signed by the subject or the representative. However, the witness shall sign both the short form and a copy of the summary, and the person actually obtaining consent shall sign a copy of the summary. A copy of the summary shall be given to the subject or the representative, in addition to a copy of the "short form."

(c) An IRB may waive the requirement for the investigator to obtain a signed consent form for some or all subjects if it finds either:

(1) That the only record linking the subject and the research would be the consent document and the principal risk would be potential harm resulting from a breach of confidentiality. Each subject will be asked whether the subject wants documentation linking the subject with the research, and the subject's wishes will govern; or

(2) That the research presents no more than minimal risk of harm to subjects and involves no procedures for which written consent is normally required outside of the research context.

In cases where the documentation requirement is waived, the IRB may require the investigator to provide subjects with a written statement regarding the research.

§ 46.118 Applications and proposals lacking definite plans for involvement of human subjects.

Certain types of applications for grants, cooperative agreements, or contracts are submitted to the Department with the knowledge that subjects may be involved within the period of funding, but definite plans would not normally be set forth in the application or proposal. These include activities such as institutional type grants (including bloc grants) where selection of specific projects is the institution's responsibility; research training grants where the activities involving subjects remain to be selected; and projects in which human subjects' involvement will depend upon completion of instruments, prior animal studies, or purification of compounds. These applications need not be reviewed by an IRB before an award may be made. However, except for research described in § 46.101(b), no human subjects may be involved in any project supported by these awards until the project has been reviewed and approved by the IRB, as provided in these regulations, and certification submitted to the Department.

§ 46.119 Research undertaken without the intention of involving human subjects.

In the event research (conducted or funded by the Department) is undertaken without the intention of involving human subjects, but it is later proposed to use human subjects in the research, the research shall first be reviewed and approved by an IRB, as provided in these regulations, a certification submitted to the Department, and final approval given to the proposed change by the Department.

§ 46.120 Evaluation and disposition of applications and proposals.

(a) The Secretary will evaluate all applications and proposals involving human subjects submitted to the Department through such officers and employees of the Department and such experts and consultants as the Secretary determines to be appropriate. This evaluation will take into consideration the risks to the subjects, the adequacy of protection against these risks, the potential benefits of the proposed research to the subjects and others, and the importance of the knowledge to be gained.

(b) On the basis of this evaluation, the Secretary may approve or disapprove the application or proposal, or enter into negotiations to develop an approvable one.

§ 46.121 Investigational new drug or device 30-day delay requirement.

When an institution is required to prepare or to submit a certification with an application or proposal under these regulations, and the application or proposal involves an investigational new drug (within the meaning of 21 U.S.C. 355(i) or 357(d) or a significant risk device (as defined in 21 CFR 812.3(m)), the institution shall identify the drug or device in the certification. The institution shall also state whether the 30-day interval required for investigational new drugs by 21 CFR 312.1(a) and for significant risk devices by 21 CFR 812.30 has elapsed, or whether the Food and Drug Administration has waived that requirement. If the 30-day interval has expired, the institution shall state whether the Food and Drug Administration has requested that the sponsor continue to withhold or restrict the use of the drug or device in human subjects. If the 30-day interval has not expired, and a waiver has not been received, the institution shall send a statement to the Department upon expiration of the interval. The Department will not consider a certification acceptable until the institution has submitted a statement that the 30-day interval has elapsed, and the Food and Drug Administration has not requested it to limit the use of the drug or device, or that the Food and Drug Administration has waived the 30-day interval.

§ 46.122 Use of Federal funds.

Federal funds administered by the Department may not be expended for research involving human subjects unless the requirement of these regulations, including all subparts of these regulations, have been satisfied.

§ 46.123 Early termination of research funding; evaluation of subsequent applications and proposals.

(a) The Secretary may require that Department funding for any project be terminated or suspended in the manner prescribed in applicable program requirements, when the Secretary finds an institution has materially failed to comply with the terms of these regulations.

(b) In making decisions about funding applications or proposals covered by these regulations the Secretary may take into account, in addition to all other eligibility requirements and program criteria, factors such as whether the applicant has been subject to a termination or suspension under paragraph (a) of this section and whether the applicant or the person who would direct the scientific and technical aspects of an activity has in the judgment of the Secretary materially failed to discharge responsibility for the protection of the rights and welfare of human subjects (whether or not Department funds were involved).

§ 46.124 Conditions.

With respect to any research project or any class of research projects the Secretary may impose additional conditions prior to or at the time of funding when in the Secretary's judgment additional conditions are necessary for the protection of human subjects.

Subpart B—Additional Protections Pertaining to Research Development, and Related Activities Involving Fetuses, Pregnant Women, and Human in Vitro Fertilization

Source: 40 FR 33528, Aug. 8, 1975, 43 FR 1758, January 11, 1978, 43 FR 51559, November 3, 1978

§ 46.201 Applicability.

(a) The regulations in this subpart are applicable to all Department of Health, Education, and Welfare grants and contract supporting research, development, and related activities involving: (1) The fetus, (2) pregnant women, and (3) human *in vitro* fertilization.

(b) Nothing in this subpart shall be construed as indicating that compliance with the procedures set forth herein will in any way render inapplicable pertinent State or local laws bearing upon activities covered by this subpart.

(c) The requirements of this subpart are in addition to those imposed under the other subparts of this part.

§ 46.202 Purpose.

It is the purpose of this subpart to provide additional safeguards in reviewing activities to which this subpart is applicable to assure that they conform to appropriate ethical standards and relate to important societal needs.

§ 46.203 Definitions.

As used in this subpart:

(a) "Secretary" means the Secretary of Health, Education, and Welfare and any other officer or employee of the Department of Health, Education, and Welfare to whom authority has been delegated.

(b) "Pregnancy" encompasses the period of time from confirmation of implantation (through any of the presumptive signs of pregnancy, such as missed menses, or by a medically acceptable pregnancy test), until expulsion or extraction of the fetus.

(c) "Fetus" means the product of conception from the time of implantation (as evidenced by any of the presumptive signs of pregnancy, such as missed menses, or a medically acceptable pregnancy test), until a determination is made, following expulsion or extraction of the fetus, that it is viable.

(d) "Viable" as it pertains to the fetus means being able, after either spontaneous or induced delivery, to survive (given the benefit of available medical therapy) to the point of independently maintaining heart beat and respiration. The Secretary may from time to time, taking into account medical advances, publish in the *Federal Register* guidelines to assist in determining whether a fetus is viable for purposes of this subpart. If a fetus is viable after delivery, it is a premature infant.

(e) "Nonviable fetus" means a fetus *ex utero* which, although living, is not viable.

(f) "Dead fetus" means a fetus *ex utero* which exhibits neither heartbeat, spontaneous respiratory activity, spontaneous movement of voluntary muscles, nor pulsation of the umbilical cord (if still attached).

(g) "*In vitro* fertilization" means any fertilization of human ova which occurs outside the body of a female, either through admixture of donor human sperm and ova or by any other means.

§ 46.204 Ethical Advisory Boards.

(a) One or more Ethical Advisory Boards shall be established by the Secretary. Members of these board(s) shall be so selected that the board(s) will be competent to deal with medical, legal, social, ethical, and related issues and may include, for example, research scientists, physicians, psychologists, sociologists, educators, lawyers, and ethicists, as well as representatives of the general public. No board member may be a regular, full-time employee of the Department of Health, Education, and Welfare.

(b) At the request of the Secretary, the Ethical Advisory Board shall render advice consistent with the policies and requirements of this Part as to ethical issues, involving activities covered by this subpart, raised by individual applications or proposals. In addition, upon request by the Secretary, the Board shall render advice as to classes of applications or proposals and general policies, guidelines, and procedures.

(c) A Board may establish, with the approval of the Secretary, classes of applications or proposals which:

(1) Must be submitted to the Board, or (2) need not be submitted to the Board. Where the Board so establishes a class of applications or proposals which must be submitted, no application or proposal within the class may be funded by the Department or any component thereof until the application or proposal has been reviewed by the Board and the Board has rendered advice as to its acceptability from an ethical standpoint.

(d) No application or proposal involving human *in vitro* fertilization may be funded by the Department or any component thereof until the application or proposal has been reviewed by the Ethical Advisory Board and the Board has rendered advice as to its acceptability from an ethical standpoint.

§ 46.205 Additional duties of the Institutional Review Boards in connection with activities involving fetuses, pregnant women, or human in vitro fertilization.

(a) In addition to the responsibilities prescribed for Institutional Review Boards under Subpart A or this part, the applicant's or offeror's Board shall, with respect to activities covered by this subpart, carry out the following additional duties:

(1) Determine that all aspects of the activity meet the requirements of this subpart;

(2) Determine that adequate consideration has been given to the manner in which potential subjects will be selected, and adequate provision has been made by the applicant or offeror for monitoring the actual informed consent process (e.g., through such mechanisms, when appropriate, as participation by the Institutional Review Board or subject advocates in: (i) Overseeing the actual process by which individual consents required by this subpart are secured either by approving induction of each individual into the activity or verifying, perhaps through sampling, that approved procedures for induction of individuals into the activity are being followed, and (ii) monitoring the

progress of the activity and intervening as necessary through such steps as visits to the activity site and continuing evaluation to determine if any unanticipated risks have arisen);

(3) Carry out such other responsibilities as may be assigned by the Secretary.

(b) No award may be issued until the applicant or offeror has certified to the Secretary that the Institutional Review Board has made the determinations required under paragraph (a) of this section and the Secretary has approved these determinations, as provided in § 46.120 of Subpart A of this part.

(c) Applicants or offerors seeking support for activities covered by this subpart must provide for the designation of an Institutional Review Board, subject to approval by the Secretary, where no such Board has been established under Subpart A of this part.

§ 46.206 General limitations.

(a) No activity to which this subpart is applicable may be undertaken unless:

(1) Appropriate studies on animals and nonpregnant individuals have been completed;

(2) Except where the purpose of the activity is to meet the health needs of the mother or the particular fetus, the risk to the fetus is minimal and, in all cases, is the least possible risk for achieving the objectives of the activity.

(3) Individuals engaged in the activity will have no part in: (i) Any decisions as to the timing, method, and procedures used to terminate the pregnancy, and (ii) determining the viability of the fetus at the termination of the pregnancy; and

(4) No procedural changes which may cause greater than minimal risk to the fetus or the pregnant woman will be introduced in the procedure for terminating the pregnancy solely in the interest of the activity.

(b) No inducements, monetary or otherwise, may be offered to terminate pregnancy for purposes of the activity.

[40 FR 33528, Aug. 8, 1975, as amended at 40 FR 51638, Nov. 6, 1975]

§ 46.207 Activities directed toward pregnant women as subjects.

(a) No pregnant woman may be involved as a subject in an activity covered by this subpart unless: (1) The purpose of the activity is to meet the health needs of the mother and the fetus will be placed at risk only to the minimum extent necessary to meet such needs, or (2) the risk to the fetus is minimal.

(b) An activity permitted under paragraph (a) of this section may be conducted only if the mother and father are legally competent and have given their informed consent after having been fully informed regarding possible impact on the fetus, except that the father's informed consent need not be secured if: (1) The purpose of the activity is to meet the health needs of the mother; (2) his identity or whereabouts cannot reasonably be ascertained; (3) he is not reasonably available; or (4) the pregnancy resulted from rape.

§ 46.208 Activities directed toward fetuses in utero as subjects.

(a) No fetus *in utero* may be involved as a subject in any activity covered by this subpart unless: (1):The purpose of the activity is to meet the health needs of the particular fetus and the fetus will be placed at risk only to the minimum extent necessary to meet such needs, or (2) the risk to the fetus imposed by the research is minimal and the purpose of the activity is the development of important biomedical knowledge which cannot be obtained by other means.

(b) An activity permitted under paragraph (a) of this section may be conducted only if the mother and father are legally competent and have given their informed consent, except that the father's consent need not be secured if: (1) His identity or whereabouts cannot reasonably be ascertained, (2) he is not reasonably available, or (3) the pregnancy resulted from rape.

§ 46.209 Activities directed toward fetuses ex utero, including nonviable fetuses, as subjects.

(a) Until it has been ascertained whether or not a fetus ex utero is viable, a fetus ex utero may not be involved as a subject in an activity covered by this subpart unless:

(1) There will be no added risk to the fetus resulting from the activity, and the purpose of the activity is the development of important biomedical knowledge which cannot be obtained by other means, or

(2) The purpose of the activity is to enhance the possibility of survival of the particular fetus to the point of viability.

(b) No nonviable fetus may be involved as a subject in an activity covered by this subpart unless:

(1) Vital functions of the fetus will not be artificially maintained,

(2) Experimental activities which of themselves would terminate the heartbeat or respiration of the fetus will not be employed, and

(3) The purpose of the activity is the development of important biomedical knowledge which cannot be obtained by other means.

(c) In the event the fetus *ex utero* is found to be viable, it may be included as a subject in the activity only to the extent permitted by and in accordance with the requirements of other subparts of this part.

(d) An activity permitted under paragraph (a) or (b) of this section may be conducted only if the mother and father are legally competent and have given their informed consent, except that the father's informed consent need not be secured if: (1) his identity or whereabouts cannot reasonably be ascertained, (2) he is not reasonably available, or (3) the pregnancy resulted from rape.

§ 46.210 Activities involving the dead fetus, fetal material, or the placenta.

Activities involving the dead fetus, macerated fetal material, or cells, tissue, or organs excised from a dead fetus shall be conducted only in accordance with any applicable State or local laws regarding such activities.

§ 46.211 Modification or waiver of specific requirements.

Upon the request of an applicant or offeror (with the approval of its Institutional Review Board), the Secretary may modify or waive specific requirements of this subpart, with the approval of the Ethical Advisory Board after such opportunity for public comment as the Ethical Advisory Board considers appropriate in the particular instance. In making such decisions, the Secretary will consider whether the risks to the subject are so outweighed by the sum of the benefit to the subject and the importance of the knowledge to be gained as to warrant such modification or waiver and that such benefits cannot be gained except through a modification or waiver. Any such modifications or waivers will be published as notices in the *Federal Register*.

Subpart C—Additional Protections Pertaining to Biomedical and Behavorial Research Involving Prisoners as Subjects

Source: 43 FR 53655, Nov. 16, 1978

§ 46.301 Applicability.

(a) The regulations in this subpart are applicable to all biomedical and behavioral research conducted or supported by the Department of Health, Education, and Welfare involving prisoners as subjects.

(b) Nothing in this subpart shall be construed as indicating that compliance with the procedures set forth herein will authorize research involving prisoners as subjects, to the extent such research is limited or barred by applicable State or local law.

(c) The requirements of this subpart are in addition to those imposed under the other subparts of this part.

§ 46.302 Purpose.

Inasmuch as prisoners may be under constraints because of their incarceration which could affect their ability to make a truly voluntary and uncoerced decision whether or not to participate as subjects in research, it is the purpose of this subpart to provide additional safeguards for the protection of prisoners involved in activities to which this subpart is applicable.

§ 46.303 Definitions.

As used in this subpart:

(a) "Secretary" means the Secretary of Health, Education, and Welfare and any other officer of employee of the Department of Health, Education, and Welfare to whom authority has been delegated.

(b) "DHEW" means the Department of Health, Education, and Welfare.

(c) "Prisoner" means any individual involuntarily confined or detained in a penal institution. The term is intended to encompass individuals sentenced to such an institution under a criminal or civil statute, individuals detained in other facilities by virtue of statutes or commitment procedures which provide alternatives to criminal prosecution or incarceration in a penal institution, and individuals detained pending arraignment, trial or sentencing.

(d) "Minimal risk" is the probability and magnitude of physical or psychological harm that is normally encountered in the daily lives, or in the routine medical, dental, or psychological examination of healthy persons.

§ 46.304 Composition of Institutional Review Boards where prisoners are involved.

In addition to satisfying the requirements in § 46.107 of this part, an Institutional Review Board, carrying out responsibilities under this part with respect to research covered by this subpart, shall also meet the following specific requirements:

(a) A majority of the Board (exclusive of prisoner members) shall have no association with the prison(s) involved, apart from their membership on the Board.

(b) At least one member of the Board shall be a prisoner, or a prisoner representative with appropriate background and experience to serve in that capacity, except that where a particular research project is reviewed by more than one Board only one Board need satisfy this requirement.

§ 46.305 Additional duties of the Institutional Review Boards where prisoners are involved.

(a) In addition to all other responsibilities prescribed for Institutional Review Boards under this part, the Board shall review research covered by this subpart and approve such research only if it finds that:

(1) The research under review represents one of the categories of research permissible under § 46.306(a)(2);

(2) Any possible advantages accruing to the prisoner through his or her participation in the research, when compared to the general living conditions, medical care, quality of food, amenities and opportunity for earnings in the prison, are not of such a magnitude that his or her ability to weigh the risks of the research against the value of such advantages in the limited choice environment of the prison is impaired;

(3) The risks involved in the research are commensurate with risks that would be accepted by nonprisoner volunteers;

(4) Procedures for the selection of subjects within the prison are fair to all prisoners and immune from arbitrary intervention by prison authorities or prisoners. Unless the principal investigator provides to the Board justification in writing for following some other procedures, control subjects must be selected randomly from the group of available prisoners who meet the characteristics needed for that particular research project;

(5) The information is presented in language which is understandable to the subject population;

(6) Adequate assurance exists that parole boards will not take into account a prisoner's participation in the research in making decisions regarding parole, and each prisoner is clearly informed in advance that participation in the research will have no effect on his or her parole; and

(7) Where the Board finds there may be a need for follow-up examination or care of participants after the end of their participation, adequate provision has been made for such examination or care, taking into account the varying lengths of individual prisoners' sentences, and for informing participants of this fact.

(b) The Board shall carry out such other duties as may be assigned by the Secretary.

(c) The institution shall certify to the Secretary, in such form and manner as the Secretary may require, that the duties of the Board under this section have been fulfilled.

§ 46.306 Permitted research involving prisoners.

(a) Biomedical or behavioral research conducted or supported by DHEW may involve prisoners as subjects only if:

(1) The institution responsible for the conduct of the research has certified to the Secretary that the Institutional Review Board has approved the research under § 46.305 of this subpart; and

(2) In the judgment of the Secretary the proposed research involves solely the following:

(A) Study of the possible causes, effects, and processes of incarceration, and of criminal behavior, provided that the study presents no more than minimal risk and no more than inconvenience to the subject:

(B) Study of prisons as institutional structures or of prisoners as incarcerated persons, provided that the study presents no more than minimal risk and no more than inconvenience to the subjects;

(C) Research on conditions particularly affecting prisoners as a class (for example, vaccine trials and other research on hepatitis which is much more prevalent in prisons than elsewhere; and research on social and psychological problems such as alcoholism, drug addiction and sexual assaults) provided that the study may proceed only after the Secretary has consulted with appropriate experts including experts in penology medicine and ethics, and published notice, in the *Federal Register,* of his intent to approve such research; or

(D) Research on practices, both innovative and accepted, which have the intent and reasonable probability of improving the health or well-being of the subject. In cases in which those studies require the assignment of prisoners in a manner consistent with protocols approved by the IRB to control groups which may not benefit from the research, the study may proceed only after the Secretary has consulted with appropriate experts, including experts in penology medicine and ethics, and published notice, in the *Federal Register,* of his intent to approve such research.

(b) Except as provided in paragraph (a) of this section, biomedical or behavioral research conducted or supported by DHEW shall not involve prisoners as subjects.

Subpart D—Additional Protections for Children Involved as Subjects in Research.

Source: 48 FR 9818, March 8, 1983

§ 46.401 To what do these regulations apply?

(a) This subpart applies to all research involving children as subjects, conducted or supported by the Department of Health and Human Services.

(1) This includes research conducted by Department employees, except that each head of an Operating Division of the Department may adopt such nonsubstantive, procedural modifications as may be appropriate from an administrative standpoint.

(2) It also includes research conducted or supported by the Department of Health and Human Services outside the United States, but in appropriate circumstances, the Secretary may, under paragraph (3) of § 46.101 of Subpart A, waive the applicability of some or all of the requirements of these regulations for research of this type.

(b) Exemptions (1), (2), (5) and (6) as listed in Subpart A at § 46.101(b) are applicable to this subpart. Exemption (4), re-

search involving the observation of public behavior, listed at § 46.101(b), is applicable to this subpart where the investigator(s) does not participate in the activities being observed. Exemption (3), research involving survey or interview procedures, listed at § 46.101(b) does not apply to research covered by this subpart.

(c) The exceptions, additions, and provisions for waiver as they appear in paragraphs (c) through (i) of § 46.101 of Subpart A are applicable to this subpart.

§ 46.402 Definitions.

The definitions in § 46.102 of Subpart A shall be applicable to this subpart as well. In addition, as used in this subpart:

(a) "Children" are persons who have not attained the legal age for consent to treatments or procedures involved in the research, under the applicable law of the jurisdiction in which the research will be conducted.

(b) "Assent" means a child's affirmative agreement to participate in research. Mere failure to object should not, absent affirmative agreement, be construed as assent.

(c) "Permission" means the agreement of parent(s) or guardian to the participation of their child or ward in research.

(d) "Parent" means a child's biological or adoptive parent.

(e) "Guardian" means an individual who is authorized under applicable state or local law to consent on behalf of a child to general medical care.

§ 46.403 IRB duties.

In addition to other responsibilities assigned to IRBs under this part, each IRB shall review research covered by this subpart and approve only research which satisfies the conditions of all applicable sections of this subpart.

§46.404 Research not involving greater than minimal risk.

HHS will conduct or fund research in

which the IRB finds that no greater than minimal risk to children is presented, only if the IRB finds that adequate provisions are made for soliciting the assent of the children and the permission of their parents or guardians, as set forth in § 46.606.

§46.405 Research involving greater than minimal risk but presenting the prospect of direct benefit to the individual subjects.

HHS will conduct or fund research in which the IRB finds that more than minimal risk to children is presented by an intervention or procedure that holds out the prospect of direct benefit for the individual subject, or by a monitoring procedure that is likely to contribute to the subject's well-being only if the IRB finds that:

(a) The risk is justified by the anticipated benefit to the subjects;

(b) The relation of the anticipated benefit to the risk is at least as favorable to the subjects as that presented by available alternative approaches; and

(c) Adequate provisions are made for soliciting the assent of the children and permission of their parents or guardians, as set forth in § 46.408.

§ 46.406 Research involving greater than minimal risk and no prospect of direct benefit to individual subjects, but likely to yield generalizable knowledge about the subject's disorder or condition.

HHS will conduct or fund research in which the IRB finds that more than minimal risk to children is presented by an intervention or procedure that does not hold out the prospect of direct benefit for the individual subject, or by a monitoring procedure which is not likely to contribute to the well-being of the subject, only if the IRB finds that:

(a) The risk represents a minor increase over minimal risk;

(b) The intervention or procedure presents experiences to subjects that are reasonably commensurate with those inherent in their actual or expected medical, dental, psychological, social or educational situations;

(c) The intervention or procedure is likely to yield generalizable knowledge about the subjects' disorder or condition which is of vital importance for the understanding or amelioration of the subjects' disorder or condition; and

(d) Adequate provisions are made for soliciting assent of the children and permission of their parents or guardians, as set forth in § 46.408.

§ 46.407 Research not otherwise approvable which presents an opportunity to understand, prevent, or alleviate a serious problem affecting the health or welfare of children.

HHS will conduct or fund research that the IRB does not believe meets the requirements of § § 46.404, 46.405, or 46.406 only if:

(a) The IRB finds that the research presents a reasonable opportunity to further understanding, prevention, or alleviation of a serious problem affecting the health or welfare of children; and

(b) The Secretary, after consultation with a panel of experts in pertinent disciplines (for example: science, medicine, education, ethics, law) and following opportunity for public review and comment, has determined either: (1) That the research in fact satisfies the conditions of § § 46.404, 46.405, or 46.406, as applicable, or (2) the following:

(i) The research presents a reasonable opportunity to further the understanding, prevention, or alleviation of a serious problem affecting the health or welfare of children;

(ii) The research will be conducted in accordance with sound ethical principles;

(iii) Adequate provisions are made for soliciting the assent of children and the permission of their parents or guardians, as set forth in § 46.408.

§ 46.408 Requirements for permission by parents or guardians and for assent by children.

(a) In addition to the determinations required under other applicable sections of this subpart, the IRB shall determine that adequate provisions are made for soliciting the assent of the children, when in the judgment of the IRB the children are capable of providing assent. In determining whether children are capable of assenting, the IRB shall take into account the ages, maturity, and psychological state of the children involved. This judgment may be made for all children to be involved in research under a particular protocol, or for each child, as the IRB deems appropriate. If the IRB determines that the capability of some or all of the children is so limited that they cannot reasonably be consulted or that the intervention or procedure involved in the research holds out a prospect of direct benefit that is important to the health or well-being of the children and is available only in the context of the research, the assent of the children is not a necessary condition for proceeding with the research. Even where the IRB determines that the subjects are capable of assenting, the IRB may still waive the assent requirement under circumstances in which consent may be waived in accord with § 46.116 of Subpart A.

(b) In addition to the determinations required under other applicable sections of this subpart, the IRB shall determine, in accordance with and to the extent that consent is required by § 46.116 of Subpart A, that adequate provisions are made for soliciting the permission of each child's parents or guardian. Where parental permission is to be obtained, the IRB may find that the permission of one parent is sufficient for research to be conducted under § § 46.404 or 46.405. Where research is covered by § § 46.406 and 46.407 and permission is to be obtained from parents, both parents must give their permission unless one parent is deceased, unknown, incompetent, or not reasonably available, or when only one parent has legal responsibility for the care and custody of the child.

(c) In addition to the provisions for waiver contained in § 46.116 of Subpart A, if the IRB determines that a research protocol is designed for conditions or for a subject population for which parental or guardian permission is not a reasonable requirement to protect the subjects (for example, neglected or abused children), it may waive the consent requirements in Subpart A of this part and paragraph (b) of this section, provided an appropriate mechanism for protecting the children who will participate as subjects in the research is substituted, and provided further that the waiver is not inconsistent with federal state or local law. The choice of an appropriate mechanism would depend upon the nature and purpose of the activities described in the protocol, the risk and anticipated benefit to the research subjects, and their age, maturity, status, and condition.

(d) Permission by parents or guardians shall be documented in accordance with and to the extent required by § 46.117 of Subpart A.

(e) When the IRB determines that assent is required, it shall also determine whether and how assent must be documented.

§ 46.409 Wards.

(a) Children who are wards of the state or any other agency, institution, or entity can be included in research approved under § § 46.406 or 46.407 only if such research is:

(1) Related to their status as wards; or

(2) Conducted in schools, camps, hospitals, institutions, or similar settings in which the majority of children involved as subjects are not wards.

(b) If the research is approved under paragraph (a) of this section, the IRB shall require appointment of an advocate for each child who is a ward, in addition to any other individual acting on behalf of the child as guardian or in loco parentis. One individual may serve as advocate for more than one child. The advocate shall be an individual who has the background and experience to act in, and agrees to act in, the best interests of the child for the duration of the child's participation in the research and who is not associated in any way (except in the role as advocate or member of the IRB) with the research, the investigator(s), or the guardian organization.

Appendix 2

Food and Drug Administration Rules and Regulations[1]

Part 56—Institutional Review Boards[2]

Subpart A—General Provisions

Sec.

56.101 Scope.

56.102 Definitions.

56.103 Circumstances in which IRB review is required.

56.104 Exemptions from IRB requirement.

56.105 Waiver of IRB requirement.

Subpart B—Organization and Personnel

Sec.

56.107 IRB membership.

Subpart C—IRB Functions and Operations

Sec.

56.108 IRB functions and operations.

56.109 IRB review of research.

56.110 Expedited review procedures for certain kinds of research involving no more than minimal risk, and for minor changes in approved research.

56.111 Criteria for IRB approval of research.

56.112 Review by institution.

56.113 Suspension or termination of IRB approval or research.

56.114 Cooperative research.

Subpart D—Records and Reports

Sec.

56.115 IRB records.

Subpart E—Administrative Action for Noncompliance

Sec.

56.120 Lesser administrative actions.

56.121 Disqualification of an IRB or an institution.

56.122 Public disclosure of information regarding revocation.

56.123 Reinstatement of an IRB or an institution.

---

[1]There are many other Federal Regulations applicable to research regulated by the FDA. When the research is conducted or supported by DHHS, their regulations also apply. In addition, 21 CFR has many parts other than 50 and 56. Several of these that are relevant to research involving human subjects are cited throughout this book.

[2]Reprinted from: Federal Register, Vol. 46, No. 17, Tuesday, January 27, 1981.

56.124 Actions alternative or additional to disqualification.

Subpart A—General Provisions

§ 56.101 Scope.

(a) This part contains the general standards for the composition, operation, and responsibility of an Institutional Review Board (IRB) that reviews clinical investigations regulated by the Food and Drug Administration under sections 505(i), 507(d), and 520(g) of the act, as well as clinical investigations that support applications for research or marketing permits for products regulated by the Food and Drug Administration, including food and color additives, drugs for human use, medical devices for human use, biological products for human use, and electronic products. Compliance with this part is intended to protect the rights and welfare of human subjects involved in such investigations.

(b) References in this part to regulatory sections of the Code of Federal Regulations are to Chapter I of Title 21, unless otherwise noted.

§ 56.102 Definitions.

As used in this part:

(a) "Act" means the Federal Food, Drug, and Cosmetic Act, as amended (secs. 201–902, 52 Stat. 1040 et seq., as amended (21 U.S.C. 321–392)).

(b) "Application for research or marketing permit" includes:

(1) A color additive petition, described in Part 71.

(2) Data and information regarding a substance submitted as part of the procedures for establishing that a substance is generally recognized as safe for a use which results or may reasonably be expected to result, directly or indirectly, in its becoming a component or otherwise affecting the characterization of any food, described in § 170.35.

(3) A food additive petition, described in Part 171.

(4) Data and information regarding a food additive submitted as part of the procedures regarding food additives permitted to be used on an interim basis pending additional study, described in § 180.1.

(5) Data and information regarding a substance submitted as part of the procedures for establishing a tolerance for unavoidable contaminants in food and food-packaging materials, described in section 406 of the act.

(6) A "Notice of Claimed Investigational Exemption for a New Drug" described in Part 312.

(7) A new drug application, described in Part 314.

(8) Data and information regarding the bioavailability or bioequivalence of drugs for human use submitted as part of the procedures for issuing, amending, or repealing a bioequivalence requirement, described in Part 320.

(9) Data and information regarding an over-the-counter drug for human use submitted as part of the procedures for classifying such drugs as generally recognized as safe and effective and not misbranded, described in Part 330.

(10) Data and information regarding an antibiotic drug submitted as part of the procedures for issuing, amending, or repealing regulations for such drugs, described in Part 430.

(11) An application for a biological product license, described in Part 601.

(12) Data and information regarding a biological product submitted as part of the procedures for determining that licensed biological products are safe and effective and not misbranded, as described in Part 601.

(13) An "Application for an Investigational Device Exemption," described in Parts 812 and 813.

(14) Data and information regarding a medical device for human use submitted as part of the procedures for classifying such devices, described in Part 860.

(15) Data and information regarding a medical device for human use submitted as

part of the procedures for establishing, amending, or repealing a standard for such device, described in Part 861.

(16) An application for premarket approval of a medical device for human use, described in section 515 of the act.

(17) A product development protocol for a medical device for human use, described in section 515 of the act.

(18) Data and information regarding an electronic product submitted as part of the procedures for establishing, amending, or repealing a standard for such products, described in section 358 of the Public Health Service Act.

(19) Data and information regarding an electronic product submitted as part of the procedures for obtaining a variance from any electronic product performance standard, as described in § 1010.4.

(20) Data and information regarding an electronic product submitted as part of the procedures for granting, amending, or extending an exemption from a radiation safety performance standard, as described in § 1010.5.

(21) Data and information regarding an electronic product submitted as part of the procedures for obtaining an exemption from notification of a radiation safety defect or failure of compliance with radiation safety performance standard, described in Subpart D of Part 1003.

(c) "Clinical investigation" means any experiment that involves a test article and one or more human subjects, and that either must meet the requirements for prior submission to the Food and Drug Administration under section 505(i), 507(d), or 520(g) of the act, or need not meet the requirements for prior submission to the Food and Drug Administration under these sections of the act, but the results of which are intended to be later submitted to, or held for inspection by, the Food and Drug Administration as part of an application for a research or marketing permit. The term does not include experiments that must meet the provisions of Part 58, regarding nonclinical laboratory studies. The terms "research,"

"clinical research," "clinical study," "study," and "clinical investigation" are deemed to be synonymous for purposes of this part.

(d) "Emergency use" means the use of a test article on a human subject in a life-threatening situation in which no standard acceptable treatment is available, and in which there is not sufficient time to obtain IRB approval.

(e) "Human subject" means an individual who is or becomes a participant in research, either as a recipient of the test article or as a control. A subject may be either a healthy individual or a patient.

(f) "Institution" means any public or private entity or agency (including Federal, State, and other agencies). The term "facility" as used in section 520(g) of the act is deemed to be synonymous with the term "institution" for purposes of this part.

(g) "Institutional Review Board (IRB)" means any board, committee, or other group formally designated by an institution, to conduct periodic review of, biomedical research involving human subjects. The primary purpose of such review is to assure the protection of the rights and welfare of the human subjects. The term has the same meaning as the phrase "institutional review committee" as used in section 520(g) of the act.

(h) "Investigator" means an individual who actually conducts a clinical investigation (i.e., under whose immediate direction the test article is administered or dispensed to, or used involving, a subject) or, in the event of an investigation conducted by a team of individuals, is the responsible leader of that team.

(i) "Minimal risk" means that the risks of harm anticipated in the proposed research are not greater, considering probability and magnitude, than those ordinarily encountered in daily life or during the performance of routine physical or psychological examinations or tests.

(j) "Sponsor" means a person or other entity that initiates a clinical investigation, but that does not actually conduct the investigation, i.e., the test article is administered or dispensed to, or used involving, a subject under the immediate direction of another individual. A person other than an individual (e.g., a corporation or agency) that uses one or more of its own employees to conduct an investigation that it has initiated is considered to be a sponsor (not a sponsor-investigator), and the employees are considered to be investigators.

(k) "Sponsor-investigator" means an individual who both initiates and actually conducts, alone or with others, a clinical investigation, i.e., under whose immediate direction the test article is administered or dispensed to, or used involving, a subject. The term does not include any person other than an individual, e.g., it does not include a corporation or agency. The obligations of a sponsor-investigator under this part include both those of a sponsor and those of an investigator.

(l) "Test article" means any drug for human use, biological product for human use, medical device for human use, human food additive, color additive, electronic product, or any other article subject to regulation under the act or under sections 351 or 354–360F of the Public Health Service Act.

§ 56.103 Circumstances in which IRB review is required.

(a) Except as provided in § § 56.104 and 56.105, any clinical investigation which must meet the requirements for prior submission (as required in Parts 312, 812, and 813) to the Food and Drug Administration shall not be initiated unless that investigation has been reviewed and approved by, and remains subject to continuing review by, an IRB meeting the requirements of this part.

(b) Except as provided in § § 56.104 and 56.105, the Food and Drug Administration may decide not to consider in support of an application for a research or marketing permit any data or information that has been derived from a clinical investigation that has not been approved by, and that was not subject to initial and continuing review by, an IRB meeting the requirements of this part. The determination that a clinical in-

vestigation may not be considered in support of an application for a research or marketing permit does not, however, relieve the applicant for such a permit of any obligation under any other applicable regulations to submit the results of the investigation to the Food and Drug Administration.

(c) Compliance with these regulations will in no way render inapplicable pertinent Federal, State, or local laws or regulations.

§ 56.104 Exemptions from IRB requirement.

The following categories of clinical investigations are exempt from the requirements of this part for IRB review:

(a) Any investigation which commenced before July 27, 1981, and was subject to requirements for IRB review under FDA regulations before that date, provided that the investigation remains subject to review of an IRB which meets the FDA requirements in effect before July 27, 1981.

(b) Any investigation commenced before July 27, 1981, and was not otherwise subject to requirements for IRB review under Food and Drug Administration regulations before that date.

(c) Emergency use of a test article, provided that such emergency use is reported to the IRB within 5 working days. Any subsequent use of the test article at the institution is subject to IRB review.

§ 56.105 Waiver of IRB requirement.

On the application of a sponsor or sponsor-investigator, the Food and Drug Administration may waive any of the requirements contained in these regulations, including the requirements for IRB review, for specific research activities or for classes of research activities, otherwise covered by these regulations.

Subpart B—Organization and Personnel

§ 56.107 IRB membership.

(a) Each IRB shall have at least five members, with varying backgrounds to promote complete and adequate review of research activities commonly conducted by the institution. The IRB shall be sufficiently qualified through the experience and expertise of its members, and the diversity of the members' backgrounds including consideration of the racial and cultural backgrounds of members and sensitivity to such issues as community attitudes, to promote respect for its advice and counsel in safeguarding the rights and welfare of human subjects. In addition to possessing the professional competence necessary to review specific research activities, the IRB shall be able to ascertain the acceptability of proposed research in terms of institutional commitments and regulations, applicable law, and standards of professional conduct and practice. The IRB shall therefore include persons knowledgeable in these areas. If an IRB regularly reviews research that involves a vulnerable category of subjects, including but not limited to subjects covered by other parts of this chapter, the IRB should include one or more individuals who are primarily concerned with the welfare of these subjects.

(b) No IRB many consist entirely of men, or entirely of women, or entirely of members of one profession.

(c) Each IRB shall include at least one member whose primary concerns are in nonscientific areas; for example: lawyers, ethicists, members of the clergy.

(d) Each IRB shall include at least one member who is not otherwise affiliated with the institution and who is not part of the immediate family of a person who is affiliated with the institution.

(e) No IRB may have a member participate in the IRB's initial or continuing review of any project in which the member has a conflicting interest, except to provide information requested by the IRB.

(f) An IRB may, in its discretion, involve individuals with competence in special areas to assist in the review of complex issues which require expertise beyond or in addition to that available on the IRB. These individuals may not vote with the IRB.

Subpart C—IRB Functions and Operations

§ 56.108 IRB functions and operations.

In order to fulfill the requirements of these regulations, each IRB shall:

(a) Follow written procedures (1) for conducting its initial and continuing review of research and for reporting its findings and actions to the investigator and the institution, (2) for determining which projects require review more often than annually and which projects need verification from sources other than the investigators that no material changes have occurred since previous IRB review, (3) for insuring prompt reporting to the IRB of changes in a research activity, (4) for insuring that changes in approved research, during the period for which IRB approval has already been given, may not be initiated without IRB review and approval except where necessary to eliminate apparent immediate hazards to the human subjects; and (5) for insuring prompt reporting to the IRB of unanticipated problems involving risks to subjects or others.

(b) Except when an expedited review procedure is used (see § 56.110), review proposed research at convened meetings at which majority of the members of the IRB are present, including at least one member whose primary concerns are in nonscientific areas. In order for the research to be approved, it shall receive the approval of a majority of those members present at the meeting.

(c) Be responsible for reporting to the appropriate institutional officials and the Food and Drug Administration any serious or continuing noncompliance by investigators with the requirements and determinations of the IRB.

§ 56.109 IRB review of research.

(a) An IRB shall review and have authority to approve, require modifications in (to secure approval), or disapprove all research activities covered by these regulations.

(b) An IRB shall require that information given to subjects as part of informed consent is in accordance with § 50.25. The IRB may require that information, in addition to that specifically mentioned in § 50.25, be given to the subject when in the IRB's judgment the information would meaningfully add to the protection of the rights and welfare of subjects.

(c) An IRB shall require documentation of informed consent in accordance with § 50.27, except that the IRB may, for some or all subjects, waive the requirement that the subject or the subject's legally authorized representative sign a written consent form if it finds that the research presents no more than minimal risk of harm to subjects and involves no procedures for which written consent is normally required outside the research context. In cases where the documentation requirement is waived, the IRB may require the investigator to provide subjects with a written statement regarding the research.

(d) An IRB shall notify investigators and the institution in writing of its decision to approve or disapprove the proposed research activity. If the IRB decides to disapprove a research activity, it shall include in its written notification a statement of the reasons for its decisions and give the investigator an opportunity to respond in person or in writing.

(e) An IRB shall conduct continuing review of research covered by these regulations at intervals appropriate to the degree of risk, but not less than once per year, and shall have authority to observe or have a third party observe the consent process and the research.

§ 56.110 Expedited review procedures for certain kinds of research involving no more than minimal risk, and for minor changes in approved research.

(a) The Food and Drug Administration has established, and published in the *Federal Register,* a list of categories of research that may be reviewed by the IRB through an expedited review procedure. The list will be amended, as appropriate, through periodic republication in the *Federal Register.*

(b) An IRB may review some or all of the research appearing on the list through an expedited review procedure, if the research involves no more than minimal risk. The IRB may also use the expedited review procedure to review minor changes in previously approved research during the period for which approval is authorized. Under an expedited review procedure, the review may be carried out by the IRB chairperson or by one or more experienced reviewers designated by the chairperson from among members of the IRB. In reviewing the research, the reviewers may exercise all of the authorities of the IRB except that the reviewers may not disapprove the research. A research activity may be disapproved only after review in accordance with the non-expedited procedure set forth in § 56.108(b).

(c) Each IRB which uses an expedited review procedure shall adopt a method for keeping all members advised of research proposals which have been approved under the procedure.

(d) The Food and Drug Administration may restrict, suspend, or terminate an institution's or IRB's use of the expedited review procedure when necessary to protect the rights or welfare of subjects.

§ 56.111 Criteria for IRB approval of research.

(a) In order to approve research covered by these regulations the IRB shall determine that all of the following requirements are satisfied:

(1) Risks to subjects are minimized: (i) by using procedures which are consistent with sound research design and which do not unnecessarily expose subjects to risk, and (ii) whenever appropriate, by using procedures already being performed on the subjects for diagnostic or treatment purposes.

(2) Risks to subjects are reasonable in relation to anticipated benefits, if any, to subjects, and the importance of the knowledge that may be expected to result. In evaluating risks and benefits, the IRB should consider only those risks and benefits that may result from the research (as distin-

guished from risks and benefits of therapies that subjects would receive even if not participating in the research). The IRB should not consider possible long-range effects of applying knowledge gained in the research (for example, the possible effects of the research on public policy) as among those research risks that fall within the purview of its responsibility.

(3) Selection of subjects is equitable. In making this assessment, the IRB should take into account the purposes of the research and the setting in which the research will be conducted.

(4) Informed consent will be sought from each prospective subject or the subject's legally authorized representative, in accordance with and to the extent required by Part 50.

(5) Informed consent will be appropriately documented, in accordance with and to the extent required by § 50.27.

(6) Where appropriate, the research plan makes adequate provision for monitoring the data collected to ensure the safety of subjects.

(7) Where appropriate, there are adequate provisions to protect the privacy of subjects and to maintain the confidentiality of data.

(b) Where some or all of the subjects are likely to be vulnerable to coercion or undue influence, such as persons with acute or severe physical or mental illness, or persons who are economically or educationally disadvantaged, appropriate additional safeguards have been included in the study to protect the rights and welfare of these subjects.

§ 56.112 Review by institution.

Research covered by these regulations that has been approved by an IRB may be subject to further appropriate review and approval or disapproval by officials of the institution. However, those officials may not approve the research if it has not been approved by an IRB.

§ 56.113 Suspension or termination of IRB approval of research.

An IRB shall have authority to suspend or terminate approval of research that is not being conducted in accordance with the IRB's requirements or that has been associated with unexpected serious harm to subjects. Any suspension or termination of approval shall include a statement of the reasons for the IRB's action and shall be reported promptly to the investigator, appropriate institutional officials, and the Food and Drug Administration.

§ 56.114 Cooperative research.

In complying with these regulations, institutions involved in multi-institutional studies may use joint review, reliance upon the review of another qualified IRB, or similar arrangements aimed at avoidance of duplication of effort.

Subpart D—Records and Reports

§ 56.115 IRB records.

(a) An institution, or where appropriate an IRB, shall prepare and maintain adequate documentation of IRB activities, including the following:

(1) Copies of all research proposals reviewed, scientific evaluations, if any, that accompany the proposals, approved sample consent documents, progress reports submitted by investigators, and reports of injuries to subjects.

(2) Minutes of IRB meetings which shall be in sufficient detail to show attendance at the meetings; actions taken by the IRB; the vote on these actions including the number of members voting for, against, and abstaining; the basis for requiring changes in or disapproving research; and a written summary of the discussion of controverted issues and their resolution.

(3) Records of continuing review activities.

(4) Copies of all correspondence between the IRB and the investigators.

(5) A list of IRB members identified by name; earned degrees; representative ca-

pacity; indications of experience such as board certifications, licenses, etc., sufficient to describe each member's chief anticipated contributions to IRB deliberations; and any employment or other relationship between each member and the institution; for example: full-time employee, part-time employee, a member of governing panel or board, stockholder, paid or unpaid consultant.

(6) Written procedures for the IRB as required by § 56.108(a).

(7) Statements of significant new findings provided to subjects, as required by § 50.25.

(b) The records required by this regulation shall be retained for at least 3 years after completion of the research, and the records shall be accessible for inspection and copying by authorized representatives of the Food and Drug Administration at reasonable times and in a reasonable manner.

(c) The Food and Drug Administration may refuse to consider a clinical investigation in support of an application for a research or marketing permit if the institution or the IRB that reviewed the investigation refuses to allow an inspection under this section.

Subpart E—Administrative Actions for Noncompliance

§ 56.120 Lesser administrative actions.

(a) If apparent noncompliance with these regulations in the operation of an IRB is observed by an FDA investigator during an inspection, the inspector will present an oral or written summary of observations to an appropriate representative of the IRB. The Food and Drug Administration may subsequently send a letter describing the noncompliance to the IRB and to the parent institution. The agency will require that the IRB or the parent institution respond to this letter within a time period specified by FDA and describe the corrective actions that will be taken by the IRB, the institution, or both to achieve compliance with these regulations.

(b) On the basis of the IRB's or the institution's response, FDA may schedule reinspection to confirm the adequacy of corrective actions. In addition, until the IRB or the parent institution takes appropriate corrective action, the agency may:

(1) Withhold approval of new studies subject to the requirements of this part that are conducted at the institution or reviewed by the IRB;

(2) Direct that no new subjects be added to ongoing studies subject to this part;

(3) Terminate ongoing studies subject to this part when doing so would not endanger the subjects; or

(4) When the apparent noncompliance creates a significant threat to the rights and welfare of human subjects, notify relevant State and Federal regulatory agencies and other parties with a direct interest in the agency's action of the deficiencies in the operation of the IRB.

(c) The parent institution is presumed to be responsible for the operation of an IRB, and the Food and Drug Administration will ordinarily direct any administrative action under this subpart against the institution. However, depending on the evidence of responsibility for deficiencies, determined during the investigation, the Food and Drug Administration may restrict its administrative actions to the IRB or to a component of the parent institution determined to be responsible for formal designation of the IRB.

§ 56.121 Disqualification of an IRB or an institution.

(a) Whenever the IRB or the institution has failed to take adequate steps to correct the noncompliance stated in the letter sent by the agency under § 56.120(a), and the Commissioner of Food and Drugs determines that this noncompliance may justify the disqualification of the IRB or of the parent institution, the Commissioner will institute proceedings in accordance with the requirements for a regulatory hearing set forth in Part 16.

(b) The Commissioner may disqualify an IRB or the parent institution if the Commissioner determines that:

(1) The IRB has refused or repeatedly failed to comply with any of the regulations set forth in this part, and

(2) The noncompliance adversely affects the rights or welfare of the human subjects in a clinical investigation.

(c) If the Commissioner determines that disqualification is appropriate, the Commissioner will issue an order that explains the basis for the determination and that prescribes any actions to be taken with regard to ongoing clinical research conducted under the review of the IRB. The Food and Drug Administration will send notice of the disqualification to the IRB and the parent institution. Other parties with a direct interest, such as sponsors and clinical investigators, may also be sent a notice of disqualification. In addition, the agency may elect to publish a notice of its action in the *Federal Register*.

(d) The Food and Drug Administration will not approve an application for a research permit for a clinical investigation that is to be under the review of a disqualified IRB or that is to be conducted at a disqualified institution, and it may refuse to consider in support of a marketing permit the data from a clinical investigation that was reviewed by a disqualified IRB or conducted at a disqualified institution, unless the IRB or the parent institution is reinstated as provided in § 56.123.

§ 56.122 Public disclosure of information regarding revocation.

A determination that the Food and Drug Administration has disqualified an institution and the administrative record regarding that determination are disclosable to the public under Part 20.

§ 56.123 Reinstatement of an IRB or an institution.

An IRB or an institution may be reinstated if the Commissioner determines, upon an evaluation of a written submission from the IRB or institution that explains the corrective action that the institution or IRB plans to take, that the IRB or institution has provided adequate assurance that it will operate in compliance with the standards set forth in this part. Notification of reinstatement shall be provided to all person notified under § 56.121(c).

§ 56.124 Actions alternative or additional to disqualification.

Disqualification of an IRB or of an institution is independent of, and neither in lieu of nor a precondition to, other proceedings or actions authorized by the act. The Food and Drug Administration may, at any time, through the Department of Justice institute any appropriate judicial proceedings (civil or criminal) and any other appropriate regulatory action, in addition to or in lieu of, and before, at the time of, or after, disqualification. The agency may also refer pertinent matters to another Federal, State, or local government agency for any action that the agency determines to be appropriate.

Effective date. This regulation shall become effective July 27, 1981.

Part 50—Protection of Human Subjects*

Subpart B—Informed Consent of Human Subjects

Sec.
50.20 General requirements for informed consent.
50.21 Effective date.
50.23 Exception from general requirements.
50.25 Elements of informed consent.
50.27 Documentation of informed consent.

§ 50.20 General requirements for informed consent.

Except as provided in § 50.23, no investigator may involve a human being as a

*Reprinted from: Federal Register, Vol. 46, No. 17, Tuesday, January 27, 1981.

subject in research covered by these regulations unless the investigator has obtained the legally effective informed consent of the subject or the subject's legally authorized representative. An investigator shall seek such consent only under circumstances that provide the prospective subject or the representative sufficient opportunity to consider whether or not to participate and that minimize the possibility of coercion or undue influence. The information that is given to the subject or the representative shall be in language understandable to the subject or the representative. No informed consent, whether oral or written, may include any exculpatory language through which the subject or the representative is made to waive or appear to waive any of the subject's legal rights, or releases or appears to release the investigator, the sponsor, the institution, or its agents from liability for negligence.

§ 50.21 Effective date.

The requirements for informed consent set out in this part apply to all human subjects entering a clinical investigation that commences on or after July 27, 1981.

§ 50.23 Exception from general requirements.

(a) The obtaining of informed consent shall be deemed feasible unless, before use of the test article (except as provided in paragraph (b) of this section), both the investigator and a physician who is not otherwise participating in the clinical investigation certify in writing all of the following:

(1) The human subject is confronted by a life-threatening situation necessitating the use of the test article.

(2) Informed consent cannot be obtained from the subject because of an inability to communicate with, or obtain legally effective consent from, the subject.

(3) Time is not sufficient to obtain consent from the subject's legal representative.

(4) There is available no alternative method of approved or generally recognized therapy that provides an equal or greater likelihood of saving the life of the subject.

(b) If immediate use of the test articles is, in the investigator's opinion, required to preserve the life of the subject, and time is not sufficient to obtain the independent determination required in paragraph (a) of this section in advance of using the test article, the determinations of the clinical investigator shall be made and, within 5 working days after the use of the article, be reviewed and evaluated in writing by a physician who is not participating in the clinical investigation.

(c) The documentation required in paragraph (a) or (b) of this section shall be submitted to the IRB within 5 working days after the use of the test article.

§ 50.25 Elements of informed consent.

(a) *Basic elements of informed consent.* In seeking informed consent, the following information shall be provided to each subject:

(1) A statement that the study involves research, an explanation of the purposes of the research and the expected duration of the subject's participation, a description of the procedures to be followed, and identification of any procedures which are experimental.

(2) A description of any reasonably foreseeable risks or discomforts to the subject.

(3) A description of any benefits to the subject or to others which may reasonably be expected from the research.

(4) A disclosure of appropriate alternative procedures or courses of treatment, if any, that might be advantageous to the subject.

(5) A statement describing the extent, if any, to which confidentiality of records identifying the subject will be maintained and that notes the possibility that the Food and Drug Administration may inspect the records.

(6) For research involving more than minimal risk, an explanation as to whether

any compensation and an explanation as to whether any medical treatments are available if injury occurs and, if so, what they consist of, or where further information may be obtained.

(7) An explanation of whom to contact for answers to pertinent questions about the research and research subjects' rights, and whom to contact in the event of a research-related injury to the subject.

(8) A statement that participation is voluntary, that refusal to participate will involve no penalty or loss of benefits to which the subject is otherwise entitled, and that the subject may discontinue participation at any time without penalty or loss of benefits to which the subject is otherwise entitled.

(b) *Additional elements of informed consent.* When appropriate, one or more of the following elements of information shall also be provided to each subject:

(1) A statement that the particular treatment or procedure may involve risks to the subject (or to the embryo or fetus, if the subject is or may become pregnant) which are currently unforeseeable.

(2) Anticipated circumstances under which the subject's participation may be terminated by the investigator without regard to the subject's consent.

(3) Any additional costs to the subject that may result from participation in the research.

(4) The consequences of a subject's decision to withdraw from the research and procedures for orderly termination of participation by the subject.

(5) A statement that significant new findings developed during the course of the research which may relate to the subject's willingness to continue participation will be provided to the subject.

(6) The approximate number of subjects involved in the study.

(c) The informed consent requirements in these regulations are not intended to preempt any applicable Federal, State, or local laws which require additional information to be disclosed for informed consent to be legally effective.

(d) Nothing in these regulations is intended to limit the authority of a physician to provide emergency medical care to the extent the physician is permitted to do so under applicable Federal, State, or local law.

§ 50.27 Documentation of informed consent.

(a) Except as provided in § 56.109(c), informed consent shall be documented by the use of a written consent form approved by the IRB and signed by the subject or the subject's legally authorized representative. A copy shall be given to the person signing the form.

(b) Except as provided in § 56.109(c), the consent form may be either of the following:

(1) A written consent document that embodies the elements of informed consent required by § 50.25. This form may be read to the subject or the subject's legally authorized representative, but, in any event, the investigator shall give either the subject or the representative adequate opportunity to read it before it is signed.

(2) A "short form" written consent document stating that the elements of informed consent required by § 50.25 have been presented orally to the subject or the subject's legally authorized representative. When this method is used, there shall be a witness to the oral presentation. Also, the IRB shall approve a written summary of what is to be said to the subject or the representative. Only the short form itself is to be signed by the subject or the representative. However, the witness shall sign both the short form and a copy of the summary, and the person actually obtaining the consent shall sign a copy of the summary. A copy of the summary shall be given to the subject or the representative in addition to a copy of the short form.

Appendix 3

The Nuremberg Code[1]

The Proof as to War Crimes and Crimes against Humanity

Judged by any standard of proof the record clearly shows the commission of war crimes and crimes against humanity substantially as alleged in counts two and three of the indictment. Beginning with the outbreak of World War II criminal medical experiments on non-German nationals, both prisoners of war and civilians, including Jews and "asocial" persons, were carried out on a large scale in Germany and the occupied countries. These experiments were not the isolated and casual acts of individual doctors and scientists working solely on their own responsibility, but were the product of coordinated policy-making and planning at high governmental, military, and Nazi Party levels, conducted as an integral part of the total war effort. They were ordered, sanctioned, permitted, or approved by persons in positions of authority who under all principles of law were under the duty to know about things and to take steps to terminate or prevent them.

Permissible Medical Experiments

The great weight of evidence before us is to the effect that certain types of medical experiments on human beings, when kept within reasonable well-defined bounds, conform to the ethics of the medical profession generally. The protagonists of the practice of human experimentation justify their views on the basis that such experiments yield results for the good of society that are unprocurable by other methods or means of study. All agree, however, that certain basic principles must be observed in order to satisfy moral, ethical and legal concepts:

1. The voluntary consent of the human subject is absolutely essential.

This means that the person involved should have legal capacity to give consent; should be so situated as to be able to exercise free power of choice, without the intervention of any element of force, fraud, deceit, duress, overreaching, or other ulterior form of constraint or coercion; and should have sufficient knowledge and comprehension of the elements of the subject matter involved as to enable him to make an understanding and enlightened decision. This latter element requires that before the acceptance of an affirmative decision by the experimental subject there should be made known to him the nature, duration, and purpose of the experiment; the method and means by which it is to be conducted; all inconveniences and hazards reasonably to be expected; and the effects upon his health or person which may

[1]Reprinted from *Trials of War Criminals before the Nuremberg Military Tribunals under Control Council Law No. 10,* Vol. 2 (Washington, D.C.: U.S. Government Printing Office, 1949), pp. 181–182.

possibly come from his participation in the experiment.

The duty anu responsibility for ascertaining the quality of the consent rests upon each individual who initiates, directs or engages in the experiment. It is a personal duty and responsibility which may not be delegated to another with impunity.

2. The experiment should be such as to yield fruitful results for the good of society, unprocurable by other methods or means of study, and not random and unnecessary in nature.

3. The experiment should be designed and based on the results of animal experimentation and a knowledge of the natural history of the disease or other problem under study that the anticipated results will justify the performance of the experiment.

4. The experiment should be so conducted as to avoid all unnecessary physical and mental suffering and injury.

5. No experiment should be conducted where there is an *a priori* reason to believe that death or disabling injury will occur except, perhaps, in those experiments where the experimental physicians also serve as subjects.

6. The degree of risk to be taken should never exceed that determined by the hu-

manitarian importance of the problem to be solved by the experiment.

7. Proper preparations should be made and adequate facilities provided to protect the experimental subject against even remote possibilities of injury, disability, or death.

8. The experiment should be conducted only by scientifically qualified persons. The highest degree of skill and care should be required through all stages of the experiment of those who conduct or engage in the experiment.

9. During the course of the experiment the human subject should be at liberty to bring the experiment to an end if he has reached the physical or mental state where continuation of the experiment seems to him to be impossible.

10. During the course of the experiment the scientist in charge must be prepared to terminate the experiment at any stage, if he has probable cause to believe, in the exercise of the good faith, superior skill and careful judgment required of him that a continuation of the experiment is likely to result in injury, disability, or death to the experimental subject. . . .

Appendix 4

World Medical Association Declaration of Helsinki: Recommendations Guiding Medical Doctors in Biomedical Research Involving Human Subjects[1]

Introduction

It is the mission of the medical doctor to safeguard the health of the people. His or her knowledge and conscience are dedicated to the fulfillment of this mission.

The Declaration of Geneva of the World Medical Association binds the doctor with the words, "The health of my patient will be my first consideration," and the International Code of Medical Ethics declares that, "Any act or advice which could weaken physical or mental resistance of a human being may be used only in his interest."

The purpose of biomedical research involving human subjects must be to improve diagnostic, therapeutic and prophylactic procedures and the understanding of the aetiology and pathogenesis of disease.

In current medical practice most diagnostic, therapeutic or prophylactic procedures involve hazards. This applies *a fortiori* to biomedical research.

Medical progress is based on research which ultimately must rest in part on experimentation involving human subjects.

In the field of biomedical research a fundamental distinction must be recognized between medical research in which the aim is essentially diagnostic or therapeutic for a patient, and medical research, the essential object of which is purely scientific and without direct diagnostic or therapeutic value to the person subjected to the research.

Special caution must be exercised in the conduct of research which may affect the environment, and the welfare of animals used for research must be respected.

Because it is essential that the results of laboratory experiments be applied to human beings to further scientific knowledge and to help suffering humanity, the World Medical Association has prepared the following recommendations as a guide to every doctor

---

[1]Adopted by the 18th World Medical Assembly, Helsinki, Finland, 1964, and as revised by the 29th World Medical Assembly, Tokyo, Japan, 1975.

in biomedical research involving human subjects. They should be kept under review in the future. It must be stressed that the standards as drafted are only a guide to physicians all over the world. Doctors are not relieved from criminal, civil and ethical responsibilities under the laws of their own countries.

I. Basic Principles

1. Biomedical research involving human subjects must conform to generally accepted scientific principles and should be based on adequately performed laboratory and animal experimentation and on a thorough knowledge of the scientific literature.

2. The design and performance of each experimental procedure involving human subjects should be clearly formulated in an experimental protocol which should be transmitted to a specially appointed independent committee for consideration, comment and guidance.

3. Biomedical research involving human subjects should be conducted only by scientifically qualified persons and under the supervision of a clinically competent medical person. The responsibility for the human subject must always rest with a medically qualified person and never rest on the subject of the research, even though the subject has given his or her consent.

4. Biomedical research involving human subjects cannot legitimately be carried out unless the importance of the objectives is in proportion to the inherent risk to the subject.

5. Every biomedical research project involving human subjects should be preceded by careful assessment of predictable risks in comparison with foreseeable benefits to the subject or to others. Concern for the interests of the subject must always prevail over the interest of science and society.

6. The right of the research subject to safeguard his or her integrity must always be respected. Every precaution should be taken to respect the privacy of the subject and to minimize the impact of the study on the subject's physical and mental integrity and on the personality of the subject.

7. Doctors should abstain from engaging in research projects involving human subjects unless they are satisfied that the hazards involved are believed to be predictable. Doctors should cease any investigation if the hazards are found to outweigh the potential benefits.

8. In publication of the results of his or her research, the doctor is obliged to preserve the accuracy of the results. Reports of experimentation not in accordance with the principles laid down in this Declaration should not be accepted for publication.

9. In any research on human beings, each potential subject must be adequately informed of the aims, methods, anticipated benefits and potential hazards of the study and the discomfort it may entail. He or she should be informed that he or she is at liberty to abstain from participation in the study and that he or she is free to withdraw his or her consent to participation at any time. The doctor should then obtain the subject's freely given informed consent, preferably in writing.

10. When obtaining informed consent for the research project the doctor should be particularly cautious if the subject is in a dependent relationship to him or her or may consent under duress. In that case the informed consent should be obtained by a doctor who is not engaged in the investigation and who is completely independent of this official relationship.

11. In case of legal incompetence, informed consent should be obtained from the legal guardian in accordance with national legislation. Where physical or mental incapacity makes it impossible to obtain informed consent, or when the subject is a minor, permission from the responsible relative replaces that of the subject in accordance with national legislation.

12. The research protocol should always contain a statement of the ethical considerations involved and should indicate that the principles enunciated in the present Declaration are complied with.

II. Medical Research Combined with Professional Care (Clinical Research)

1. In the treatment of the sick person, the doctor must be free to use a new diagnostic and therapeutic measure, if in his or her judgment it offers hope of saving life, reestablishing health or alleviating suffering.

2. The potential benefits, hazards and discomforts of a new method should be weighed against the advantages of the best current diagnostic and therapeutic methods.

3. In any medical study, every patient— including those of a control group, if any— should be assured of the best proven diagnostic and therapeutic method.

4. The refusal of the patient to participate in a study must never interfere with the doctor-patient relationship.

5. If the doctor considers it essential not to obtain informed consent, the specific reasons for this proposal should be stated in the experimental protocol for transmission to the independent committee (I, 2).

6. The doctor can combine medical research with professional care, the objective being the acquisition of new medical knowledge, only to the extent that medical research is justified by its potential diagnostic or therapeutic value for the patient.

III. Nontherapeutic Biomedical Research Involving Human Subjects (Nonclinical Biomedical Research)

1. In the purely scientific application of medical research carried out on a human being, it is the duty of the doctor to remain the protector of the life and health of that person on whom biomedical research is being carried out.

2. The subjects should be volunteers— either healthy persons or patients for whom the experimental design is not related to the patient's illness.

3. The investigator or the investigating team should discontinue the research if in his/her or their judgment it may, if continued, be harmful to the individual.

4. In research on man, the interest of science and society should never take precedence over considerations related to the well-being of the subject.

Appendix 5

Leo Szilard's Ten Commandments

1. Recognize the connections of things and the laws of conduct of men, so that you may know what you are doing.

2. Let your acts be directed towards a worthy goal, but do not ask if they will reach it; they are to be models and examples, not means to an end.

3. Speak to all men as you do to yourself, with no concern for the effect you make, so that you do not shut them out from your world; lest in isolation the meaning of life slips out of sight and you lose the belief in the perfection of the creation.

4. Do not destroy what you cannot create.

5. Touch no dish, except that you are hungry.

6. Do not covet what you cannot have.

7. Do not lie without need.

8. Honor children. Listen reverently to their words and speak to them with infinite love.

9. Do your work for six years; but in the seventh, go into solitude or among strangers, so that the memory of your friends does not hinder you from being what you have become.

10. Lead your life with a gentle hand and be ready to leave whenever you are called.

Circulated by Mrs. Szilard in July 1964, in a letter to their friends (translated by Dr. Jacob Bronowski).

Index

433